2012

INTERNATIONAL GREEN CONSTRUCTION CODE™

A Member of the International Code Family®

IgCC™

2012 International Green Construction Code™

First Printing: March 2012

ISBN: 978-1-60983-059-5

With the cooperating sponsorships of the American Institute of Architects, ASTM International, Amercian Society of Heating, Refrigerating and Air-Conditioning Engineers, United States Green Building Council and the Illuminating Engineering Society.

PRINTED IN THE U.S.A.

PREFACE

Introduction

Internationally, code officials recognize the need for a modern, up-to-date code governing the impact of buildings and structures on the environment. This first edition, the 2012 edition, of the *International Green Construction Code*™ (IgCC™), is designed to meet this need through model code regulations that contain clear and specific requirements with provisions that promote safe and sustainable construction in an integrated fashion with the ICC Family of Codes.

This comprehensive green code establishes minimum regulations for building systems and site considerations using prescriptive and performance-related provisions. It is intended to be an overlay code to be used with, and is fully compatible with, all of the *International Codes*® (I-Codes®) published by the International Code Council (ICC)®, including the *International Building Code*®, *International Energy Conservation Code*®, *International Existing Building Code*®, *International Fire Code*®, *International Fuel Gas Code*®, *International Mechanical Code*®, ICC *Performance Code*®, *International Plumbing Code*®, *International Private Sewage Disposal Code*®, *International Property Maintenance Code*®, *International Residential Code*®, *International Swimming Pool and Spa Code*™, *International Wildland-Urban Interface Code*® and *International Zoning Code*®.

The *International Green Construction Code*™ provisions provide many benefits, among which is the model code development process that offers an international forum for building professionals to discuss performance and prescriptive code requirements. This forum provides an excellent arena to debate proposed revisions. This model code also encourages international consistency in the application of provisions.

This code has been developed in collaboration with the following Cooperating Sponsors: The American Institute of Architects (AIA); ASTM International; the American Society of Heating, Refrigerating and Air-Conditioning Engineers (ASHRAE), the Illuminating Engineering Society (IES); and the U.S. Green Building Council (USGBC). ICC wishes to thank these Cooperating Sponsors for recognizing the need for the development of a comprehensive set of green regulations that are enforceable, usable and adoptable.

Development

This first edition of the 2012 *International Green Construction Code* is the culmination of an effort that started in 2010 with the drafting of Public Version 1.0 (PV 1.0) by the Sustainable Building Technology Committee (SBTC) established by the ICC Board of Directors. Following that, Public Version 2.0 was created, based upon public comments submitted to PV 1.0 and approved by the IgCC Public Comment Committee. Following the issuance of PV 2.0, a full cycle of code development in accordance with ICC's Code Development Procedures was held in 2011. This included the submission of code change proposals followed by a Code Development Hearing, the submission of public comments and a Final Action Hearing. A new edition of the code is promulgated every three years.

This code is founded on principles intended to establish provisions consistent with the scope of a green construction code that adequately protects public health, safety and welfare; provisions that do not unnecessarily increase construction costs; provisions that do not restrict the use of new materials, products or methods of construction; and provisions that do not give preferential treatment to particular types or classes of materials, products or methods of construction. This is achieved by comprehensive provisions which are enforceable, useable and adoptable.

Adoption

The *International Green Construction Code* is available for adoption and use by jurisdictions internationally. Its use within a governmental jurisdiction is intended to be accomplished through adoption by reference in accordance with proceedings established in the jurisdiction's laws. At the time of adoption, jurisdictions should insert the appropriate information in provisions requiring specific local information, such as the name of the adopting jurisdiction. These locations are shown in bracketed words in small capital letters in the code and in the sample ordinance. The sample adoption ordinance on page xiii addresses several key elements of a code adoption ordinance, including the information required for insertion into the code text.

Maintenance

The *International Green Construction Code* is kept up to date through the review of proposed changes submitted by code enforcing officials, industry representatives, design professionals and other interested parties. Proposed changes are carefully considered through an open code development process in which all interested and affected parties may participate.

The contents of this work are subject to change both through the Code Development Cycles and the governmental body that enacts the code into law. For more information regarding the code development process, contact the Codes and Standards Development Department of the International Code Council.

While the development procedure of the *International Green Construction Code* assures the highest degree of care, the ICC, AIA, ASHRAE, ASTM International, IES and the USGBC and their members and those participating in the development of this code do not accept any liability resulting from compliance or noncompliance with the provisions given herein, for any restrictions imposed on materials or processes, or for the completeness of the text. ICC, AIA, ASHRAE, ASTM International, IES and the USGBC do not have power or authority to police or enforce compliance with the contents of this code. Only the governmental body that enacts the code into law has such authority.

Note that, for the development of the 2015 edition of the I-Codes, there will be two groups of code development committees and they will meet in separate years. The groupings are as follows:

Group A Codes (Heard in 2012, Code Change Proposals Deadline: January 3, 2012)	**Group B Codes (Heard in 2013, Code Change Proposals Deadline: January 3, 2013)**
International Building Code	Administrative Provisions (Chapter 1 all codes except IRC and ICCPC, administrative updates to currently referenced standards, and designated definitions)
International Fuel Gas Code	International Energy Conservation Code
International Mechanical Code	International Existing Building Code
International Plumbing Code	International Fire Code
International Private Sewage Disposal Code	International Green Construction Code
	ICC Performance Code
	International Property Maintenance Code
	International Residential Code
	International Swimming Pool and Spa Code
	International Wildland-Urban Interface Code
	International Zoning Code

Italicized Terms

Selected terms set forth in Chapter 2, Definitions, are italicized where they appear in code text. Such terms are not italicized where the definition set forth in Chapter 2 does not impart the intended meaning in the use of the term. The terms selected have definitions which the user should read carefully to facilitate better understanding of the code.

Effective Use of the International Green Construction Code

The *International Green Construction Code*™ (IgCC™) is a model code that provides minimum requirements to safeguard the environment, public health, safety and general welfare through the establishment of requirements that are intended to reduce the negative impacts and increase the positive impacts of the built environment on the natural environment and building occupants. The IgCC is fully compatible with the ICC family of codes, including the *International Building Code*® (IBC®), the *International Code Council Performance Code*® (ICCPC®), the *International Energy Conservation Code*® (IECC®), the *International Existing Building Code*® (IEBC®), the *International Fire Code*® (IFC®), the *International Fuel Gas Code*® (IFGC®), the *International Mechanical Code*® (IMC®), the *International Plumbing Code*® (IPC®), the *International Private Sewage Disposal Code*® (IPSDC®), the *International Property Maintenance Code*® (IPMC®), the *International Residential Code*® (IRC®), the *International Swimming Pool and Spa Code*™ (ISPSC™), the *International Wildland-Urban Interface Code*® (IWUIC®), and the *International Zoning Code*® (IZC®).

The IgCC addresses natural resource, material, water and energy conservation, as well as indoor environmental quality and comfort, building commissioning, operations and maintenance for new and existing buildings, building sites and building materials, components, equipment and systems. The code will be promulgated on a 3-year cycle to allow for new construction methods and technologies to be incorporated into the code. Innovative approaches and alternative materials, designs, and methods not specifically addressed in the code can be approved by the code official where the proposed innovative approaches or materials, designs or methods comply with the intent of the provisions of the code (see Section 105.4).

The IgCC applies to all occupancies other than temporary structures approved under Section 3103 of the *International Building Code*, except that application to the following is subject to jurisdictional choices in Table 302.1: one- and two-family dwellings and townhouses that are within the scope of the *International Residential Code*; Group R-3 occupancies; and Group R-2 and R-4 residential occupancies that are four stories or less in height.

Arrangement and Format of the 2012 IgCC

Before applying the requirements of the IgCC, it is beneficial to understand its arrangement and format.

Chapters	Subjects
1-2	Administration and definitions
3	Jurisdictional requirements and life cycle assessment
4	Site development and land use
5	Material resource conservation and efficiency
6	Energy conservation, efficiency and CO_2e emission reduction
7	Water resource conservation, quality and efficiency
8	Indoor environmental quality and comfort
9	Commissioning, operation and maintenance
10	Existing buildings
11	Existing building site development
12	Referenced standards
Appendix A	Project electives
Appendix B	Radon mitigation
Appendix C	Optional ordinance
Appendix D	Enforcement procedures

The following is a chapter-by-chapter synopsis of the scope and intent of the provisions of the *International Green Construction Code*:

Chapter 1 Scope and Administration. Chapter 1 of the IgCC establishes the limits of applicability of the code and describes the manner in which the code is to be applied and enforced. Chapter 1 is divided into two parts: Part 1 – Scope and Application (Sections 101 and 102); and Part 2 – Administration and Enforcement (Sections 103 – 109).

Section 101 identifies which buildings and structures come under its purview and Section 102 references other ICC codes as applicable. Section 103 establishes the duties and powers of the code official, requires that compliance and enforcement be part of the enforcement of other ICC codes listed in Section 102.4, and grants authority to the code official to make inspections. Section 105 provides guidance to the code official in the approval of materials, methods of construction, designs, systems and innovative approaches where they are not specifically prescribed in the IgCC. Section 106, in conjunction with Section 101.2 as an overlay code, requires that permits be issued under other ICC codes.

The provisions of Chapter 1 also establish the rights and privileges of the design professional, contractor and property owner.

It is important to note that by reference to Section 301.1.1, Section 101.3 allows ASHRAE 189.1, *Standard for the Design of High-Performance Green Buildings*, to be used. In addition, Exception 1 to Section 101.3 notes that the code is not applicable to low-rise residential structures unless the jurisdiction selects ICC 700 in Table 302.1 for application to various types of residential buildings and occupancies. Further, ICC 700 is noted in Section 101.3.1 as being a "deemed to comply document" for mid- and high-rise R-2 and R-4 occupancies.

The green building code is intended to be adopted as a legally enforceable document and it cannot be effective without adequate provisions for its administration and enforcement.

Chapter 2 Definitions. All terms that are defined in the code are listed alphabetically in Chapter 2. Terms are defined in Chapter 2. Codes are technical documents and every word, term and punctuation mark can impact the meaning of the code text and the intended results. The code often uses terms that have a unique meaning in the code and that code meaning can differ substantially from the ordinarily understood meaning of the term as used outside of the code. Where understanding of a term's definition is especially key to or necessary for understanding a particular code provision, the term is shown in *italics* wherever it appears in the code. However, this is true only for those terms that have a meaning that is unique to the code. In other words, the generally understood meaning of a term or phrase might not be sufficient or consistent with the meaning prescribed by the code; therefore, it is essential that the code-defined meaning be known.

Definitions are deemed to be of prime importance in establishing the meaning and intent of the code text that uses code-defined terms. The user of the code should be familiar with and consult this chapter because the definitions are essential to the correct interpretation of the code and because the user may not be aware that a term is defined in a manner that is not commonly understood.

Chapter 3 Jurisdictional Requirements and Life Cycle Assessment. As indicated earlier, Section 301.1.1 allows ASHRAE 189.1, *Standard for the Design of High-Performance Green Buildings*, to be used. Similarly, ICC 700 may be applicable to specific types of residential construction in accordance with the decisions made by the jurisdiction in the portions of Table 302.1 related to Section 101.3.

The jurisdictional requirements contained in Section 302 are formatted to afford jurisdictions the flexibility to adapt the code in a manner that is best suited to meet their unique environmental and regional goals and needs. The section numbers and optional enhanced performance features listed in Table 302.1 do not become enforceable unless they are specifically selected in the table by the jurisdiction and the appropriate "Yes" box is checked or otherwise specifically indicated in the jurisdiction's adopting ordinance. Those provisions selected by the jurisdiction in Table 302.1 become enforceable for all buildings constructed in the jurisdiction. The text of all section numbers listed in Table 302.1 also contains a reference to Table 302.1, reinforcing the fact that they are not enforceable unless they are specifically adopted. Furthermore, the sample ordinance provided in the IgCC references Table 302.1 and requires that the jurisdiction indicate those provisions from the list that it intends to enforce.

Jurisdictions must take great care when making their choices in Table 302.1. While various requirements listed in Table 302.1 may be environmentally beneficial in many jurisdictions, some may be inappropriate in other jurisdictions. If these practices were appropriate for all jurisdictions, they would have been included in the baseline requirements of the IgCC, not in Table 302.1.

Where jurisdictions find the concept of jurisdictional requirements to be unnecessary, they are able to opt out by simply checking the "No" boxes for all provisions listed in Table 302.1. Because relatively few of the code's provisions are listed in Table 302.1, even where jurisdictions do not choose any of the provisions or enhanced performance options listed in Table 302.1, the IgCC remains a strong and effective green and sustainable building tool. That said, many jurisdictions will appreciate the flexibility that the jurisdictional requirements provide in their efforts to address specific green and sustainable building concerns. Where jurisdictions begin to specifically adopt more of the items listed in Table 302.1 in future years, they will also appreciate the opportunities that the IgCC provides to grow and to produce a more sustainable built environment with each future adoption of the IgCC.

Section 303 contains provisions for whole building life cycle assessment. The IgCC does not require that whole building life cycle assessment be performed. However, where these provisions are complied with, compliance with the material selection provisions of Section 505 is not required. In this manner, whole building life cycle assessment is encouraged, though not required.

Chapter 4 Site Development and Land Use. Chapter 4 is intended to minimize the negative environmental impacts on and protect, restore and enhance the natural features and environmental quality of building sites.

Section 401.2 requires predesign site inventory and assessment. Where indicated by the jurisdiction in Table 302.1, Section 402 limits building construction near surface water, in conservation and flood hazard areas and on greenfield sites, park land or agricultural land. Section 403 requires stormwater management. Section 404 limits potable water uses related to landscape irrigation and outdoor fountains. Section 405 addresses vegetation, soil and water quality protection.

Section 406 requires that a plan be developed to ensure that least 75 percent of land-clearing debris and excavated soils is diverted from disposal.

Section 407.1 requires that at least one walkway or bicycle path connect building entrances to streets or other paths. Buildings with a total floor area of over 10,000 square feet (929 m^2) must also provide changing and shower facilities. Bicycle parking and storage requirements are contained in Sections 407.3 through 407.3.2 and Table 407.3.

Where indicated by the jurisdiction in Table 302.1, and where the total building floor area is greater than 10,000 square feet (929 m^2), preferred parking is required for high-occupancy and low-emission, hybrid and electric vehicles. These requirements, however, do not take precedence over the accessible parking requirements of the *International Building Code*.

Section 408 mitigates heat island effects through requirements related to site hardscape materials, shading and roof surfaces and coverings. Where indicated to be enforceable in the jurisdiction in Table 302.1, light pollution from building sites must be controlled in accordance with Section 409.

Chapter 5 Material Resource Conservation and Efficiency. Chapter 5 addresses material resource conservation and efficiency by means of provisions related to material selection, recycling, reuse, renewability, toxicity and durability, including resistance to damage caused by moisture.

Section 502 addresses material storage and handling during the construction phase. Section 503 requires that a construction material and waste management plan be prepared and allows the jurisdiction to increase the percentage of waste that must be recycled in Table 302.1. Section 504 requires areas be designed and constructed to facilitate the recycling of waste generated post certificate of occupancy.

Section 505 requires that at least 55 percent of constructed materials selected for each project be any combination of the following material types: used, recycled, recyclable, bio-based, or indigenous. However, where a whole building life cycle analysis is performed in accordance with Section 303, compliance with Section 505 is not required.

Section 506 regulates the mercury content of fluorescent lamps. Section 507 contains requirements for inspections that are tied to Table 903.1 and are intended to control moisture intrusion in the building envelope.

Chapter 6 Energy Conservation, Efficiency and CO_2e Emission Reduction. Chapter 6 is intended to provide flexibility and permit the use of innovative approaches to achieve the effective use of energy.

All buildings that consume energy must comply with the requirements of Sections 603 (Energy Metering, Monitoring and Reporting), 609 (Specific Appliances and Equipment), 610 (Building Renewable Energy Systems), 611 (Energy Systems Commissioning and Completion) and, where indicated by the jurisdiction in Table 302.1, must also comply with Section 604 (Automated Demand Response Infrastructure).

In addition to the preceding, buildings designed on a performance basis must comply with Sections 602 (Modeled Performance Pathway Requirements) and 608.6 (Plug load controls), while buildings designed on a prescriptive basis must comply with the prescriptive requirements of Sections 605 (Building Envelope Systems), 606 (Building Mechanical Systems), 607 (Building Service Water Heating Systems) and 608 (Building Electrical Power and Lighting Systems).

Section 602.1.1 requires that performance based designs demonstrate a zEPI of not more than 51, as determined in accordance with Equation 6-1. This equation contains a reference to EUI (energy use index), which must be calculated in accordance Appendix G of ASHRAE 90.1, as modified by Sections 602.1.2.2 and 602.1.2.3 of the IgCC. Section 602.1.1 requires that performance based designs also demonstrate CO_2e emissions reduction in accordance with Section 602.2 and Equation 6-2.

Section 603 addresses energy metering, monitoring and reporting and is applicable to all buildings that consume energy. Section 603.2 requires that energy distribution systems be designed to provide separate metering of the energy use categories listed in Table 603.2. For buildings greater than 25,000 square feet in gross floor area, meters must be installed. For buildings less than 25,000 square feet in gross floor area, the system must be designed to accommodate the installation of future meters. Section 603.3 requires that building energy metering be capable of determining energy use and peak demand for the types of energy indicated in Sections 603.3.1 through 603.3.7.

Where the jurisdiction has indicated in Table 302.1 that Section 604.1 is enforceable, an automated demand-response infrastructure must be provided. This requires that building energy, HVAC and lighting systems and specific building energy-using components be provided with controls that respond to changes in energy demand by means of automated preprogrammed strategies.

Section 605 provides building envelope system requirements for buildings that are designed on a prescriptive basis. Section 605.1.1 requires that insulation and fenestration exceed the requirements of the *International Energy Conservation Code* by at least 10 percent. Section 605.1.2.2 requires testing of the building thermal envelope for air tightness.

Section 610 establishes minimum renewable energy source requirements for all buildings that consume energy. It requires that buildings use renewable energy sources to provide either 2 percent of total calculated annual energy use by means of solar photovoltaic or wind, or 10 percent of annual estimated hot water energy by means of solar hot water heating.

Section 611 is applicable to all buildings that consume energy. It requires the commissioning and completion of mechanical, lighting, electrical and building envelope systems. These systems are also listed in Table 903.1, Commissioning Plan.

There are also provisions outside of Chapter 6 that have significant impacts on energy: Table 302.1 allows jurisdictions to require lower zEPI values, or require more stringent levels of efficiency, by occupancy; where indicated to be enforceable in Table 302.1, the project electives of Section A106 in Appendix A require additional energy conserving practices be implemented and recognize and encourage energy performance that exceeds the baseline minimum requirements of Chapters 3 and 6; Section 1003.2 addresses energy use where existing buildings are altered; and, where indicated to be enforceable in Table 302.1, Section 1007.2 requires that owners of existing buildings report post certificate of occupancy zEPI, energy demand and CO_2e emissions.

Chapter 7 Water Resource Conservation, Quality and Efficiency. Chapter 7 provides requirements that are intended to conserve water, protect water quality and provide for safe water consumption.

Section 702 regulates water consumption through limitations of fixture and fitting flow rates and by means of requirements related to specific equipment and appliances. It also requires that municipal reclaimed water, where available and required by the jurisdiction in Table 302.1, be supplied to water-supplied toilets, urinals, trap primers and applicable industrial systems. Hot water distribution systems must be designed to reduce the volume of water between fixtures and sources of hot or tempered water in accordance with Section 702.8.

Section 703 regulates water used in HVAC systems and equipment including hydronic closed systems, humidification systems, condensate coolers, condensate drainage recovery, once through heat exchangers, humidifier discharge, cooling towers, evaporative condensers, fluid cooers, wet-hood exhaust scrubber systems and evaporative cooling systems.

Section 704 regulates water treatment devices and equipment including water softeners, reverse osmosis water treatment systems and onsite reclaimed water treatment systems.

Section 705 contains specific water conservation measures for indoor ornamental fountains and other water features. It also requires the separate metering of water consumed from any source associated with the building or its site.

Section 706 contains signage and water quality requirements related to nonpotable water. Sections 707, 708 and 709 contain requirements related to rainwater collection and distribution systems, gray water systems, and reclaimed water systems, respectively. Section 710 contains provisions for other alternative onsite sources of nonpotable water.

Chapter 8 Indoor Environmental Quality and Comfort. Chapter 8 is intended to ensure that the building's interior environment is conducive to the health of building occupants.

Section 801.2.requires that an indoor air quality management plan be developed to ensure compliance with Sections 802 through 805. Section 802 addresses air-handling system access for cleaning and repair, as well as air-handling filter rack design. Section 803 contains requirements for the ventilation of buildings during the construction phase, prohibits smoking within buildings, limits pollutant sources in print, copy and janitorial rooms, and provides filters requirements for air-conditioning systems. Section 804 contains specific indoor air quality and pollutant control requirements for fireplaces, solid fuel-burning appliances, vented decorative gas appliances, vented gas fireplace heaters and decorative gas appliances. Where the jurisdiction has indicated in Table 302.1 that Section 804.2 is enforceable, baseline indoor air quality testing is required. Section 805 prohibits the use of urea-formaldehyde foam insulation and materials that contain asbestos.

Section 806 regulates emissions from wood products, adhesives, sealants, paints, coatings, flooring, acoustical ceiling tiles, wall systems and insulation.

Where the jurisdiction has indicated in Table 302.1 that Section 807.1 is to be enforceable, sound transmission levels must be limited in accordance with Sections 807.2 through 807.5.2.

Section 808 requires that fenestration be provided to ensure that interior spaces in the specified occupancies benefit from exposure to natural light.

Chapter 9 Commissioning, Operation and Maintenance. Chapter 9 addresses building commissioning, operation and maintenance. It requires inspections as specifically listed in Table 903.1. Chapter 9 also requires that construction documents contain information related to building operation and maintenance in accordance with Section 904.3.

Many of the provisions of Chapter 9, and in particular those in Sections 902 and 903, are essentially based on the requirements for special inspections contained in the *International Building Code*. Both Table 903.1 and Section 904 also contain ties to, and are coordinate with, various provisions in Chapters 4 through 8 of the IgCC. The building operation and maintenance documents required by Section 904.3 are intended to help and encourage building owners and facility management staff to operate and maintain buildings in a manner, and a performance level, as originally intended by design professionals as they strove to configure building systems in a manner that satisfied the requirements of the IgCC.

Chapter 10 Existing Buildings. Conceptually, the requirements of Chapter 10 of the IgCC are based upon the requirements of Chapter 34 of the *International Building Code* for existing buildings. These provisions are not retroactive. They apply only where buildings are altered or added to.

Additions are essentially handled as new construction.

Alterations must meet the requirements of other applicable chapters of the code for those portions or elements of the building that are being altered. However, similar to the means by which the *International Building Code* addresses accessibility in existing buildings, Section 1003.2 requires that at least 10 percent of the cost of alterations be dedicated to improvements related to water and energy conservation and efficiency. Water and energy conservation and efficiency requirements that are intended to apply specifically to existing buildings are listed in Sections 1003.2.1 through 1003.2.7. These sections address the following: metering devices; heating, ventilation and air conditioning; service water systems; lighting; swimming pools and spas; insulation of unconditioned attics; and roof replacement insulation.

Section 1005 provides relief for historic buildings under certain conditions. Where buildings are decommissioned, Section 1006 requires that a material and waste management plan be developed to ensure that such buildings are d and demolished in such a manner that at least 50 percent of materials are diverted from landfills.

Where indicated to be enforceable in the jurisdiction in Table 302.1, Section 1007.2 requires post certificate of occupancy zEPI, energy demand and CO_2e emissions reporting.

Chapter 11 Existing Building Site Development. While Chapter 10 is applicable to existing buildings, Chapter 11 is applicable to additions to and to the alteration, repair, maintenance and operation of the sites upon which those buildings are located. Conceptually, much like Chapter 10, the requirements of Chapter 11 of the IgCC are based upon the requirements of Chapter 34 of the *International Building Code* for existing buildings. These provisions are not retroactive. They apply only where buildings are altered or added to.

Additions are essentially handled as new construction. Alterations must meet the requirements of other applicable chapters of the code for those portions or elements of the building that are being altered.

Section 1105 provides relief for historic buildings under certain conditions.

Chapter 12 Referenced Standards. The code contains numerous references to standards that are used to regulate materials and methods of construction. Chapter 12 contains a comprehensive list of all standards that are referenced in the code. The standards are part of the code to the extent of the reference to the standard (see Sections 102.4 and 102.4.1). Compliance with the referenced standard is necessary for compliance with this code. By providing specifically adopted standards, the construction and installation requirements necessary for compliance with the code can be readily determined. The basis for code compliance is, therefore, established and available on an equal basis to the code official, contractor, designer and owner.

Chapter 12 is organized in a manner that makes it easy to locate specific standards. It lists all of the referenced standards, alphabetically, by acronym of the promulgating agency of the standard. Each agency's standards are then listed in either alphabetical or numeric order based upon the standard identification. The list also contains the title of the standard; the edition (date) of the standard referenced; any addenda included as part of the ICC adoption; and the section or sections of this code that reference the standard.

Appendices. Appendices are provided in the IgCC to offer optional or supplemental criteria to the provisions in the main chapters of the code. Appendices provide additional information and standards not typically administered by all building departments. Appendices have the same force and effect as the first 12 chapters of the IgCC only when they are explicitly adopted by the jurisdiction.

Appendix A Project Electives. Where Appendix A is adopted, it mandates buildings which are "greener" and "more sustainable" than those that meet only the baseline minimum requirements found in the body of the IgCC.

Project electives were created to encourage performance which exceeds the minimum requirements of the IgCC and to encourage, but not mandate, the implementation of green and sustainable practices that are otherwise difficult or impossible to mandate. For example, it would not be realistic to require that all buildings be constructed on brownfield sites. It is, however, environmentally beneficial to encourage construction on brownfield sites. Thus, Appendix A contains a project elective related to brownfield sites.

Project electives encourage the consideration of, but do not require the implementation of, all green and sustainable practices contained in Appendix A. Where green and sustainable practices and provisions are generally suitable as mandatory requirements, they have typically been placed in the body of the IgCC. Green and sustainable practices that are seldom or never appropriate as mandatory requirements for all projects in all regions, or where they are intended to encourage and recognize, but not necessarily require, higher building performance, are typically more appropriately integrated in the code as project electives.

Sections A104 through A108 of Appendix A are arranged by major sections that correspond with the fundamental principles addressed in Chapters 4 through 8 of the IgCC: site; material resource conservation and efficiency; energy conservation, efficiency and earth atmospheric quality; water resource conservation and efficiency; and indoor environmental quality and comfort. In each of these major sections, jurisdictions that intend to enforce Appendix A must determine the number of project electives that must be complied with from the list of project electives tables associated with each of those major sections. Jurisdictions must exercise discretion when determining these minimum values, as it may be difficult or impossible for some projects to comply with various provisions. In addition, if jurisdictions have chosen to enforce certain provisions listed in Table 302.1, they may be unknowingly reducing the number of project electives available in Appendix A. The text of some project electives indicates that they are not available if the jurisdiction has made the practice mandatory in Table 302.1. Thus the specific text of the project electives should be reviewed and coordinated with the jurisdictional requirements from Table 302.1 that are enforced in the jurisdiction. Where jurisdictions have not chosen to enforce various provisions listed in Table 302.1, project electives encourage the implementation of many of the same green and sustainable practices that Table 302.1 addressed.

Although the jurisdiction determines the number of project electives that must be satisfied in each table in Appendix A, the specific project electives to be implemented on each project are selected by the owner and design professional. It is because the specific electives selected can vary from project to project that they are deemed "project" electives, and it is the fact that these provisions are not mandatory until they are selected by the owner and design professional that they are deemed project "electives."

Appendix B Radon Mitigation. Radon comes from the natural radioactive decay of the element radium in soil, rock and water and finds its way into the air. Appendix B contains requirements for the design and construction of systems that mitigate the transfer of radon gases from the soil to building interior spaces.

Appendix C Optional Ordinance. The optional ordinance contained in Appendix C addresses key elements of an evidentiary-based adoption structure that includes performance-bonding requirements. These bonding requirements are tied to the issuance of building permits, certificates of occupancy and the process of compliance verification.

Appendix D Enforcement Procedures. Appendix D is intended to ensure that buildings constructed in accordance with the IgCC are maintained in a manner that is compliant with the code. Appendix D requires that existing buildings that do not comply with these code requirements be altered or repaired to restore compliance with the IgCC.

LEGISLATION

The *International Codes* are designed and promulgated to be adopted by reference by legislative action. Jurisdictions wishing to adopt the 2012 *International Green Construction Code*™ as an enforceable regulation governing structures and premises should ensure that certain factual information is included in the adopting legislation at the time adoption is being considered by the appropriate governmental body. The following sample adoption legislation addresses several key elements, including the information required for insertion into the code text.

SAMPLE LEGISLATION FOR ADOPTION OF THE *INTERNATIONAL GREEN CONSTRUCTION CODE* ORDINANCE NO.______

A[N] **[ORDINANCE/STATUTE/REGULATION]** of the **[JURISDICTION]** adopting the 2012 edition of the *International Green Construction Code*™, regulating and governing the impact of buildings and structures on the environment in the **[JURISDICTION]**; providing for the issuance of permits and collection of fees therefor; repealing **[ORDINANCE/STATUTE/REGULATION]** No. ______ of the **[JURISDICTION]** and all other ordinances or parts of laws in conflict therewith.

The **[GOVERNING BODY]** of the **[JURISDICTION]** does ordain as follows:

Section 1. That a certain document, three (3) copies of which are on file in the office of the **[TITLE OF JURISDICTION'S KEEPER OF RECORDS]** of **[NAME OF JURISDICTION]**, being marked and designated as the *International Green Construction Code*, 2012 edition, including Appendix Chapters **[FILL IN THE APPENDIX CHAPTERS BEING ADOPTED]**, as published by the International Code Council, be and is hereby adopted as the Green Construction Code of the **[JURISDICTION]**, in the State of **[STATE NAME]** regulating and governing the conditions and maintenance of structures and premises as herein provided; the standards for physical things and conditions essential to safeguard the environment, public health, safety and general welfare through the establishment of requirements to reduce the negative impacts and increase the positive impacts of the built environment on the natural environment and building occupants; and each and all of the regulations, provisions, penalties, conditions and terms of said Green Construction Code on file in the office of the **[JURISDICTION]** are hereby referred to, adopted, and made a part hereof, as if fully set out in this legislation, with the additions, insertions, deletions and changes, if any, prescribed in Section 2 of this ordinance.

Section 2. The following sections are hereby revised:

Section 101.1. Insert: **[NAME OF JURISDICTION]**

Table 302.1. Insert: **[JURISDICTIONAL REQUIREMENTS]**

Section 1007.3.3.1. Insert: **[AGENCY RESPONSIBLE]** where Section 1007.3 is selected in Table 302.1.

Section 1007.3.3.2. Insert: **[AGENCY RESPONSIBLE]** where Section 1007.3 is selected in Table 302.1.

Section 1007.3.3.3. Insert: **[AGENCY RESPONSIBLE]** where Section 1007.3 is selected in Table 302.1.

Section 3. That **[ORDINANCE/STATUTE/REGULATION]** No. ______ of **[JURISDICTION]** entitled **[FILL IN HERE THE COMPLETE TITLE OF THE LEGISLATION OR LAWS IN EFFECT AT THE PRESENT TIME SO THAT THEY WILL BE REPEALED BY DEFINITE MENTION]** and all other ordinances or parts of laws in conflict herewith are hereby repealed.

Section 4. That if any section, subsection, sentence, clause or phrase of this legislation is, for any reason, held to be unconstitutional, such decision shall not affect the validity of the remaining portions of this ordinance. The **[GOVERNING BODY]** hereby declares that it would have passed this law, and each section, subsection, clause or phrase thereof, irrespective of the fact that any one or more sections, subsections, sentences, clauses and phrases be declared unconstitutional.

Section 5. That nothing in this legislation or in the Green Construction Code hereby adopted shall be construed to affect any suit or proceeding impending in any court, or any rights acquired, or liability incurred, or any cause or causes of action acquired or existing, under any act or ordinance hereby repealed as cited in Section 3 of this law; nor shall any just or legal right or remedy of any character be lost, impaired or affected by this legislation.

Section 6. That the **[JURISDICTION'S KEEPER OF RECORDS]** is hereby ordered and directed to cause this legislation to be published. (An additional provision may be required to direct the number of times the legislation is to be published and to specify that it is to be in a newspaper in general circulation. Posting may also be required.)

Section 7. That this law and the rules, regulations, provisions, requirements, orders and matters established and adopted hereby shall take effect and be in full force and effect **[TIME PERIOD]** from and after the date of its final passage and adoption.

TABLE OF CONTENTS

CHAPTER 1
SCOPE AND ADMINISTRATION

PART 1—SCOPE AND APPLICATION

SECTION 101
GENERAL

[A] 101.1 Title. These regulations shall be known as the Green Construction Code of [NAME OF JURISDICTION] hereinafter referred to as "this code."

101.2 General. This code is an overlay document to be used in conjunction with the other codes and standards adopted by the jurisdiction. This code is not intended to be used as a standalone construction regulation document and permits are not to be issued under this code. This code is not intended to abridge or supersede safety, health or environmental requirements under other applicable codes or ordinances.

101.3 Scope. The provisions of this code shall apply to the design, construction, addition, alteration, change of occupancy, relocation, replacement, repair, equipment, building site, maintenance, removal and demolition of every building or structure or any appurtenances connected or attached to such buildings or structures and to the site on which the building is located. Occupancy classifications shall be determined in accordance with the *International Building Code*® (IBC®).

Exceptions:

1. The code shall not apply to items 1.1, 1.2 and 1.3 except where the jurisdiction adopts the jurisdictional requirements of Section 302.1, Item 1, for residential buildings.
    1.1. Detached one- and two-family dwellings and multiple single-family dwellings (townhouses) not more than three stories in height above grade plane with a separate means of egress, their accessory structures, and the site or lot upon which these buildings are located.
    1.2. Group R-3 residential buildings, their accessory structures, and the site or lot upon which these buildings are located.
    1.3. Group R-2 and R-4 residential buildings four stories or less in height above grade plane, their accessory structures, and the site or lot upon which these buildings are located.
2. The code shall not apply to equipment or systems that are used primarily for industrial or manufacturing.
3. The code shall not apply to temporary structures *approved* under Section 3103 of the *International Building Code.*
4. Where ASHRAE 189.1 is selected in accordance with Section 301.1.1, ASHRAE 189.1 shall not apply to buildings identified in Exceptions 1 through 3.

101.3.1 Residential construction. In lieu of the requirements of this code the following shall be deemed-to-comply with this code:

1. Group R-2 and R-4 residential buildings five stories or more in height above grade plane, their accessory structures, and the site or lot upon which these buildings are located that comply with ICC 700, with a minimum energy efficiency category requirements of the Silver performance level or equivalent.
2. Group R-2 and R-4 portions of mixed use buildings that comply with ICC 700, with a minimum energy efficiency category requirements of the Silver performance level or equivalent. The remainder of the building and the site upon which the building is located shall comply with the provisions of this code.

101.4 Appendices. Provisions in the appendices shall not apply unless specifically adopted.

101.5 Intent. This code is intended to safeguard the environment, public health, safety and general welfare through the establishment of requirements to reduce the negative impacts and increase the positive impacts of the built environment on the natural environment and building occupants. This code is not intended to abridge or supersede safety, health or environmental requirements under other applicable codes or ordinances.

SECTION 102
APPLICABILITY

102.1 Code conflicts. Where there is a conflict between a general requirement and a specific requirement of this code, the specific requirement shall be applicable. Where, in any specific case, different sections of the code specify different materials, methods of construction or other requirements, the most practical requirement to meet the intent of the code shall govern.

102.2 Other laws. The provisions of this code shall not be deemed to nullify any provisions of local, state or federal law.

102.3 Application of references. References to chapter or section numbers, or to provisions not specifically identified by number, shall be construed to refer to such chapter, section or provision of this code.

102.4 Referenced codes and standards. The following codes shall be considered part of the requirements of this code: the *International Building Code*, the *International Code Council Performance Code*® (ICCPC®), the *International Energy Conservation Code*® (IECC®), the *Interna-*

tional Existing Building Code® (IEBC®), the *International Fire Code®* (IFC®), the *International Fuel Gas Code®* (IFGC®), the *International Mechanical Code®* (IMC®), the *International Plumbing Code®* (IPC®), *International Property Maintenance Code®* (IPMC®), and the *International Residential Code®* (IRC®).

102.4.1 Conflicting provisions. Where the extent of the reference to a referenced code or standard includes subject matter that is within the scope of this code or the International Codes listed in Section 102.4, the provisions of this code or the International Codes listed in Section 102.4, as applicable, shall take precedence over the provisions in the referenced code or standard.

102.5 Partial invalidity. In the event that any part or provision of this code is held to be illegal or void, this shall not have the effect of making void or illegal any of the other parts or provisions.

102.6 Existing structures. The legal occupancy of any structure existing on the date of adoption of this code shall be permitted to continue without change, except as is specifically covered in this code, the *International Building Code*, the *International Existing Building Code*, the *International Property Maintenance Code* or the *International Fire Code*, or as is deemed necessary by the *code official* for the general safety and welfare of building occupants and the public.

102.7 Mixed occupancy buildings. In mixed occupancy buildings, each portion of a building shall comply with the specific requirements of this code applicable to each specific occupancy.

PART 2—ADMINISTRATION AND ENFORCEMENT

SECTION 103
DUTIES AND POWERS OF THE CODE OFFICIAL

103.1 General. The *code official* established in the *International Building Code* is hereby authorized and directed to enforce the provisions of this code. The *code official* shall have the authority to render interpretations of this code and to adopt policies and procedures in order to clarify the application of its provisions and how this code relates to other applicable codes and ordinances. Such interpretations, policies and procedures shall be in compliance with the intent and purpose of this code and other applicable codes and ordinances. Such policies and procedures shall not have the effect of waiving requirements specifically provided for in this code or other applicable codes and ordinances.

103.2 Applications and permits. The *code official* shall enforce compliance with the provisions of this code as part of the enforcement of other applicable codes and regulations, including the referenced codes listed in Section 102.4.

103.3 Notices and orders. The *code official* shall issue all necessary notices or orders to ensure compliance with this code.

103.4 Inspections. The *code official* shall make inspections, as required, to determine code compliance, or the *code official* shall have the authority to accept reports of inspection by approved agencies or individuals. The *code official* is authorized to engage such expert opinion as deemed necessary to report on unusual technical issues that arise, subject to the approval of the appointing authority.

SECTION 104
CONSTRUCTION DOCUMENTS

104.1 Information on construction documents. The content and format of construction documents shall comply with the *International Building Code.*

SECTION 105
APPROVAL

105.1 General. This code is not intended to prevent the use of any material, method of construction, design, system, or innovative approach not specifically prescribed herein, provided that such construction, design, system or innovative approach has been *approved* by the *code official* as meeting the intent of this code and all other applicable laws, codes and ordinances.

105.2 Approved materials and equipment. Materials, equipment, devices and innovative approaches *approved* by the *code official* shall be constructed, installed and maintained in accordance with such approval.

105.2.1 Used materials, products and equipment. The use of used materials, products and equipment that meet the requirements of this code for new materials is permitted. Used equipment and devices shall be permitted to be reused subject to the approval of the *code official.*

105.3 Modifications. Wherever there are practical difficulties involved in carrying out the provisions of this code, the *code official* shall have the authority to grant modifications for individual cases, upon application of the owner or owner's representative, provided the *code official* shall first find that special individual reason makes the strict letter of this code impractical and that the modification is in compliance with the intent and purpose of this code and that such modification does not lessen the minimum requirements of this code. The details of granting modifications shall be recorded and entered in the files of the department.

105.4 Innovative approaches and alternative materials, design, and methods of construction and equipment. The provisions of this code are not intended to prevent the installation of any material or to prohibit any design, innovative approach, or method of construction not specifically prescribed by this code, provided that any such alternative has been *approved.* An alternative material, design, innovative approach or method of construction shall be reviewed and *approved* where the *code official* finds that the proposed design is satisfactory and complies with the intent of the provisions of this code, and that the material, design, method or work offered is, for the purpose intended, at least the equivalent of that prescribed in this code The details of granting the use of alternative materials, designs, innovative approach and methods of construction shall be recorded and entered in the files of the department.

105.4.1 Research reports. Supporting data, where necessary to assist in the approval of materials or assemblies not specifically provided for in this code, shall consist of valid research reports from *approved* sources.

105.4.2 Tests. Wherever there is insufficient evidence of compliance with the provisions of this code, or evidence that a material or method does not conform to the requirements of this code, or in order to substantiate claims for alternative materials or methods, the *code official* shall have the authority to require tests as evidence of compliance to be made at no expense to the jurisdiction. Test methods shall be as specified in this code or by other recognized test standards. In the absence of recognized and accepted test methods, the *code official* shall approve the testing procedures. Tests shall be performed by an approved agency. Reports of such tests shall be retained by the *code official* for the period required for retention of public records.

105.5 Compliance materials. The *code official* shall be permitted to approve specific computer software, worksheets, compliance manuals and other similar materials that meet the intent of this code.

105.6 Approved programs. The *code official* or other authority having jurisdiction shall be permitted to deem a national, state or local program to meet or exceed this code. Buildings *approved* in writing by such a program shall be considered to be in compliance with this code.

105.6.1 Specific approval. The *code official* or authority having jurisdiction shall be permitted to approve programs or compliance tools for a specified application, limited scope or specific locale. For example, a specific approval shall be permitted to apply to a specific section or chapter of this code.

SECTION 106
PERMITS

106.1 Required. Any owner or authorized agent who intends to construct, enlarge, alter, repair, move, demolish, or change the occupancy of a building or structure, or to erect, install, enlarge, alter, repair, remove, convert or replace any energy, electrical, gas, mechanical or plumbing system, the installation of which is regulated by this code, or to cause any such work to be done, shall first make application to the *code official* and obtain the required permit under the applicable code or regulation relevant to the intended work. Separate permits shall not be issued under this code. Exemptions from permit requirements shall not be deemed to grant authorization for any work to be done in any manner in violation of the provisions of this code or any other applicable laws, codes or ordinances of this jurisdiction.

SECTION 107
FEES

107.1 Fees. Fees for permits shall be paid as required, in accordance with the schedule as established by the applicable governing authority for the intended work prescribed in an application.

SECTION 108
BOARD OF APPEALS

108.1 General. Appeals of orders, decisions or determinations made by the *code official* relative to the application and interpretation of this code shall be made to the Board of Appeals created under the applicable *International Code®*.

108.2 Limitations on authority. An application for appeal shall be based on a claim that the true intent of this code or the rules legally adopted there under have been incorrectly interpreted, the provisions of this code do not fully apply or an equivalent or better form of construction is proposed. The board shall have no authority to waive requirements of this code.

108.3 Qualifications. The members of the board of appeals related to interpretation of this code shall be qualified by experience and training in the matters covered by this code and shall not be employees of the jurisdiction.

SECTION 109
CERTIFICATE OF OCCUPANCY

109.1 Violations. Issuance of a certificate of occupancy shall not be construed as an approval of a violation of the provisions of this code or of other ordinances of the jurisdiction.

CHAPTER 2

DEFINITIONS

SECTION 201
GENERAL

201.1 Scope. Unless otherwise expressly stated, the following words and terms shall, for the purposes of this code, have the meanings shown in this chapter.

201.2 Interchangeability. Words used in the present tense include the future; words stated in the masculine gender include the feminine and neuter; the singular number includes the plural and the plural, the singular.

201.3 Terms defined in other codes. Where terms are not defined in this code and are defined in the *International Building Code*® (IBC®), *International Energy Conservation Code*® (IECC®), *International Fire Code*® (IFC®), *International Fuel Gas Code*® (IFGC®), *International Mechanical Code*® (IMC®), *International Plumbing Code*® (IPC®) or *International Residential Code*® (IRC®), such terms shall have the meanings ascribed to them as in those codes.

201.4 Terms not defined. Where terms are not defined through the methods authorized by this section, such terms shall have ordinarily accepted meanings such as the context implies.

SECTION 202
DEFINITIONS

95th-PERCENTILE RAINFALL EVENT. The rainfall event having a precipitation total greater than or equal to 95 percent of all rainfall events during a 24-hour period on an annual basis.

A-WEIGHTED SOUND LEVEL. Sound pressure level in decibels measured with a sound level meter using an A-weighted network.

ADDITION. An extension or increase in floor area or height of a building or structure.

AIR CURTAIN. A device that generates and discharges a laminar air stream installed at the building entrance intended to prevent the infiltration of external, unconditioned air into the conditioned spaces, or the loss of interior, conditioned air to the outside.

ALTERATION. Any construction or renovation to an existing structure other than repair or addition.

ALTERNATE ON–SITE NONPOTABLE WATER. Nonpotable water from other than public utilities, onsite surface sources and subsurface natural freshwater sources. Examples of such water are gray water, onsite reclaimed water, collected rainwater, captured condensate, and rejected water from reverse osmosis systems.

APPROVED. Acceptable to the *code official* or authority having jurisdiction.

APPROVED AGENCY. An established and recognized agency regularly engaged in conducting tests or furnishing inspection services or commissioning services, where such agency has been *approved.*

APPROVED SOURCE. An independent person, firm or corporation, *approved* by the *code official*, who is competent and experienced in the application of engineering principles to materials, methods or systems analyses.

AREA, TOTAL BUILDING FLOOR. The total of the *total floor areas* on all stories of the building.

AREA, TOTAL FLOOR. The total area of a story as measured from the interior side of the exterior walls.

ASBESTOS-CONTAINING PRODUCTS. Building materials containing one or more of the following mineral fibers in any detectable amount that have been intentionally added or are present as a contaminant: chrysotile, amosite, crocidolite, tremolite, actinolite, anthophyllite and any fibrous amphibole.

AUTOMATIC. Self-acting, operating by its own mechanism when actuated by some impersonal influence, such as a change in current strength, pressure, temperature or mechanical configuration (see "Manual").

AUTOMATIC TIME SWITCH CONTROL. A device or system that automatically controls lighting or other loads, including switching ON or OFF, based on time schedules.

BACKWATER VALVE. A device or valve installed in the system drain piping which prevents drainage or waste from backing up into the system and causing contamination or flooding.

BICYCLE PARKING, LONG TERM. Bicycle racks or storage lockers provided for bicycle riders including, but not limited to, employees and students, anticipated to be at a building site for four or more hours.

BICYCLE PARKING, SHORT TERM. Bicycle racks or storage lockers provided for bicycle riders including, but not limited to, customers, visitors, and event audiences, anticipated to be at a building site for less than four hours.

BIO-BASED MATERIAL. A commercial or industrial material or product, other than food or feed, that is composed of, or derived from, in whole or in significant part, biological products or renewable domestic agricultural materials, including plant, animal, and marine materials, or forestry materials.

BRANCH CIRCUIT. All circuit conductors between the final branch-circuit overcurrent device and the load.

BROWNFIELD. A site in which the expansion, redevelopment or reuse of would be required to address the presence or potential presence of a hazardous substance, pollutant or contaminant. *Brownfield* sites include:

1. EPA-recognized *brownfield* sites as defined in Public Law 107-118 (H.R. 2869) "Small Business Liability Relief and Brownfields Revitalization Act," 40 CFR, Part 300; and
2. Sites determined to be contaminated according to local or state regulation.

BTU. Abbreviation for British thermal unit, which is the quantity of heat required to raise the temperature of 1 pound (454 g) of water 1 °F (0.56 °C) (1 Btu = 1055 J).

BUFFER. The number of feet of setback from a wetland or water body determined by a jurisdiction to be necessary to protect a specific wetland or water body. The width of the buffer varies based on characteristics of the wetland and surrounding areas including, but not limited to, the type and function of the wetland, soils, slopes, land uses, habitats, and needs for wildlife or water quality protection.

BUILDING. Any structure used or intended for supporting or sheltering any use or occupancy, including the energy using systems and site subsystems powered through the building's electrical service.

BUILDING COMMISSIONING (See "Commissioning").

BUILDING SITE. A lot, or a combination of adjoining lots, that are being developed and maintained subject to the provisions of this code. A building site shall be permitted to include public ways, private roadways, bikeways and pedestrian ways that are developed as an element of the total development.

BUILDING THERMAL ENVELOPE. The basement walls, exterior walls, floor, roof, and any other building elements that enclose conditioned space. This boundary also includes the boundary between conditioned space and any exempt or unconditioned space.

CAPTIVE KEY CONTROL. An automatic control device or system that energizes circuits when the key that unlocks the sleeping unit is inserted into the device and that de-energizes those circuits when the key is removed.

CARBON DIOXIDE EQUIVALENT (CO_2e) EMISSIONS. A measure used to compare the emissions from various greenhouse gases based upon their 100-year time horizon global warming potential (GWP). CO_2e emissions from carbon dioxide (CO_2), methane (CH_4), and nitrous oxide (N_2O) are included. The carbon dioxide equivalent for a gas is derived by multiplying the weight of the gas by the associated GWP.

CHANGE OF OCCUPANCY. A change in the purpose or level of activity within a building that involves a change in application of the requirements of this code.

CODE OFFICIAL. The officer or other designated authority charged with the administration and enforcement of this code, or a duly authorized representative.

COLLECTION PIPING. Unpressurized piping used within the collection system that drains rainwater or gray water to the storage tank by gravity.

COMBINATION OVEN/STEAMER. A chamber designed for heating, roasting, or baking food by a combination of conduction, convection, radiation, electromagnetic energy or steam.

COMMISSIONING. A process that verifies and documents that the selected building and site systems have been designed, installed, and function in accordance with the owner's project requirements and construction documents, and minimum code requirements

COMPOSITE WOOD PRODUCTS. Hardwood plywood, particleboard, and medium-density fiberboard.

Composite wood products do not include the following:

1. Hardboard and structural plywood as specified in DOC PS-1;
2. Structural panels as specified in DOC PS-2;
3. Structural composite lumber as specified in ASTM D 5456;
4. Oriented strand board and glued laminated timber as specified in ANSI A190.1;
5. Prefabricated wood I-joists as specified in ASTM D 5055; and
6. Finger-jointed lumber.

CONSERVATION AREA. Land designated by the jurisdiction or by state or federal government, as appropriate for conservation from development because of the land possessing natural values important to the community including, but not limited to, wildlife habitat, forest or other significant vegetation, steep slopes, ground water recharge area, riparian corridor or wetland.

CONSTRUCTION-COMPACTED SUBSOIL. Subsoils that are compacted through any of the following: clearing, grading, smearing and topsoil removal such that the infiltrative capacity of the soils or the bulk density of the soils is significantly altered in comparison to the reference soil properties.

CONSTRUCTION DOCUMENTS. Written, graphic and pictorial documents prepared or assembled for describing the design, location and physical characteristics of the elements of a project necessary for obtaining a building permit.

CONTROL. A specialized automatic or manual device or system used to regulate the operation of lighting, equipment or appliances.

CO_2e. Weight of each gas emitted when consuming a specific energy type in the building per unit of the specific energy type provided to the building at the utility meter multiplied by the global warming potential (GWP) of the specific gas, and then summed over all three gases emitted.

where:

GWP (CO_2) = 1

GWP (CH_4) = 25

GWP (N_2O) = 298.

COURT. An open, uncovered space, unobstructed to the sky, bounded on three or more sides by exterior building walls or other enclosing devices.

DAYLIGHT CONTROL. A device or system that provides automatic control of electric light levels based on the amount of daylight in a space.

DAYLIGHT SATURATION. The percentage of daylight hours throughout the year when not less than 28 foot candles (300 lux) of natural light is provided at a height of 30 inches (760 mm) above the floor.

DAYLIT AREA. That portion of a building's interior floor area that is regularly illuminated by natural light.

DECIBELS (dB). Term used to identify ten times the common logarithm of the ratio of two like quantities proportional to the power of energy.

DECONSTRUCTION. The process of systematically disassembling a building, structure, or portion thereof, so that the materials, products, components, assemblies and modules can be salvaged for repurpose, reuse or recycling.

DEMAND LIMIT. The shedding of loads when pre-determined peak demand limits are about to be exceeded.

DEMAND RESPONSE (DR). The ability of a building system to reduce the energy consumption for a specified time period after receipt of demand response signal typically from the power company or demand response provider. Signals requesting demand response are activated at times of peak usage or when power reliability is at risk.

DEMAND RESPONSE, AUTOMATED (AUTO-DR). Fully automated demand response initiated by a signal from a utility or other appropriate entity, providing fully automated connectivity to customer energy end-use control strategies.

DEMAND RESPONSE AUTOMATION INTERNET SOFTWARE. Software that resides in a building energy management control system that can receive a demand response signal and automatically reduce heating, ventilation, air-conditioning (HVAC) and lighting system loads.

DEMOLITION. The process of razing, relocation, or removal of an existing building or structure, or a portion thereof.

DETENTION. The short-term storage of stormwater on a site in order to regulate the runoff from a given rainfall event and to control discharge rates to reduce the impact on downstream stormwater systems.

DISHWASHER.

Dishwasher, door type. A machine designed to clean and sanitize plates, glasses, cups, bowls, utensils, and trays by applying sprays of detergent solution and a sanitizing final rinse, that is designed to accept a standard 20-inch by 20-inch (508 mm by 508 mm) dish rack which requires the raising of a door to place the rack into the wash/rinse chamber.

Dishwasher, multiple tank conveyor. A machine designed to clean and sanitize plates, glasses, cups, bowls, utensils, and trays by applying sprays of detergent solution and a sanitizing final rinse, using a conveyor or similar mechanism to carry dishes through a series of wash and rinse sprays utilizing one or more tanks within the machine. This type of machine may include a prewashing section before the washing section and an auxiliary rinse section between the power rinse and final rinse section.

Dishwasher, pot pan and utensil. A machine designed to clean and sanitize pots, pans, and kitchen utensils by applying sprays of detergent solutions and a sanitizing final rinse.

Dishwasher, rackless conveyor. A machine designed to clean and sanitize plates, glasses, cups, bowls, utensils, and trays by applying sprays of detergent solution and a sanitizing final rinse, using a conveyor or similar mechanism to carry dishes through a series of wash and rinse sprays within the machine. Rackless conveyor machines utilize permanently installed, vertical pegs to carry dishware through the wash and rinse cycles.

Dishwasher, single tank conveyor. A machine designed to clean and sanitize plates, glasses, cups, bowls, utensils, and trays by applying sprays of detergent solution and a sanitizing final rinse, using a conveyor or similar mechanism to carry dishes through a series of wash and rinse sprays within the machine. This type of machine does not have a pumped rinse tank but may include a prewashing section ahead of the washing section.

Dishwasher, under counter. A machine designed to clean and sanitize plates, glasses, cups, bowls, utensils, and trays by applying sprays of detergent solution and a sanitizing final rinse, that has an overall height 38 inches (965 mm) or less, designed to be installed under food preparation workspaces.

DISTRIBUTION PIPE. Pressurized or nonpressure piping used within the plumbing system.

DIVERSE USE CATEGORIES. Categories of occupancies and land uses which are designated as either community, retail or service facilities:

Community facilities. The community facilities category includes: child care; civic or community center; a building containing a place of worship; police or fire station; post office, public library, public park, school, senior care facility, homeless shelter, and similar social services facilities.

Retail uses. The retail use category includes: convenience store, florist, hardware store, pharmacy, grocery or supermarket and similar retail uses.

Service uses. The service use category includes: bank, coffee shop or restaurant; hair care; health club or fitness center; laundry or dry cleaner, medical or dental office and similar service uses.

DWELLING UNIT. A single unit providing complete, independent living facilities for one or more persons, including permanent provisions for living, sleeping, eating, cooking and sanitation.

ENERGY MANAGEMENT AND CONTROL SYSTEM, BUILDING (EMCS). A computerized, intelligent network of electronic devices, designed to automatically monitor and control the energy using systems in a building.

ENERGY STAR. A joint program of the U.S. Environmental Protection Agency (EPA) and the U.S. Department of Energy (DOE) designed to identify and promote energy-efficient products and practices.

ENERGY STAR QUALIFIED. Appliances or equipment that have been found to comply with ENERGY STAR requirements by a third-party organization recognized by the U.S. Environmental Protection Agency (EPA) and the U.S. Department of Energy (DOE).

EQUIPMENT. All piping, ducts, vents, control devices and other components of systems other than appliances which are permanently installed and integrated to provide control of environmental conditions for buildings. This definition shall also include other systems specifically regulated in this code.

EVAPORATIVE COOLING SYSTEM. A system for cooling the air in a building or space by removing heat from the outdoor air by means of the evaporation of water. The system forces air through wet porous pads, causing the latent heat of evaporation to cool the air. Water is continuously circulated over the pads to replenish the evaporated water. Where the cooled air is sent directly into the building, the system is referred to as "direct evaporative cooling." Where the cooled air is sent through heat exchangers re-circulating indoor air, the system is referred to as "indirect evaporative cooling."

EXISTING BUILDING. A building erected prior to the date of adoption of the appropriate code, or one for which a legal building permit has been issued.

EXISTING STRUCTURE. A structure erected prior to the date of adoption of the appropriate code, or one for which a legal building permit has been issued.

EXTERIOR WALL, OBSTRUCTED. That portion of an exterior wall with limited access to natural light due to shading from buildings, structures, or geological formations,

FACILITY OPERATIONS. A facility is operational during the time when the primary activity that facility is designed for is taking place. For Group A and Group M occupancies, this is the time during which the facility is open to the public.

FAN EFFICIENCY GRADE (FEG). A numerical rating identifier that specifies the fan's aerodynamic ability to convert shaft power, or impeller power in the case of a direct driven fan, to air power. FEGs are based on fan peak (optimum) energy efficiency that indicates the quality of the fan energy usage and the potential for minimizing the fan energy usage.

FARMLAND.

Farmlands of statewide significance. Land, in addition to prime and unique farmlands, that is of statewide importance for the production of food, feed, fiber, forage and oil seed crops. Criteria for delineating this land is determined by the appropriate state agency.

Prime farmland. Land that has the best combination of physical and chemical characteristics for producing food, fiber, feed, forage, and oil seed crops and that is also available for these uses, including cropland, pastureland, forest land, range land and similar lands which are not water areas or urban or built-up land areas.

Unique farmland. Land other than prime farmland that is used for the production of specific high-value food or fiber crops. The land has the special combination of soil quality, location, growing season and moisture supply needed to economically produce sustained high-quality crops or high yields of a specific crop where the lands are treated and managed according to acceptable farming methods.

FEEDER CONDUCTORS. The circuit conductors between the service equipment, the source of a separately derived system, or other power supply source and the final branch-circuit overcurrent device.

FENESTRATION. Skylights, roof windows, vertical windows (fixed or moveable), opaque doors, glazed doors, glazed block, and combination opaque/glazed doors. Fenestration includes products with glass and nonglass glazing materials.

FIBER PROCUREMENT SYSTEM. A system that ensures that fiber procured for the manufacture of wood and wood-based products comes from responsible or certified sources in accordance with ASTM D 7612.

FIREPLACE. An assembly consisting of a hearth and fire chamber of noncombustible material and provided with a chimney for use with solid fuels.

Factory-built fireplace. A listed and labeled fireplace and chimney system composed of factory-made components, and assembled in the field in accordance with the manufacturer's instructions and the conditions of the listing.

Masonry fireplace. A field-constructed fireplace composed of solid masonry units, bricks, stones or concrete.

FLOOD HAZARD AREA. The greater of the following two areas:

1. The area within a *floodplain* subject to a 1-percent or greater chance of flooding in any given year;
2. The area designated as a *flood hazard area* on a community's flood hazard map, or otherwise legally designated.

FLOOD OR FLOODING. A general and temporary condition of partial or complete inundation of normally dry land from:

1. The overflow of inland or tidal waters.
2. The unusual and rapid accumulation of runoff of surface waters from any source.

FLOODPLAIN. An area of land at risk of being inundated with water during high flows. *Floodplains* are associated with both water courses, such as rivers and streams, and bodies of water, such as oceans and lakes.

FLOOR AREA, NET. The actual occupied area not including unoccupied accessory areas such as corridors, stairways, toilet rooms, mechanical rooms and closets.

FREEZER. Equipment designed to enclose a space of mechanically cooled and temperature-controlled air used to maintain prescribed frozen food holding temperatures.

FRYER, DEEP FAT. A unit with a width between 12 and 18 inches (305 and 457 mm) that cooks food by immersion in a tank of oil or fat more than 25 pounds (11 kg) and less than 50 pounds (23 kg).

FRYER, LARGE VAT. A unit with a width greater than 18 inches (457 mm) that cooks food by immersion in a tank of oil or fat more than 50 pounds (23 kg).

GLOBAL WARMING POTENTIAL (GWP). The cumulative radiative forcing effects of a gas over a 100-year time horizon resulting from the emission of a unit mass of gas relative to a reference gas. The GWP-weighted emissions of direct greenhouse gases in the U.S. Inventory are presented in terms of equivalent emissions of carbon dioxide (CO_2), using units of teragrams of carbon dioxide equivalents ($TgCO_2$ Eq.). conversion: $Tg=10^9$ kg = 10^6 metric tons = 1 million metric tons.

GRAY WATER. Untreated waste water that has not come into contact with waste water from water closets, urinals, kitchen sinks, or dishwashers. Gray water includes, but is not limited to, waste water from bathtubs, showers, lavatories, clothes washers, and laundry trays.

GREENFIELD. Land that has not been previously developed or has a history of only agricultural use.

GREENHOUSE GAS. A gas in the atmosphere that absorbs and emits radiation within the thermal infrared range.

GRIDDLE, DOUBLE-SIDED. Equipment used to cook food between flat, smooth, or grooved horizontal plates heated from above and underneath.

GRIDDLE, SINGLE-SIDED. Equipment used to cook food directly on a flat, smooth, or grooved horizontal plate heated from underneath.

GROUND SOURCE OR GEOEXCHANGE. Where the earth is used as a heat sink in air conditioning or heat pump island systems. This also applies to systems utilizing subsurface water. Ground source heating and cooling uses the relatively constant temperature of the earth below the frost line. This steady temperature profile allows the earth to be used as a heat source in the winter and as a heat sink in the summer.

HARDSCAPE. Areas of a building site covered by manmade materials.

HIGH-OCCUPANCY VEHICLE. A vehicle which is occupied by two or more people, including carpools, vanpools, and buses.

HISTORIC BUILDINGS. Buildings that are listed in or eligible for listing in the National Register of Historic Places, or designated as historic under an appropriate state or local law.

ICE MACHINE.

Ice machine, ice-making head. A factory-made assembly consisting of a condensing unit and ice-making section operating as an integrated unit, with means for making and harvesting ice, that combines the ice-making mechanism and the condensing unit in a single package, but requires a separate ice storage bin.

Ice machine, remote-condensing unit. A factory-made assembly consisting of a condensing unit and ice-making section operating as an integrated unit, with means for making and harvesting ice, where the ice-making mechanism and condenser or condensing unit are in separate sections.

Ice machine, self-contained unit. A factory-made assembly consisting of a condensing unit and ice-making section operating as an integrated unit, with means for making and harvesting ice and where the ice-making mechanism and storage compartment are combined into an integral cabinet.

IMPERVIOUS SURFACE. Paved concrete or asphalt and other similar surfaces that readily accommodate the flow of water with relatively little absorption, as typically used at exterior horizontal areas including, but not limited to, parking lots, bikeways, walkways, plazas and fire lanes.

INDEPENDENT SYSTEM OPERATOR (ISO). The electric system's operator.

INFEASIBLE. An alteration of a building, site feature, or system that has little likelihood of being accomplished because existing physical or site constraints prohibit modification or addition of elements, spaces or features which are in full and strict compliance with the minimum requirements for new construction.

INFILL SITE. Infill sites are one of the following:

1. A vacant lot, or collection of adjoining lots, located in an established, developed area that is already served by existing infrastructure;
2. A previously developed lot or a collection of previously developed adjoining lots, that is being redeveloped or is designated for redevelopment.

INFRASTRUCTURE. Facilities within a jurisdiction that provide community services and networks for travel and communication including: transportation services such as, but not limited to roads, bikeways and pedestrian ways and transit services; utility systems such as, but not limited to, water, sanitary sewage, stormwater management, telecommunications, power distribution and waste management; and community services such as, but not limited to, public safety, parks, schools and libraries.

INFRASTRUCTURE, ADEQUATE. The capacity of infrastructure systems, as determined by the jurisdiction, to serve the demands imposed by a new development on building sites without negatively impacting services to existing users of the infrastructure and without negatively impacting the overall functionality of the infrastructure. Adequacy can be determined based on existing infrastructure or on the infrastructure as augmented by a development project.

INVASIVE PLANT SPECIES. Species that are not native to the ecosystem under consideration and that cause, or are likely to cause, economic or environmental harm or harm to human, animal or plant health, defined by using the best scientific knowledge of that region. Consideration for inclusion as an invasive species shall include, but shall not be limited to, those species identified on:

1. *Approved* city, county or regional lists.
2. State noxious weeds laws,
3. Federal noxious weeds laws.

JURISDICTION. The governmental unit that has adopted this code under due legislative authority.

LABEL. An identification applied on a product by the manufacturer that contains the name of the manufacturer, the function and performance characteristics of the product or material, and the name and identification of an approved agency and that indicates that the representative sample of the product or material has been tested and evaluated by an approved agency.

LABELED. Equipment, materials or products to which has been affixed a label, seal, symbol or other identifying mark of a nationally recognized testing laboratory, inspection agency or other organization concerned with product evaluation that maintains periodic inspection of the production of the above-labeled items and whose labeling indicates either that the equipment, material or product meets identified standards or has been tested and found suitable for a specified purpose.

LIFE CYCLE ASSESSMENT (LCA). A technique to evaluate the relevant energy and material consumed and environmental emissions associated with the entire life of a building, product, process, material, component, assembly, activity or service.

LIGHTING BOUNDARY. Where the lot line abuts a public walkway, bikeway, plaza, or parking lot, the *lighting boundary* shall be a line 5 feet (1524 mm) from the lot line and located on the public property. Where the lot line abuts a public roadway or public transit corridor, the *lighting boundary* shall be the centerline of the public roadway or public transit corridor. In all other circumstances, the *lighting boundary* shall be at the lot line.

LISTED. Equipment, materials, products or services included in a list published by an organization acceptable to the *code official* and concerned with evaluation of products or services that maintains periodic inspection of production of listed equipment or materials or periodic evaluation of services and whose listing states either that the equipment, material, product or service meets identified standards or has been tested and found suitable for a specified purpose.

LOT. A portion or parcel of land considered as a unit.

LOT LINE. A line dividing one lot from another, or from a street or any public place.

LOW EMISSION, HYBRID AND ELECTRIC VEHICLE. Vehicles that achieve EPA Tier 2, California LEV-II, or a minimum of EPA LEV standards, whether by means of hybrid, alternative fuel, or electric power.

LOW VOLTAGE DRY-TYPE DISTRIBUTION TRANSFORMER. A NEMA 'Class 1' transformer that is air-cooled, does not use oil as a coolant, has an input voltage $\leq$ 600 volts, and is rated for operation at a frequency of 60 hertz.

MANUAL. Capable of being operated by personal intervention (see "Automatic").

MINIMUM EFFICIENCY REPORTING VALUE (MERV). Minimum efficiency-rated value for the effectiveness of air filters.

METER. A measuring device used to collect data and indicate usage.

MODIFIED ENERGY FACTOR (MEF). The capacity in cubic feet of the clothes container of a clothes washing machine, C, divided by the clothes washing total energy consumption in kWh per cycle. Total energy consumption per cycle is the sum of the machine electrical energy consumption per cycle, M; the hot water energy consumption per cycle, E; and the energy required for removal of the remaining moisture in the wash load per cycle, D. The equation is:

$MEF = C/(M + E + D)$

MUNICIPAL RECLAIMED WATER. Reclaimed water treated by a municipality.

NATIVE PLANT SPECIES. Species that are native to the ecosystem under consideration, defined by using the best scientific knowledge of that region. Consideration for inclusion as a native species shall include, but is not limited to, those species identified in any of the following:

1. *Approved* city, county and regional lists.
2. State laws.
3. Federal laws.

NONPOTABLE WATER. Water not safe for drinking, personal or culinary utilization.

OCCUPANT LOAD. The occupant load as calculated in accordance with the requirements of Chapter 10 of the *International Building Code.*

OCCUPANT SENSOR CONTROL. A device or system that detects the presence or absence of people within an area and causes lighting, equipment, or appliances to be regulated accordingly.

ONCE-THROUGH COOLING. The use of water as a cooling medium where the water is passed through a heat exchanger one time and then discharged to the drainage system. This also includes the use of water to reduce the temperature of condensate or process water before discharging it to the drainage system.

ORGANIC MATTER. Carbon-containing material composed of both living organisms and formerly living, decomposing plant and animal matter. Soil organic matter content is either naturally occurring or is a result of supplementation with compost or other partially decomposed plant and animal material.

OUTDOOR ORNAMENTAL FOUNTAIN. An outdoor fixture whose dominant use is aesthetic consisting of a catch basin, reservoir or chamber from which one or more jets or streams of water is emitted.

OVEN, CONVECTION. A chamber designed for heating, roasting, or baking food by conduction, convection, radiation, and/or electromagnetic energy.

PERMIT. An official document or certificate issued by the jurisdiction which authorizes performance of a specified activity.

PERVIOUS CONCRETE. Hydraulic cement concrete with distributed, interconnected macroscopic voids that allows water to pass through the material with little resistance.

POST-CONSUMER RECYCLED CONTENT. The proportion of recycled material in a product generated by households or by commercial, industrial, and institutional facilities in their role as end users of the product that can no longer be used for its intended purpose. This includes returns of material from the distribution chain.

POTABLE WATER. Water free from impurities present in amounts sufficient to cause disease or harmful physiological effects and conforming to the bacteriological and chemical quality requirements of the Public Health Service Drinking Water Standards or the regulations of the public health authority having jurisdiction.

POWER CONVERSION SYSTEM. The equipment used to convert incoming electrical power, to the force causing vertical motion of the elevator. In a traction system, this would include the electrical drive, motor, and transmission.

PRECONSUMER (POST-INDUSTRIAL) RECYCLED CONTENT. The proportion of recycled material in a product diverted from the waste stream during the manufacturing process. Preconsumer recycled content does not include reutilization of material such as rework, regrind, or scrap generated in a process and capable of being reclaimed within the same process that generated it.

PRIMARY ENERGY USE. The total fuel-cycle energy embedded within building materials and all forms of energy required for building operation. Units of energy are reported in total Btu's for building materials and total Btu's per unit of energy (e.g., kWh, therms and gallons) consumed in the operation of building mechanical systems (HVAC and lighting). Total fuel-cycle energy includes energy required from the point of initial extraction, through processing and delivery to the final point of consumption into building materials or building operation.

PROCESS LOADS. Building energy loads that are not related to building space conditioning, lighting, service water heating or ventilation for human comfort.

PROJECTION FACTOR. A ratio that describes the geometry of a horizontal projection, as determined in accordance with Equation 4-2 of Section C402.3.3 of the *International Energy Conservation Code*.

PROPOSED DESIGN. A description of the proposed building used to estimate annual energy use for determining compliance based on total building performance including improvements in design such as the use of passive solar energy design concepts and technologies, improved *building thermal envelope* strategies, increased equipment and systems efficiency, increased use of daylighting, improved control strategies and improved lighting sources that will result in a decrease in annual energy.

R-VALUE (THERMAL RESISTANCE). The inverse of the time rate of heat flow through a body from one of its bounding surfaces to the other surface for a unit temperature difference between the two surfaces, under steady state conditions, per unit area ($h \times ft^2 \times °F/Btu$) [$(m^2 \times K)/W$].

RAINWATER. Water from natural precipitation.

RAINWATER COLLECTION AND CONVEYANCE SYSTEM. Rainwater collection system components extending between the collection surface and the storage tank that convey collected rainwater, usually through a gravity system.

REBOUND AVOIDANCE, EXTENDED AUTO-DR CONTROL. The rebound avoidance, extended Auto-DR control strategy is essentially an extension of the rebound avoidance, slow recovery strategy. Although a slow recovery strategy is critical to maximize the benefit of an Auto-DR strategy, the building energy management and control system (EMCS) programming for just such a strategy can be very complex or might not be possible for many conventional EMCS's. A rebound avoidance, extended Auto-DR control strategy also includes logic and controls for avoiding a rebound peak when the control signal is stopped.

REBOUND AVOIDANCE, SEQUENTIAL EQUIPMENT RECOVERY. Sequential equipment recovery that disperses short duration equipment start up spikes gradually, thereby avoiding a larger whole building demand spike.

REBOUND AVOIDANCE, SLOW RECOVERY. Slow recovery strategies slowly recover the target parameter that was controlled in the demand response strategy. Where this strategy is applied, the zone setpoints are gradually restored to the normal setpoints. Where air moving systems are targeted, a limit strategy is applied to the adjustable speed drives; fan adjustable speed drive limits are gradually shifted up.

RECEIVING WATERS. Groundwater, creeks, streams, rivers, lakes or other water bodies that receive treated or untreated waste water or stormwater, including water from combined sewer systems and stormwater drains.

RECLAIMED WATER. Nonpotable water that has been derived from the treatment of waste water by a facility or system licensed or permitted to produce water meeting the jurisdiction's water requirements for its intended uses. Also known as "Recycled Water."

RECYCLABILITY. Ability of a material or product to be captured and separated from a waste stream for conversion, reprocessing or reuse.

REFRIGERATOR. Equipment designed to enclose a space of mechanically cooled and temperature-controlled air used to maintain prescribed cold food holding temperatures.

REGISTERED DESIGN PROFESSIONAL. An individual who is registered or licensed to practice their respective design profession as defined by the statutory requirements of the professional registration laws of the state or jurisdiction in which the project is to be constructed.

REGISTERED DESIGN PROFESSIONAL IN RESPONSIBLE CHARGE. A *registered design professional* engaged by the owner to review and coordinate certain aspects of the project, as determined by the building official, for compatibility with the design of the building or structure, including submittal documents prepared by others, deferred submittal documents and phased submittal documents.

RELOCATABLE (RELOCATED) MODULAR BUILDING. A partially or completely assembled building using a modular construction process and designed to be reused or repurposed multiple times and transported to different building sites.

RENEWABLE ENERGY CREDIT (REC). An REC represents the property rights to the environmental, social, and other nonpower qualities of renewable electricity generation. An REC, and its associated attributes and benefits, is sold separately from the underlying physical electricity associated with an onsite renewable energy source. REC's allow organizations to support renewable energy development and protect the environment where renewable power products are not locally available. There are two approaches to verifying REC ownership and the right to make environmental claims: (1) REC contracts from a list of *approved* providers, including an audit of the chain of custody; and (2) REC tracking systems.

RENEWABLE ENERGY SOURCE, ONSITE. Energy derived from solar radiation, wind, waves, tides, biogas, biomass, or geothermal energy. The energy system providing onsite renewable energy is located on or adjacent to the building site, and generate energy for use on the building site or to send back to the energy supply system.

REPAIR. The reconstruction or renewal of any part of an existing building or building site for the purpose of its maintenance.

RETENTION (STORMWATER). The permanent holding of stormwater on a site, preventing the water from leaving the site as surface drainage and allowing for use of the water on site, or loss of the water through percolation, evaporation or absorption by vegetation.

REUSE. To divert a material, product, component, module, or a building from the waste stream in order to use it again.

ROOF COVERING. The covering applied to the roof deck for weather resistance, fire classification or appearance.

ROOF REPLACEMENT. The process of removing the existing roof covering, repairing any damaged substrate and installing a new roof covering.

ROOF WASHER. A device or method for removal of sediment and debris from collection surface by diverting initial rainfall from entry into the storage tank. Also referred to as a First Flush Device.

SEQUENCE OF OPERATIONS (HVAC). A fully descriptive detailed account of the intended operation of HVAC systems covering the operation of systems in narrative terms, accounting for all of the equipment that makes up the systems, how the systems are designed to operate, and how they are to be controlled.

SITE DISTURBANCE. Site preparation or construction which negatively affects the native soils, native vegetation, or native animal life of the site

SKYLIGHTS AND SLOPED GLAZING. Glass or other transparent or translucent glazing material installed at a slope of less than 60 degrees (1.05 rad) from horizontal. Glazing material in skylights, including unit skylights, tubular daylighting devices, solariums, sunrooms, roofs and sloped walls, are included in this definition.

SKYLIGHT, UNIT. A factory-assembled, glazed fenestration unit, containing one panel of glazing material that allows for natural lighting through an opening in the roof assembly while preserving the weather-resistant barrier of the roof.

SLEEPING UNIT. A room or space in which people sleep, that can also include permanent provisions for living, eating, and either sanitation or kitchen facilities but not both. Such rooms and spaces that are also part of a dwelling unit are not sleeping units.

SOLAR HEAT GAIN COEFFICIENT (SHGC). The ratio of the solar heat gain entering the space through the fenestration assembly to the incident solar radiation. Solar heat gain includes directly transmitted solar heat and absorbed solar radiation which is then reradiated, conducted or convected into the space.

SOLAR PHOTOVOLTAIC SYSTEM. Devices such as photovoltaic (PV) modules and inverters that are used to transform solar radiation into energy.

SOLAR REFLECTANCE. A measure of the ability of a surface material to reflect sunlight. It is the fraction of incident sunlight reflected by a surface, expressed on a scale of 0 to 1. Solar reflectance is also referred to as "albedo."

SOLAR REFLECTANCE INDEX (SRI). A value that incorporates both solar reflectance and thermal emittance in a single measure to represent a surface's relative temperature in the sun. SRI compares a surface's temperature to those of standard black and standard white surfaces. It typically ranges from 0 for standard black to 100 for standard white, but can be less than 0 or greater than 100.

SOLAR THERMAL EQUIPMENT. A device that uses solar radiation to heat water or air for use within the facility for service water heating, process heat, space heating or space cooling.

STANDARD REFERENCE DESIGN. A building design that meets the minimum requirements of the *International Energy Conservation Code* and the additional requirements of Section 602.2.

STANDBY MODE (ELEVATOR). An operating mode during periods of inactivity in which electrical loads are reduced to conserve energy. For elevators, standby mode begins up to 5 minutes after an elevator is unoccupied and has parked and completed its last run and ends when the doors are re-opened. For escalators and moving walkways, standby mode begins after traffic has been absent for up to 5 minutes and ends when the next passenger arrives.

STEAM COOKER. Equipment in which potable steam is used for heating, cooking, and reconstituting food.

STORAGE TANK (GRAY WATER OR RAINWATER). A fixed container for holding water at atmospheric pressure for subsequent use as part of a plumbing or irrigation system.

STORY. That portion of a building included between the upper surface of a floor and the upper surface of the floor or roof next above. It is measured as the vertical distance from top to top of two successive tiers of beams or finished floor surfaces and, for the topmost story, from the top of the floor finish to the top of the ceiling joists or, where there is not a ceiling, to the top of the roof rafters.

STRUCTURE. That which is built or constructed.

SUBSTANTIAL IMPROVEMENT. Any repair, reconstruction, rehabilitation, addition or improvement of a building or structure, the cost of which equals or exceeds 50 percent of the market value of the structure before the improvement or repair is started. If the structure has sustained substantial damage, any repairs are considered *substantial improvement* regardless of the actual repair work performed. The term does not include either of the following:

1. Any project for improvement of a building required to correct existing health, sanitary or safety code violations identified by the *code official* and that are the minimum necessary to assure safe living conditions.
2. Any alteration of a historic structure provided that the alteration will not preclude the structure's continued designation as a historic structure.

THERMAL EMITTANCE. The ratio of radiative power emitted by a sample to that emitted by a black body radiator at the same temperature.

TOPSOIL. The upper, outmost layer of soil having the highest concentration of organic matter and microorganisms and where the majority of biological soil activity occurs.

TRACTION ELEVATOR. An elevator system in which the cars are suspended by ropes wrapped around a sheave that is driven by an electric motor.

TRANSIT SERVICE. A service that a public transit agency serving the area has committed to provide including, but not limited to, bus, streetcar, light or heavy rail, passenger ferry or tram service.

TUBULAR DAYLIGHTING DEVICE (TDD). A nonoperable fenestration unit primarily designed to transmit daylight from a roof surface to an interior space via a tubular conduit. The basic unit consists of an exterior glazed weathering surface, a light-transmitting tube with a reflective interior surface, and an interior-sealing device such as a translucent panel. The unit is either factory assembled, or field assembled from a manufacturing kit.

***U*-FACTOR (THERMAL TRANSMITTANCE).** The coefficient of heat transmission (air to air) through a building component or assembly, equal to the time rate of heat flow per unit area and unit temperature difference between the warm side and cold side air films (Btu/h $\times$ ft^2 $\times$ °F) [W/(m^2 $\times$ K)].

VEGETATIVE ROOF. An assembly of interacting components designed to waterproof and normally insulate a building's top surface that includes, by design, vegetation and related landscaping elements.

VENTILATION. The natural or mechanical process of supplying conditioned or unconditioned air to, or removing such air from, any space.

VISIBLE TRANSMITTANCE (VT). The ratio of visible light entering the space through the fenestration product assembly to the incident visible light. VT includes the effects of glazing material and frame and is expressed as a number between 0 and 1.

VOLATILE ORGANIC COMPOUND (VOC). A volatile chemical compound based on carbon chains or rings that typically contain hydrogen and sometimes contain oxygen, nitrogen and other elements, and that has a vapor pressure of greater than 0.1 mm of mercury at room temperature.

VOLTAGE DROP. A decrease in voltage caused by losses in the circuit conductors connecting the power source to the load.

WATER FACTOR (WF). The quantity of water, in gallons per cycle (Q), divided by a clothes washing machine clothes container capacity in cubic feet (C). The equation is:

$WF = Q/C$

WATER FEATURE. An outdoor open water installation or natural open water way within a built landscape to retain water supplied from source other than rainwater naturally flowing into the feature.

WATERSENSE. A program of the U.S. Environmental Protection Agency (EPA) designed to identify and promote water-efficient products and practices.

WETLAND. Areas that are inundated or saturated by surface or groundwater at a frequency and duration sufficient to support, and that under normal circumstances do support, a prevalence of vegetation typically adapted for life in saturated soil conditions.

ZERO ENERGY PERFORMANCE INDEX (zEPI). A scalar representing the ratio of energy performance of the proposed design compared to the average energy performance of buildings relative to a benchmark year.

CHAPTER 3

JURISDICTIONAL REQUIREMENTS AND LIFE CYCLE ASSESSMENT

SECTION 301 GENERAL

301.1 Scope. This chapter contains requirements that are specific to and selected by the jurisdiction and provisions for whole building life cycle assessment.

301.1.1 Application. The requirements contained in this code are applicable to buildings, or portions of buildings. As indicated in Section 101.3, these buildings shall meet either the requirements of ASHRAE 189.1 or the requirements contained in this code.

301.2 Jurisdictional requirements. This chapter requires that the jurisdiction indicate in Table 302.1 whether specific provisions are mandatory for all buildings regulated by this code and, where applicable, the level of compliance required. All other provisions of this code shall be mandatory as applicable.

SECTION 302 JURISDICTIONAL REQUIREMENTS

302.1 Requirements determined by the jurisdiction. The jurisdiction shall indicate the following information in Table 302.1 for inclusion in its code adopting ordinance:

1. The jurisdiction shall indicate whether requirements for residential buildings, as indicated in Exception 1 to Section 101.3, are applicable by selecting "Yes" or "No" in Table 302.1. Where "Yes" is selected, the provisions of ICC 700 shall apply and the remainder of this code shall not apply.
2. Where the jurisdiction requires enhanced energy performance for buildings designed on a performance basis, the jurisdiction shall indicate a zEPI of 46 or less in Table 302.1 for each occupancy required to have enhanced energy performance.
2. Where "Yes" or "No" boxes are provided, the jurisdiction shall check the box to indicate "Yes" where that section is to be enforced as a mandatory requirement in the jurisdiction, or "No" where that section is not to be enforced as a mandatory requirement in the jurisdiction.

302.1.1 zEPI of 46 or less. Where a zEPI of 46 or less is indicated by the jurisdiction in Table 302.1, buildings shall comply on a performance-basis in accordance with Section 601.3.1.

Exception: Buildings less than 25,000 square feet (2323 m^2) in *total building floor area* pursuing compliance on a prescriptive basis shall be deemed to have a zEPI of 51 and shall not be required to comply with the zEPI of Jurisdictional Choice indicated by the jurisdiction in Table 302.1.

SECTION 303 WHOLE BUILDING LIFE CYCLE ASSESSMENT

303.1 Whole building life cycle assessment. Where a whole building life cycle assessment is performed in accordance with Section 303.1, compliance with Section 505 shall not be required. The requirements for the execution of a whole building life cycle assessment shall be performed in accordance with the following:

1. The assessment shall demonstrate that the building project achieves not less than a 20-percent improvement in environmental performance for global warming potential and at least two of the following impact measures, as compared to a reference design of similar usable floor area, function and configuration that meets the minimum energy requirements of this code and the structural requirements of the *International Building Code*. For relocatable buildings, the reference design shall be comprised of the number of reference buildings equal to the estimated number of uses of the relocatable building.
    - 1.1. Primary energy use.
    - 1.2. Acidification potential.
    - 1.3. Eutrophication potential.
    - 1.4. Ozone depletion potential.
    - 1.5. Smog potential.
2. The reference and project buildings shall utilize the same life cycle assessment tool.
3. The life cycle assessment tool shall be *approved* by the *code official*.
4. Building operational energy shall be included. For relocatable buildings, an average building operational energy shall be estimated to reflect potential changes in location, siting, and configuration by adding or subtracting modules, or function.
5. Building process loads shall be permitted to be included.
6. Maintenance and replacement schedules and actions for components shall be included in the assessment. For relocatable buildings, average transportation energy, material and waste generation associated with reuse of relocatable buildings shall be included in the assessment.
7. The full life cycle, from resource extraction to demolition and disposal, including but not limited to, onsite

construction, maintenance and replacement, relocation and reconfiguration, and material and product embodied acquisition, process and transportation energy, shall be assessed.

Exception: Electrical and mechanical equipment and controls, plumbing products, fire detection and alarm systems, elevators and conveying systems shall not be included in the assessment.

8. The complete building envelope, structural elements, inclusive of footings and foundations, and interior walls, floors and ceilings, including interior and exterior finishes, shall be assessed to the extent that data are available for the materials being analyzed in the selected life cycle assessment tool.
9. The life cycle assessment shall conform to the requirements of ISO 14044.

TABLE 302.1
REQUIREMENTS DETERMINED BY THE JURISDICTION

Section	Section Title or Description and Directives	Jurisdictional Requirements	
CHAPTER 1. SCOPE			
101.3 Exception 1.1	Detached one- and two-family dwellings and multiple single-family dwellings (townhouses) not more than three stories in height above grade plane with a separate means of egress, their accessory structures, and the site or lot upon which these buildings are located, shall comply with ICC 700.	☐ Yes	☐ No
101.3 Exception 1.2	Group R-3 residential buildings, their accessory structures, and the site or lot upon which these buildings are located, shall comply with ICC 700.	☐ Yes	☐ No
101.3 Exception 1.3	Group R-2 and R-4 residential buildings four stories or less in height above grade plane, their accessory structures, and the site or lot upon which these buildings are located, shall comply with ICC 700.	☐ Yes	☐ No
CHAPTER 4. SITE DEVELOPMENT AND LAND USE			
402.2.1	Flood hazard area preservation, general	☐ Yes	☐ No
402.2.2	Flood hazard area preservation, specific	☐ Yes	☐ No
402.3	Surface water protection	☐ Yes	☐ No
402.5	Conservation area	☐ Yes	☐ No
402.7	Agricultural land	☐ Yes	☐ No
402.8	Greenfield sites	☐ Yes	☐ No
407.4.1	High-occupancy vehicle parking	☐ Yes	☐ No
407.4.2	Low-emission, hybrid and electric vehicle parking	☐ Yes	☐ No
409.1	Light pollution control	☐ Yes	☐ No
CHAPTER 5. MATERIAL RESOURCE CONSERVATION AND EFFICIENCY			
503.1	Minimum percentage of waste material diverted from landfills	☐ 50% ☐ 65% ☐ 75%	
CHAPTER 6. ENERGY CONSERVATION, EFFICIENCY AND CO_2e EMISSION REDUCTION			
302.1, 302.1.1, 602.1	zEPI of Jurisdictional Choice – The jurisdiction shall indicate a zEPI of 46 or less in each occupancy for which it intends to require enhanced energy performance.	Occupancy: ______ zEPI: __________	
604.1	Automated demand response infrastructure	☐ Yes	☐ No
CHAPTER 7. WATER RESOURCE CONSERVATION, QUALITY AND EFFICIENCY			
702.7	Municipal reclaimed water	☐ Yes	☐ No
CHAPTER 8. INDOOR ENVIRONMENTAL QUALITY AND COMFORT			
804.2	Post-Construction Pre-Occupancy Baseline IAQ Testing	☐ Yes	☐ No
807.1	Sound transmission and sound levels	☐ Yes	☐ No
CHAPTER 10. EXISTING BUILDINGS			
1007.2	Evaluation of existing buildings	☐ Yes	☐ No
1007.3	Post Certificate of Occupancy zEPI, energy demand, and CO_2e emissions reporting	☐ Yes	☐ No

CHAPTER 4
SITE DEVELOPMENT AND LAND USE

SECTION 401
GENERAL

401.1 Scope and intent. This chapter provides requirements for the development and maintenance of building and building sites to minimize negative environmental impacts and to protect, restore and enhance the natural features and environmental quality of the site.

401.2 Predesign site inventory and assessment. An inventory and assessment of the natural resources and baseline conditions of the building site shall be submitted with the construction documents.

The inventory and assessment shall:

1. Determine the location of any protection areas identified in Section 402.1 that are located on, or adjacent to, the building site;
2. Determine whether, and to the degree to which, the native soils and hydrological conditions of the building site have been disturbed and altered by previous use or development;
3. Identify *invasive plant species* on the site for removal; and
4. Identify native plant species on the site.

SECTION 402
PRESERVATION OF NATURAL RESOURCES

402.1 Protection by area. Where *flood hazard areas*, surface water bodies or wetlands, conservation areas, parklands, agricultural lands or *greenfields* are located on, or adjacent to, a lot, the development of the lot as a building site shall comply with the provisions of Sections 402.2 through 402.8.

402.2. Flood hazard areas. For locations within *flood hazard areas*, unless compliance with Section 402.2.1 or Section 402.2.2 is required by Table 302.1, new buildings and structures and *substantial improvements* shall comply with Section 402.2.3.

402.2.1 Flood hazard area preservation, general. Where this section is indicated to be applicable in Table 302.1, new buildings and structures, site disturbance, and development of land shall be prohibited within *flood hazard areas*.

402.2.2 Flood hazard area preservation, specific. Where this section is indicated to be applicable in Table 302.1, new buildings and structures, site disturbance, and development of land shall be prohibited within the specific *flood hazard areas* established pursuant to local land use authority.

402.2.3 Development in flood hazard areas. New buildings, structures and *substantial improvements* constructed in *flood hazard areas* shall be in compliance with Section 1612 of the *International Building Code* provided the lowest floors are elevated or dry floodproofed to not less than 1 foot (25 mm) above the elevation required by Section 1612 of the *International Building Code*, or the elevation established by the jurisdiction, whichever is higher.

402.3 Surface water protection. Where this section is indicated to be applicable in Table 302.1, buildings and building site improvements shall not be located over, or located within a buffer as established by the jurisdiction, around or adjacent to oceans, lakes, rivers, streams and other bodies of water that support or could support fish, recreation or industrial use. The buffer shall be measured from the ordinary high-water mark of the body of water.

Exceptions:

1. Buildings and associated site improvements specifically related to the use of the water including, but not limited to, piers, docks, fish hatcheries, and habitat restoration facilities, shall be permitted where the impacts of the construction and location adjacent to or over the water on the habitat is mitigated.
2. Buildings and associated site improvements shall be permitted where a wetlands permit has been issued under a national wetlands permitting program or otherwise issued by the authority having jurisdiction.

402.4 Wetland protection. Buildings and building site improvements shall not be located within a wetland or within a buffer as established by the jurisdiction around a wetland.

Exception: Buildings and associated site improvements specifically related to the use of the wetland including, but not limited to, piers, docks, fish hatcheries, and habitat restoration facilities, shall be permitted where the impacts of the construction and location adjacent to or over the wetland on the habitat are mitigated.

402.5 Conservation area. Where this section is indicated to be applicable in Table 302.1, site disturbance or development of land in or within 50 feet (15 240 mm) of any designated conservation area shall not be permitted.

Exception: Buildings and associated site improvements located in or within 50 feet (15 240 mm) of a conservation area shall be permitted where the building and associated site improvements serve a purpose related to the conservation area as determined by the authority that designated the conservation area.

402.6 Park land. Site disturbance of development of land located within a public park shall not be permitted.

Exceptions:

1. Buildings and site improvements shall be permitted to be located within a park where the building and site improvements serve a park-related purpose.

2. Park lands owned and managed by the Federal government shall be exempt from this prohibition.
3. Privately held property located within the established boundary of a park shall be exempt from this prohibition.

402.7 Agricultural land. Where this section is indicated to be applicable in Table 302.1, buildings and associated site improvements shall not be located on land zoned for agricultural purposes.

Exception: Buildings and associated site improvements shall be permitted to be located on agriculturally zoned land where the building serves an agriculturally related purpose, including, but not limited to, primary residence, farmhouse, migrant workers housing, farm produce storage, processing and shipping.

402.8 Greenfield sites. Where this section is indicated to be applicable in Table 302.1, site disturbance or development shall not be permitted on *greenfield* sites.

Exception: The development of new buildings and associated site improvements shall be permitted on *greenfield* sites where the jurisdiction determines that adequate infrastructure exists, or will be provided, and where the sites comply with not less than one of the following:

1. The *greenfield* site is located within $^1/_4$ mile (0.4 km) of developed residential land with an average density of not less than 8 dwelling units per acre (19.8 dwelling units per hectare).
2. The *greenfield* site is located within $^1/_4$ mile (0.4 km) distance, measured over roads or designated walking surfaces, of not less than 5 diverse uses and within $^1/_2$ mile (0.8 km) walking distance of not less than 7 diverse uses. The diverse uses shall include not less than one use from each of the following categories of diverse uses: retail, service, or community facility.
3. The *greenfield* site has access to transit service. The building on the building site shall be located in compliance with one of the following:
   - 3.1. Within $^1/_4$ mile (0.4 km) distance, measured over designated walking surfaces, of existing or planned bus or streetcar stops.
   - 3.2. Within $^1/_2$ mile (0.8 km) distance, measured over designated walking surfaces, of existing or planned rapid transit stops, light or heavy passenger rail stations, ferry terminals, or tram terminals.
4. The *greenfield* site is located adjacent to areas of existing development that have connectivity of not less than 90 intersections per square mile (35 intersections per square kilometer). Not less than 25 percent of the perimeter of the building site shall adjoin, or be directly across a street, public bikeway or pedestrian pathway from the qualifying area of existing development.
   - 4.1. Intersections included for determination of connectivity shall include the following:
     - 4.1.1. Intersections of public streets with other public streets;
     - 4.1.2. Intersections of public streets with bikeways and pedestrian pathways that are not part of a public street for motor vehicles; and
     - 4.1.3. Intersections of bikeways and pedestrian pathways that are not part of a public street for motor vehicles with other bikeways and pedestrian pathways that are not part of a public street for motor vehicles.
   - 4.2. The following areas need not be included in the determination of connectivity:
     - 4.2.1. Water bodies, including, but not limited to lakes and wetlands.
     - 4.2.2. Parks larger than $^1/_2$ acre (2023 m^2), designated conservation areas and areas preserved from development by the jurisdiction or by the state or federal government.
     - 4.2.3. Large facilities including, but not limited to airports, railroad yards, college and university campuses.

402.8.1 Site disturbance limits on greenfield sites. For *greenfield* sites that are permitted to be developed, site disturbances shall be limited to the following areas:

1. Within 40 feet (18 288 mm) of the perimeter of the building;
2. Within 15 feet (4572 mm) of proposed surface walkways, roads, paved areas and utilities;
3. Within 25 feet (7620 mm) of constructed areas with permeable surfaces that require additional staging areas to limit compaction in the constructed areas.

SECTION 403
STORMWATER MANAGEMENT

403.1 Stormwater management. Stormwater management systems, including, but not limited to, infiltration, evapo-transpiration; rainwater harvest and runoff reuse; shall be provided and maintained on the building site.

403.1.1 Increased runoff. Stormwater management systems shall address the increase in runoff that would occur resulting from development on the building site and shall either:

1. Manage rainfall onsite and size the management system to retain not less than the volume of a single storm which is equal to the 95th-percentile rainfall event and all smaller storms and maintain the predevelopment natural runoff; or
2. Maintain or restore the predevelopment stable, natural runoff hydrology of the site throughout the development or redevelopment process. Postcon-

struction runoff rate, volume, and duration shall not exceed predevelopment rates. The stormwater management system design shall be based, in part, on a hydrologic analysis of the building site.

403.1.2 Adjoining lots and property. The stormwater management system shall not redirect or concentrate off-site discharge that would cause increased erosion or other drainage related damage to adjoining lots or public property.

403.1.3 Brownfields. Stormwater management systems on areas of *brownfields* where contamination is left in place shall not use infiltration. Stormwater management systems shall not penetrate, damage, or otherwise compromise remediation actions at the building site.

403.2 Coal tar sealants. Coal tar sealants shall not be used in any application exposed to stormwater, wash waters, condensates, snowmelt, icemelt or any source of water that could convey coal tar sealants into soils, surface waters or groundwaters.

SECTION 404 LANDSCAPE IRRIGATION AND OUTDOOR FOUNTAINS

404.1 Landscape irrigation systems. Irrigation of exterior landscaping shall comply with Sections 404.1.1 and 404.1.2.

404.1.1 Water for outdoor landscape irrigation. Outdoor landscape irrigation systems shall be designed and installed to reduce potable water use by 50 percent from a calculated mid-summer baseline in accordance with Section 404.1.2 or, where permitted by State regulation or local ordinances, the system shall be supplied by municipal reclaimed water or with *alternate onsite nonpotable water* complying with Chapter 7.

Exceptions: Potable water is permitted to be used as follows:

1. During the establishment phase of newly planted landscaping.
2. To irrigate food production.
3. To supplement nonpotable water irrigation of shade trees provided in accordance with Section 408.2.3.
4. Potable water is permitted for landscape irrigation where approved by local ordinance or regulation.

404.1.2 Irrigation system design and installation. Where in-ground irrigation systems are provided, the systems shall comply with all of the following:

1. The design and installation of outdoor irrigation systems shall be under the supervision of an irrigation professional accredited or certified by an appropriate local or national body.
2. Landscape irrigation systems shall not direct water onto building exterior surfaces, foundations or exterior paved surfaces. Systems shall not generate runoff.
3. Where an irrigation control system is used, the system shall be one that regulates irrigation based on weather, climatological or soil moisture status data. The controller shall have integrated or separate sensors to suspend irrigation events during rainfall.
4. Irrigation zones shall be based on plant water needs with plants of similar need grouped together. Turfgrass shall not be grouped with other plantings on the same zone.
5. Microirrigation zones shall be equipped with pressure regulators that ensure zone pressure is not greater than 40 psi (275.8 kPa), filters, and flush end assemblies.
6. Sprinklers shall:
    6.1. Have nozzles with matched precipitation rates.

    6.2. Be prohibited on landscape areas less than 4 feet (1230 mm) in any dimension.

    6.3. Be prohibited on slopes greater than 1 unit vertical to 4 units horizontal (25-percent slope).

    Exception: Where the application rate of the sprinklers is less than or equal to 0.5 inches (12.7 mm) per hour.

    6.4. Be permitted for use on turfgrass and crop areas only excepting microsprays of a flow less than 45 gallons (170 liters) per hour.

    6.5. If of the pop-up configuration, pop-up to a height of not less than 4 inches (101 mm).

    6.6. Only be installed in zones composed exclusively of sprinklers and shall be designed to achieve a lower quarter distribution uniformity of not less than 0.65.

404.2 Outdoor ornamental fountains and water features. Where available and *approved* for use by the authority having jurisdiction, alternate nonpotable onsite water sources complying with Chapter 7 shall be used for outdoor ornamental fountains and other water features constructed or installed on a building site. Where the fountain or water feature is the primary user of the building site's nonpotable water source, a potable makeup water connection is prohibited.

Exception: Outdoor ornamental fountains and water features are allowed to use potable water provided water is recirculated and there is not an automatic refill valve connection to a source of potable water, and provided that either:

1. The catch basin or reservoir is no greater than 100 gallons (379 L); or
2. Less than 20 square feet (1.86 m^2) of water surface area is exposed.

404.2.1 Treatment. The treatment required to maintain appropriate water quality shall comply with the authority having jurisdiction.

404.2.2 Recirculation. Outdoor ornamental fountains and water features shall be equipped to recirculate and reuse the supplied water.

404.2.3 Signage. Signage in accordance with Chapter 7 shall be posted at each outdoor ornamental fountain and water feature where nonpotable water is used.

SECTION 405
MANAGEMENT OF VEGETATION, SOILS AND EROSION CONTROL

405.1 Soil and water quality protection. Soil and water quality shall be protected in accordance with Sections 405.1.1 through 405.1.6.

405.1.1 Soil and water quality protection plan. A soil and water quality protection plan shall be submitted by the owner and *approved* prior to construction. The protection plan shall address the following:

1. A soils map, site plan, or grading plan that indicates designated soil management areas for all site soils, including, but not limited to:
    - 1.1. Soils that will be retained in place and designated as vegetation and soil protection areas (VSPAs).
    - 1.2. Topsoils that will be stockpiled for future reuse and the locations for the stockpiles.
    - 1.3. Soils that will be disturbed during construction and plans to restore disturbed soils and underlying subsoils to soil reference conditions.
    - 1.4. Soils that will be restored and re-vegetated.
    - 1.5. Soils disturbed by previous development that will be restored in place and re-vegetated.
    - 1.6. Locations for all laydown and storage areas, parking areas, haul roads and construction vehicle access, temporary utilities and construction trailer locations.
    - 1.7. Treatment details for each zone of soil that will be restored, including the type, source and expected volume of materials, including compost amendments, mulch and topsoil.
    - 1.8. A narrative of the measures to be taken to ensure that areas not to be disturbed and areas of restored soils are protected from compaction by vehicle traffic or storage, erosion, and contamination until project completion.
2. A written erosion, sedimentation and pollutant control program for construction activities associated with the project. The program shall describe the best management practices (BMPs) to be employed including how the BMPs accomplish the following objectives:
    - 2.1. Prevent loss of soil during construction due to stormwater runoff or wind erosion, including the protection of topsoil by stockpiling for reuse.
    - 2.2. Prevent sedimentation of stormwater conveyances or *receiving waters* or other public infrastructure.
    - 2.3. Prevent polluting the air with dust and particulate matter.
    - 2.4. Prevent runoff and infiltration of other pollutants from construction site, including, but not limited to thermal pollution, concrete wash, fuels, solvents, hazardous chemical runoff, pH and pavement sealants. Ensure proper disposal of pollutants.
    - 2.5. Protect from construction activities the designated vegetation and soil protection areas, *flood hazard areas* and other areas of vegetation that will remain on site.
3. A written periodic maintenance protocol for landscaping and stormwater management systems, including, but not limited to:
    - 3.1. A schedule for periodic watering of new planting that reflects different water needs during the establishment phase of new plantings as well as after establishment. Where development of the building site changed the amount of water reaching the preserved natural resource areas, include appropriate measures for maintaining the natural areas.
    - 3.2. A schedule for the use of fertilizers appropriate to the plants species, local climate and the preestablishment and post-establishment needs of the installed landscaping. Nonorganic fertilizers shall be discontinued following plant establishment.
    - 3.3. A requirement for a visual inspection of the site after major precipitation events to evaluate systems performance and site impacts.
    - 3.4. A schedule of maintenance activities of the stormwater management system including, but not limited to, cleaning of gutters, downspouts, inlets and outlets, removal of sediments from pretreatment sedimentation pits and wet detention ponds, vacuum sweeping followed by high-pressure hosing at porous pavement and removal of litter and debris.
    - 3.5. A schedule of maintenance activities for landscaped areas including, but not limited to, the removal of dead or unhealthy vegetation; reseeding of turf areas; mowing of grass to a height which optimizes lawn health and retention of precipitation.

405.1.2 Topsoil protection. Topsoil that could potentially be damaged by construction activities or equipment shall be removed from areas to be disturbed and stockpiled on the building site for future reuse on the building site or other *approved* location. Topsoil stockpiles shall be secured and protected throughout the project with temporary or permanent soil stabilization measures to prevent erosion or compaction.

405.1.3 Imported soils. Topsoils or soil blends imported to a building site to serve as topsoil shall not be mined from the following locations:

1. Sites that are prime farmland, unique farmland, or farmland of statewide importance.
2. *Greenfield* sites where development is prohibited by Section 402.8.

Exception: Soils shall be permitted to be imported from the locations in Items 1 and 2 where those soils are a byproduct of a building and building site development process provided that imported soils are reused for functions comparable to their original function.

405.1.4 Soil reuse and restoration. Soils that are being placed or replaced on a building site shall be prepared, amended and placed in a manner that establishes or restores the ability of the soil to support the vegetation that has been protected and that will be planted. Soil reuse and restoration shall be in accordance with Sections 405.1.4.1 and 405.1.4.2.

405.1.4.1 Preparation. Before placing stockpiled or imported topsoils, compliance with all of the following shall occur:

1. Areas shall be cleared of debris including, but not limited to, building materials, plaster, paints, road base type materials, petroleum based chemicals, and other harmful materials;
2. Areas of construction-compacted subsoil shall be scarified; and
3. The first lift of replaced soil shall be mixed into this scarification zone to improve the transition between the subsoil and overlying soil horizons.

Exception: Scarification is prohibited in all of the following locations:

1. Where scarification would damage existing tree roots.
2. On inaccessible slopes.
3. On or adjacent to trenching and drainage installations.
4. On areas intended by the design to be compacted such as abutments, footings, inslopes.
5. *Brownfields*.
6. Other locations where scarification would damage existing structures, utilities and vegetation being preserved.

405.1.4.2 Restoration. Soils disturbed during construction shall be restored in areas that will not be covered by buildings, structures or hardscapes. Soil restoration shall comply with the following:

1. Organic matter. To provide appropriate organic matter for plant growth and for water storage and infiltration, soils shall be amended with a mature, stable compost material so that not less than the top 12 inches (305 mm) of soil contains not less than 3 percent organic matter. Sphagnum peat or organic amendments that contain sphagnum peat shall not be used. Soil organic matter shall be determined in accordance with ASTM D 2974. Organic materials selected for onsite amendment or for blending of imported soils shall be renewable within a 50-year cycle.

   Exception: Where the reference soil for a building site has an organic level depth other than 12 inches (305 mm), soils shall be amended to organic matter levels and organic matter depth that are comparable to the site's reference soil.

2. Additional soil restoration criteria. In addition to compliance with Item 1, soil restoration shall comply with not less than three of the following criteria:

   2.1. Compaction. Bulk densities within the root zone shall not exceed the densities specified in Table 405.1.2 and shall be measured using a soil cone penetrometer in accordance with ASAE S313.3. The root zone shall be not less than 12 inches (305 mm) nor less than the site's reference soil, whichever results in the greater depth of measurement. Data derived from a soil cone penetrometer shall be reported in accordance with ASAE EP542.

   2.2. Infiltration rates. Infiltration rates or saturated hydraulic conductivity of the restored soils shall be comparable to the site's reference soil. Infiltration rates shall be determined in accordance with ASTM D 3385 or ASTM D 5093. For sloped areas where the methods provided in the referenced standards cannot be used successfully, alternate methods *approved* by the *code official* shall be permitted provided that the same method is used to test both reference soil and onsite soil.

   2.3. Soil biological function. Where remediated soils are used, the biological function of the soils' mineralizable nitrogen shall be permitted as a proxy assessment of biological activity.

   2.4. Soil chemical characteristics. Soil chemical characteristics appropriate for plant growth shall be restored. The pH, cation exchange capacity and nutrient profiles of the original undisturbed soil or the site's reference soil shall be matched in restored soils. Salinity suitable for regionally appropriate vegetation shall be established. Soil amendments and fertilizers shall be selected from those which minimize nutrient loading to waterways or groundwater.

TABLE 405.1.2
MAXIMUM CONE PENETROMETER READINGS

SURFACE RESISTANCE (PSI)	SUBSURFACE RESISTANCE (PSI)		
All Textures Sand	Sand (includes loamy sand, sandy loam, sandy clay loam, and sandy clay)	Silt (includes loam, silt loam, silty clay loam, and silty clay)	Clay (includes clay loam)
110	260	260	225

405.1.5 Engineered growing media. Where engineered growing media are used onsite, including, but not limited to vegetative roofs, trees located within hardscape areas, and special soils specified for wetlands and environmental restoration sites, such media shall comply with the best available science and practice standards for that engineered growing media and use.

405.1.6 Documentation. The following shall be provided to document compliance with Sections 405.1.3 through 405.1.5:

1. Documentation, such as receipts from a soil, compost and amendments supplier, to demonstrate that techniques to restore soil occurred; and
2. Soil test results to demonstrate that the selected techniques achieved the criteria of Section 405.1.4.2. Not less than two soil tests shall be conducted on the building site. For building sites where more than 8,000 square feet (744 m^2) of soil is to be disturbed during construction, there shall be not less than one report for every 4,000 square feet (372 m^2) disturbed or report frequency as determined by the *registered design professional.*

405.2 Vegetation and soil protection. Vegetation and soils shall be protected in accordance with Sections 405.2.1 and 405.2.2.

405.2.1 Vegetation and soil protection plan. Where existing soils and vegetation are to be protected, a vegetation and soil protection plan establishing designated vegetation and soil protection areas (VSPAs) shall be submitted with the construction documents and other submittal documents. The protection plan shall address the following:

1. Identification of existing vegetation located on a building site that is to be preserved and protected.
2. Identification of portions of the building site to be designated as vegetation and soil protection areas (VSPAs) that are to be protected during the construction process from being affected by construction activities.
3. Specification of methods to be used such as temporary fencing or other physical barriers to maintain the protection of the designated vegetation and soil protection areas (VSPAs).
4. Specification of protected perimeters around trees and shrubs that are to be included in the designated vegetation and soil protection areas (VSPAs). Perimeters around trees shall be identified as a circle with a radius of not less than 1 foot (305 mm) for every inch (25 mm) of tree diameter with a radius of not less than 5 feet (1524 mm). The perimeters around shrubs shall be not less than twice the radius of the shrub.

   Exception: *Approved* alternative perimeters appropriate to the location and the species of the trees and shrubs shall be permitted.
5. Specification of methods to protect the viability of the designated vegetation and soil protection areas (VSPAs) to support the remaining vegetation at the conclusion of the construction process including minimizing impacts on the existing stormwater drainage patterns associated with the VSPAs.
6. Identification of plans, methods and practices used to designate essential areas of soil and subsoil disturbance.

405.2.1.1 Tree protection zones (TPZ). Where tree protection zones are specified, the specifications and documentation shall be in accordance with Part 5 of TCIA/ANSI A300.

405.2.2 Invasive plant species. *Invasive plant species* shall not be planted on a building site. A management plan for the containment, removal and replacement of any *invasive plant species* currently on the site shall be generated based on either published recommendation for the referenced invasive plant or guidance prepared by a qualified professional. Existing vegetation that is to be retained on a building site shall be protected as required by Section 405.2.

405.3 Native plant landscaping. Where new landscaping is installed as part of a site plan or within the building site, not less than 75 percent of the newly landscaped area shall be planted with native plant species.

SECTION 406
BUILDING SITE WASTE MANAGEMENT

406.1 Building site waste management plan. A building site waste management plan shall be developed and implemented to divert not less than 75 percent of the land-clearing debris and excavated soils. Land-clearing debris includes rock, trees, stumps and associated vegetation. The plan shall include provisions that address all of the following:

1. Materials to be diverted from disposal by efficient usage, recycling or reuse on the building site shall be specified.
2. Diverted materials shall not be sent to sites that are agricultural land, *flood hazard areas* or *greenfield* sites where development is prohibited by Section 402.1 except where *approved* by the *code official.*
3. The effective destruction and disposal of *invasive plant species.*
4. Where contaminated soils are removed, the methods of removal and location where the soils are to be treated and disposed.
5. The amount of materials to be diverted shall be specified and shall be calculated by weight or volume, but not both.
6. Where the site is located in a federal or state designated quarantine zone for invasive insect species, building site vegetation management shall comply with the quarantine rules.
7. Receipts or other documentation related to diversion shall be maintained through the course of construction. When requested by the *code official*, evidence of diversion shall be provided.

406.2 Construction waste. Construction materials and waste and hardscape materials removed during site preparation shall be managed in accordance with Section 503.

SECTION 407 TRANSPORTATION IMPACT

407.1 Walkways and bicycle paths. Not less than one independent, paved walkway or bicycle path suitable for bicycles, strollers, pedestrians, and other forms of nonmotorized locomotion connecting a street or other path to a building entrance shall be provided. Walkways and bicycle paths shall connect to existing paths or sidewalks, and shall be designed to connect to any planned future paths. Paved walkways and bicycle paths shall be designed to minimize stormwater runoff. Pervious and permeable pavement shall be designed in accordance with Section 408.2.4.

407.2 Changing and shower facilities. Buildings with a *total building floor area* greater than 10,000 square feet (929 m^2) and that are required to be provided with long-term bicycle parking and storage in accordance with Section 407.3 shall be provided with onsite changing room and shower facilities. Not less than one shower shall be provided for each 20 long-term bicycle parking spaces, or fraction thereof.

Where more than one changing room and shower facility is required, separate facilities shall be provided for each sex.

407.3 Bicycle parking and storage. Long-term and short-term bicycle parking shall be designated on the site plan by a *registered design professional* and as specified in Table 407.3. The required number of spaces shall be determined based on the net floor area of each primary use or occupancy of a building except where Table 407.3 specifies otherwise. Accessory occupancy areas shall be included in the calculation of primary occupancy area.

Exceptions:

1. Long-term bicycle parking shall not be required where the *total building floor area* is less than 2,500 square feet (232 m^2).
2. Subject to the approval of the *code official*, the number of bicycle parking spaces shall be permitted to be reduced because of building site characteristics including, but not limited to, isolation from other development.

407.3.1 Short-term bicycle parking. Short-term bicycle parking shall comply with all of the following:

1. It shall be provided with illumination of not less than 1 footcandle (11 lux) at the parking surface;
2. It shall be located at the same grade as the sidewalk or at a location reachable by ramp or accessible route;
3. It shall have an area of not less than 18 inches (457 mm) by 60 inches (1524 mm) for each bicycle;
4. It shall be provided with a rack or other facility for locking or securing each bicycle; and
5. It shall be located within 100 feet (30 480 mm) of, and visible from, the main entrance.

Exception: Where directional signage is provided at the main building entrances, short-term bicycle parking shall be permitted to be provided at locations not visible from the main entrance.

407.3.2 Long-term bicycle parking. Long-term bicycle parking shall comply with all of the following:

1. It shall be located on the same site and within the building or within 300 feet (91 440 mm) of the main entrances;
2. It shall be provided with illumination of not less than 1 footcandle (11 lux) at the parking surface;
3. It shall have an area of not less than 18 inches (457 mm) by 60 inches (1524 mm) for each bicycle; and
4. It shall be provided with a rack or other facility for locking or securing each bicycle.

Not less than 50 percent of long-term bicycle parking shall be within a building or provided with a permanent cover including, but not limited to, roof overhangs, awnings, or bicycle storage lockers.

Vehicle parking spaces, other than those required by Section 407.4, local zoning requirements and accessible parking required by the *International Building Code*, shall be permitted to be used for the installation of long term bicycle parking spaces.

407.4 Preferred vehicle parking. Where either Section 407.4.1 or 407.4.2 is indicated to be applicable in Table 302.1, parking provided at a building site shall comply with this section. Preferred parking spaces required by this section shall be those in the parking facility that are located on the

TABLE 407.3
BICYCLE PARKING[a]

OCCUPANCY	SPECIFIC USE	SHORT-TERM SPACES	LONG-TERM SPACES[b]
A-1	Movie theaters	1 per 50 seats; not less than 4 spaces	2 spaces
	Concert halls, theaters other than for movies	1 per 500 seats	
A-2	Restaurants	1 per 50 seats; not less than 2 spaces	
A-3	Places of worship	1 per 500 seats	
A-3	Assembly spaces other than places of worship	1 per 25,000 square feet; not less than 2 spaces	1 per 50,000 square feet; not less than 2 spaces
A-4 – A-5	All	1 per 500 seats	2 spaces
B	All	1 per 50,000 square feet; not less than 2 spaces	1 per 25,000 square feet; not less than 2 spaces
E	Schools	None	1 per 250 square feet of classroom area
E, I-4	Day care	None	2 spaces
F, H	All	None	1 per 25,000 square feet; not less than 2 spaces
I-1	All	None	2 spaces
I-2	All	1 per 25,000 square feet; not less than 2 spaces	1 per 50,000 square feet; not less than 2 spaces
M	All	1 per 25,000 square feet; not less than 2 spaces	1 per 50,000 square feet; not less than 2 spaces
R-1	Hotels, motels, boarding houses	None	1 per 25,000 square feet; not less than 2 spaces
R-2, R-3, R-4	All	None	None
S	Transit park and ride lots	None	1 per 20 vehicle parking spaces
	Commercial parking facilities	1 per 20 vehicle parking spaces	None
	All other	None	2 spaces
Other	Outdoor recreation, parks	1 per 20 vehicle parking spaces; not less than 2 spaces	None

For SI: 1 square foot = 0.0929 m^2.

a. Requirements based on square feet shall be the net floor area of the occupancy or use.

b. When a calculation results in a fraction of space, the requirements shall be rounded to the next higher whole number.

shortest route of travel from the parking facility to a building entrance, but shall not take precedence over parking spaces that are required to be accessible in accordance with the *International Building Code*. Where buildings have multiple entrances with adjacent parking, parking spaces required by this section shall be dispersed and located near the entrances. Such parking spaces shall be provided with *approved* signage that specifies the permitted usage.

407.4.1 High-occupancy vehicle parking. Where employee parking is provided for a building that has a *total building floor area* greater than 10,000 square feet (929 m^2), a building occupant load greater than 100 and not less than 20 employees, at least 5 percent, but not less than two, of the employee parking spaces provided shall be designated as preferred parking for *high occupancy vehicles*.

407.4.2 Low-emission, hybrid, and electric vehicle parking. Where parking is provided for a building that has a *total building floor area* greater than 10,000 square feet (929 m^2) and that has an building occupant load greater than 100, at least 5 percent, but not less than two, of the parking spaces provided shall be designated as preferred parking for low emission, hybrid, and electric vehicles.

SECTION 408
HEAT ISLAND MITIGATION

408.1 General. The heat island effect of building and building site development shall be mitigated in accordance with Sections 408.2 and 404.3.

408.2 Site hardscape. In climate zones 1 through 6, as established in the *International Energy Conservation Code*, not less than 50 percent of the site hardscape shall be provided with one or any combination of options described in Sections 408.2.1 through 408.2.4. For the purposes of this section, site hardscape shall not include areas of the site covered by solar photovoltaic arrays or solar thermal collectors.

408.2.1 Site hardscape materials. Hardscape materials shall have an initial solar reflectance value of not less than 0.30 in accordance with ASTM E 1918 or ASTM C 1549.

Exception: The following materials shall be deemed to comply with this section and need not be tested:

1. Pervious and permeable concrete pavements.
2. Concrete paving without added color or stain.

408.2.2 Shading by structures. Where shading is provided by a building or structure or a building element or component, such building, structure, component or element shall comply with all of the following:

1. Where open trellis-type, free-standing structures such as, but not limited to, covered walkways, and trellises or pergolas, are covered with native plantings, the plantings shall be designed to achieve mature coverage within five years;
2. Where roofed structures are used to shade parking, those roofs shall comply with Section 408.3 in climate zones 1 through 6; and
3. Shade provided onto the hardscape by an adjacent building or structure located on the same lot shall be calculated and credited toward compliance with this section based on the projected peak sun angle on the summer solstice.

408.2.3 Shading by trees. Where shading is provided by trees, such trees shall be selected and placed in accordance with all of the following:

1. Trees selected shall be those that are native or adaptive to, the region and climate zone in which the project site is located. *Invasive plant species* shall not be selected. Plantings shall be selected and sited to produce a hardy and drought resistant vegetated area;
2. Construction documents shall be submitted that show the planting location and anticipated ten year canopy growth of trees and that show the contributions of existing tree canopies; and
3. Shading calculations shall be shown on the construction documents demonstrating compliance with this section and shall include only those hardscape areas directly beneath the trees based on a ten year growth canopy. Duplicate shading credit shall not be granted for those areas where multiple trees shade the same hardscape.

408.2.4 Pervious and permeable pavement. Pervious and permeable pavements including open grid paving systems and open-graded aggregate systems shall have a percolation rate not less than 2 gallons per minute per square foot (100 L/min $\times$ m^2). Pervious and permeable pavement shall be permitted where the use of these types of hardscapes does not interfere with fire and emergency apparatus or vehicle or personnel access and egress, utilities, or telecommunications lines. Aggregate used shall be of uniform size.

408.3 Roof surfaces. Not less than 75 percent of the roof surfaces of buildings and covered parking located in climate zones 1 through 3, as established in the *International Energy Conservation Code*, shall be a roof complying with Section 408.3.1; shall be covered with a vegetative roof complying with Section 408.3.2; or a combination of these requirements. The provisions of this section shall apply to roofs of structures providing shade to parking in accordance with Section 408.2.2 where located in climate zones 1 through 6.

Exception: Portions of roof surfaces occupied by the following shall be permitted to be deducted from the roof surface area required to comply with this section:

1. Solar thermal collectors.
2. Solar photovoltaic systems.
3. Roof penetrations and associated equipment.
4. Portions of the roof used to capture heat for building energy technologies.
5. Rooftop decks and rooftop walkways.

408.3.1 Roof coverings—solar reflectance and thermal emittance. Where roof coverings are used for compliance with Section 408.3, roof coverings shall comply with Section 408.3.1.1 or 408.3.1.2. The values for solar reflectance and thermal emittance shall be determined by an independent laboratory accredited by a nationally recognized accreditation program. Roof products shall be listed and labeled and certified by the manufacturer demonstrating compliance.

408.3.1.1 Roof products testing. Roof products shall be tested for a minimum three-year aged solar reflectance in accordance with ASTM E 1918, ASTM C 1549 or the CRRC-1 Standard and thermal emittance in accordance with ASTM C 1371, ASTM E 408 or the CRRC-1 Standard, and shall comply with the minimum values in Table 408.3.1.

408.3.1.2 Solar reflectance index. Roof products shall be permitted to use a solar reflectance index (SRI) where the calculated value is in compliance with Table 408.3.1 values for minimum aged SRI. The SRI value shall be determined using ASTM E 1980 with a convection coefficient of 2.1 Btu/h-ft^2 (12 W/m^2 $\times$ k) based on three-year aged roof samples tested in accordance with the test methods in Section 408.3.1.1.

TABLE 408.3.1
REFLECTANCE AND EMITTANCE

ROOF SLOPE	MINIMUM AGED SOLAR REFLECTANCE	MINIMUM AGED THERMAL EMITTANCE	MINIMUM AGED SRI
2:12 or less	0.55	0.75	60
Greater than 2:12	0.30	0.75	25

408.3.2 Vegetative roofs. Vegetative roofs, where provided in accordance with Section 408.3, shall comply with the following:

1. All plantings shall be selected based on their hardiness zone classifications in accordance with USDA MP1475 and shall be capable of withstanding the climate conditions of the jurisdiction and the micro climate conditions of the building site including, but

not limited to, wind, precipitation and temperature. Planting density shall provide foliage coverage, in the warm months, of not less than 80 percent within two years of the date of installation unless a different time period is established in the *approved* design. Plants shall be distributed to meet the coverage requirements. *Invasive plant species* shall not be planted.

2. The engineered soil medium shall be designed for the physical conditions and local climate to support the plants and shall consist of nonsynthetic materials. The planting design shall include measures to protect the engineered soil medium until the plants are established. Protection measures include, but are not limited to, installation of pregrown vegetated mats or modules, tackifying agents, fiber blankets and reinforcing mesh. The maximum wet weight and water holding capacity of an engineered soil medium shall be determined in accordance with ASTM E 2399.
3. Where access to the building facades is provided from locations on the perimeter of the roof, nonvegetated buffers adequate to support associated equipment and to protect the roof shall be provided.
4. Nonvegetated clearances as required for fire classification of vegetative roof systems shall be provided in accordance with the *International Fire Code*.
5. Plantings shall be capable of being managed to maintain the function of the vegetative roof as provided in the documents required by Section 904.3.

SECTION 409
SITE LIGHTING

409.1 Light pollution control. Where this section is indicated to be applicable in Table 302.1, uplight, light trespass, and glare shall be limited for all exterior lighting equipment as described in Sections 409.2 and 409.3.

Exception: Lighting used for the following exterior applications is exempt where equipped with a control device independent of the control of the nonexempt lighting:

1. Specialized signal, directional, and marker lighting associated with transportation.
2. Advertising signage or directional signage.
3. Lighting integral to equipment or instrumentation and installed by its manufacturer.
4. Theatrical purposes, including performance, stage, film production, and video production.
5. Athletic playing areas where lighting is equipped with hoods or louvers for glare control.
6. Temporary lighting.
7. Lighting for industrial production, material handling, transportation sites, and associated storage areas where lighting is equipped with hoods or louvers for glare control.
8. Theme elements in theme and amusement parks.
9. Roadway lighting required by governmental authorities.
10. Lighting used to highlight features of public monuments and registered landmark structures.
11. Lighting classified for and used in hazardous areas.
12. Lighting for swimming pools and water features.

409.1.1 Exterior lighting zones. The lighting zone for the building site shall be determined from Table 409.1.1 unless otherwise specified by the jurisdiction.

[E] TABLE 409.1.1
EXTERIOR LIGHTING ZONES

LIGHTING ZONE	DESCRIPTION
1	Developed areas of national parks, state parks, forest land and rural areas
2	Areas predominantly consisting of residential zoning, neighborhood business districts, light industrial with limited nighttime use and residential mixed use areas
3	All other areas
4	High-activity commercial districts in major metropolitan areas as designated by the local jurisdiction

409.2 Uplight. Exterior lighting shall comply with the requirements of Table 409.2 for the exterior lighting zones (LZ) appropriate to the building site.

Exception: Lighting used for the following exterior applications shall be exempt from the requirements of Table 409.2.

1. Lighting for building facades, landscape features, and public monuments in exterior lighting zones 3 and 4.
2. Lighting for building facades in exterior lighting zone 2.

TABLE 409.2
UPLIGHT RATINGS[a, b]

	LIGHTING ZONE (LZ)			
	1	2	3	4
Maximum Luminaire Uplight Rating	U1	U2	U3	U4

a. Uplight ratings (U) are defined by IESNA TM-15-07 Addendum A.
b. The rating shall be determined by the actual photometric geometry in the specified mounting orientation.

409.3 Light trespass and glare. Where luminaires are mounted on buildings with their backlight oriented towards the building, such luminaires shall not exceed the applicable glare ratings specified in Table 409.3(1). Other exterior luminaires shall not exceed the applicable backlight and glare ratings specified in Table 409.3(2).

Table 409.3(1)
MAXIMUM GLARE RATINGS FOR BUILDING MOUNTED LUMINAIRES WITH THE BACKLIGHT ORIENTED TOWARDS THE BUILDING[a, b]

HORIZONTAL DISTANCE TO LIGHTING BOUNDARY (H_{LB})	LIGHTING ZONE (LZ)			
	1	2	3	4
$H_{LB} > 2h_m$	G1	G2	G3	G4
$h_m < H_{LB} \leq 2\ h_m$	G0	G1	G1	G2
$0.5\ h_m \leq H_{LB} \leq h_m$	G0	G0	G1	G1
$H_{LB} < 0.5\ h_m$	G0	G0	G0	G1

h_m = Mounting height: The distance above finished grade at which a luminaire is mounted, measured to the midpoint of the luminaire.

a. Glare (G) ratings are defined by IESNA TM-15-07 Addendum A.

b. The rating shall be determined by the actual photometric geometry in the specified mounting orientation.

Table 409.3(2)
MAXIMUM ALLOWABLE BACKLIGHT AND GLARE RATINGS[a, b, c]

HORIZONTAL DISTANCE TO LIGHTING BOUNDARY (H_{LB})	LIGHTING ZONE (LZ)			
	1	2	3	4
$H_{LB} > 2h_m$	B3 G1	B4 G2	B5 G3	B5 G4
$h_m < H_{LB} \leq 2\ h_m$	B2 G1	B3 G2	B4 G3	B4 G4
$0.5\ h_m \leq H_{LB} \leq h_m$	B1 G1	B2 G2	B3 G3	B3 G4
$H_{LB} < 0.5\ h_m$	B0 G1	B0 G2	B1 G3	B2 G4

h_m = Mounting height: The distance above finished grade at which a luminaire is mounted, measured to the midpoint of the luminaire.

a. Backlight (B) and glare (G) ratings are defined by IESNA TM-15-07 Addendum A.

b. Luminaires located two mounting heights or less from the *lighting boundary* shall be installed with backlight towards the nearest *lighting boundary*, unless lighting a roadway, bikeway or walkway that intersects a public roadway.

c. The rating shall be determined by the actual photometric geometry in the specified mounting orientation.

CHAPTER 5

MATERIAL RESOURCE CONSERVATION AND EFFICIENCY

SECTION 501 GENERAL

501.1 Scope. The provisions of this chapter shall govern matters related to building material conservation, resource efficiency and environmental performance.

SECTION 502 CONSTRUCTION MATERIAL MANAGEMENT

502.1 Construction material management. Construction material management shall comply with Sections 502.1.1 and 502.1.2.

502.1.1 Storage and handling of materials. Materials stored and handled onsite during construction phases shall comply with the applicable manufacturer's printed instructions. Where manufacturer's printed instructions are not available, *approved* standards or guidelines shall be followed.

502.1.2 Construction phase moisture control. Porous or fibrous materials and other materials subject to moisture damage shall be protected from moisture during the construction phase. Material damaged by moisture or that are visibly colonized by fungi either prior to delivery or during the construction phase shall be cleaned and dried or, where damage cannot be corrected by such means, shall be removed and replaced.

SECTION 503 CONSTRUCTION WASTE MANAGEMENT

503.1 Construction material and waste management plan. Not less than 50 percent of nonhazardous construction waste shall be diverted from disposal, except where other percentages are indicated in Table 302.1. A Construction Material and Waste Management Plan shall be developed and implemented to recycle or salvage construction materials and waste. The Construction Material and Waste Management Plan shall comply with all of the following:

1. The location for collection, separation and storage of recyclable construction waste shall be indicated.
2. Materials to be diverted from disposal by efficient usage, recycling, reuse, manufacturer's reclamation, or salvage for future use, donation or sale shall be specified.
3. The percentage of materials to be diverted shall be specified and shall be calculated by weight or volume, but not both.
4. Receipts or other documentation related to diversion shall be maintained through the course of construction. Where requested by the *code official*, evidence of diversion shall be provided.

For the purposes of this section, construction materials and waste shall include all materials delivered to the site and intended for installation prior to the issuance of the certificate of occupancy, including related packaging. Construction and waste materials shall not include land-clearing debris, excavated soils and fill and base materials such as, but not limited to, topsoil, sand and gravel. Land-clearing debris shall include trees, stumps, rocks, and vegetation. Excavated soil, fill material and land-clearing debris shall be managed in accordance with Section 406.1.

SECTION 504 WASTE MANAGEMENT AND RECYCLING

504.1 Recycling areas for waste generated post certificate of occupancy. Waste recycling areas for use by building occupants shall be provided in accordance with one of the following:

1. Waste recycling areas shall be designed and constructed in accordance with the jurisdiction's laws or regulations;
2. Where laws or regulations do not exist or where limited recycling services are available, waste recycling areas shall be designed and constructed to accommodate recyclable materials based on the availability of recycling services; or
3. Where recycling services are not available, waste recycling areas shall be designed and constructed to accommodate the future recycling of materials in accordance with an *approved* design. The *approved* design shall meet one of the following:
   3.1. The *approved* waste recycling area design shall be based on analysis of other regional recycling services, laws or regulations.
   3.2. The *approved* waste recycling area shall be designed to meet the needs of the occupancy, facilitate efficient pick-up, and shall be available to occupants and haulers.

504.2 Storage of lamps, batteries and electronics. Storage space shall be provided for fluorescent lamps, high-intensity discharge (HID) lamps, batteries, electronics, and other discarded items requiring special disposal by the jurisdiction.

SECTION 505 MATERIAL SELECTION

505.1 Material selection and properties. Building materials shall conform to Section 505.2.

Exceptions:

1. Electrical, mechanical, plumbing, security and fire detection, and alarm equipment and controls, automatic fire sprinkler systems, elevators and convey-

ing systems shall not be required to comply with Section 505.2.

2. Where a whole building life cycle assessment is performed in accordance with Section 303.1, compliance with Section 505.2 shall not be required.

505.2 Material selection. Not less than 55 percent of the total building materials used in the project, based on mass, volume or cost, shall comply with Section 505.2.1, 505.2.2, 505.2.3, 505.2.4 or 505.2.5. Where a material complies with more than one section, the material value shall be multiplied by the number of sections that it complies with. The value of total building material mass, volume or cost shall remain constant regardless of whether materials are tabulated in more than one section.

505.2.1 Used materials and components. Used materials and components shall comply with the provisions for such materials in accordance with the applicable code referenced in Section 102.4 and the applicable requirements of this code.

505.2.2 Recycled content building materials. Recycled content building materials shall comply with one of the following:

1. Contain not less than 25 percent combined post-consumer and preconsumer recovered material, and shall comply with Section 505.2.3.
2. Contain not less than 50 percent combined post-consumer and preconsumer recovered material.

505.2.3 Recyclable building materials and building components. Building materials and building components that can be recycled into the same material or another material with a minimum recovery rate of not less than 30 percent through recycling and reprocessing or reuse, or building materials shall be recyclable through an established, nationally available closed loop manufacturer's take-back program.

505.2.4 Bio-based materials. Bio-based materials shall be those materials that comply with one or more of the following:

1. The bio-based content is not less than 75 percent as determined by testing in accordance with ASTM D 6866.
2. Wood and wood products used to comply with this section, other than salvaged or reused wood products, shall be labeled in accordance with the SFI Standard, FSC STD-40-004 V2-1 EN, PEFC Council Technical Document or equivalent *fiber procurement system.* As an alternative to an on-product label, a Certificate of Compliance indicating compliance with the *fiber procurement system* shall be permitted. Manufacturer's *fiber procurement systems* shall be audited by an accredited third-party.
3. The requirements of USDA 7CFR Part 2902.

505.2.5 Indigenous materials. Indigenous materials or components shall be composed of resources that are recovered, harvested, extracted and manufactured within a 500 mile (800 km) radius of the building site. Where only a portion of a material or product is recovered, harvested, extracted and manufactured within 500 miles (800 km), only that portion shall be included. Where resources are transported by water or rail, the distance to the building site shall be determined by multiplying the distance that the resources are transported by water or rail by 0.25, and adding that number to the distance transported by means other than water or rail.

SECTION 506
LAMPS

506.1 Low mercury lamps. The mercury content in lamps shall comply with Section 506.2 or 506.3.

Exception: Appliance, black light, bug, colored, germicidal, plant, shatter-resistant/shatterproof/shatterprotected, showcase, UV, T-8 and T-12 lamps with a color rendering index of 87 or higher, lamps with RDC bases, and lamps used for special-needs lighting for individuals with exceptional needs.

506.2 Straight fluorescent lamps. Straight, double-ended fluorescent lamps less than 6 feet (1829 mm) in nominal length and with bi-pin bases shall contain not more than 5 milligrams of mercury per lamp.

Exception: Lamps with a rated lifetime greater than 22,000 hours at 3 hours per start operated on an ANSI reference ballast shall not exceed 8 milligrams of mercury per lamp.

506.3 Compact fluorescent lamps. Single-ended pin-base and screw-base compact fluorescent lamps shall contain not more than 5 milligrams of mercury per lamp, and shall be listed and labeled in accordance with UL 1993.

Exception: Lamps rated at 25 watts or greater shall contain not more than 6 milligrams of mercury per lamp.

SECTION 507
BUILDING ENVELOPE MOISTURE CONTROL

507.1 Moisture control preventative measures. Moisture preventative measures shall be inspected in accordance with Sections 902 and 903 for the categories listed in Items 1 through 7. Inspections shall be executed in a method and at a frequency as listed in Table 903.1.

1. Foundation sub-soil drainage system.
2. Foundation waterproofing.
3. Foundation dampproofing.
4. Under slab water vapor protection.
5. Flashings: Windows, exterior doors, skylights, wall flashing and drainage systems.
6. Exterior wall coverings.
7. Roof coverings, roof drainage, and flashings.

CHAPTER 6

ENERGY CONSERVATION, EFFICIENCY AND CO_2e EMISSION REDUCTION

SECTION 601 GENERAL

601.1 Scope. The provisions of this chapter regulate the design, construction, commissioning, and operation of buildings and their associated building sites for the effective use of energy.

601.2 Intent. This chapter is intended to provide flexibility to permit the use of innovative approaches and techniques to achieve the effective use of energy.

601.3 Application. Buildings and their associated building sites shall comply with Section 601.3.1 or Section 601.3.2.

601.3.1 Performance-based compliance. Buildings designed on a performance basis shall comply with Sections 602, 608.6, 609, 610 and 611.

601.3.2 Prescriptive-based compliance. Buildings designed on a prescriptive basis shall comply with the requirements of Sections 605, 606, 607, 608, 609, 610 and 611.

601.4 Minimum requirements. Buildings shall be provided with metering complying with Section 603, and commissioning complying with Section 611. Where required in accordance with Section 604.1, building shall be provided with automated-demand response complying with Section 604.

601.5 Multiple buildings on a site and mixed use buildings. Where there is more than one building on a site and where a building has more than one use in the building, each building or each portion of a building associated with a particular use shall comply with Sections 601.5.1 or 601.5.2 or a combination of both.

601.5.1 Multiple buildings on a site. For building sites with multiple buildings, the energy use associated with the building site shall be assigned on a proportional basis to each building based on total gross floor area of each building in relation to the total gross floor area of all buildings on the building site.

Where energy is derived from either renewable or waste energy, or both sources located on the building site, within individual buildings, or on individual buildings and delivered to multiple buildings, the energy so derived shall be assigned on a proportional basis to the buildings served based on building gross floor area. Energy delivered from renewable and waste energy sources located on or within a building shall be assigned to that building.

Exception: Where it can be shown that energy to be used at the building site is associated with a specific building, that energy use shall be assigned to that specific building.

601.5.2 Mixed use buildings. Where buildings have more than one use, the energy use requirements shall be based on each individual occupancy.

SECTION 602 MODELED PERFORMANCE PATHWAY REQUIREMENTS

602.1 Performance-based compliance. Compliance for buildings and their sites to be designed on a performance basis shall be determined by predictive modeling. Predictive modeling shall use source energy kBtu/sf-y unit measure based on compliance with Section 602.1.1 and CO_2e emissions in Section 602.3. Where a building has mixed uses, all uses shall be included in the performance-based compliance.

602.1.1 zEPI. Performance-based designs shall demonstrate a zEPI of not more than 51 as determined in accordance with Equation 6-1 for energy use reduction and shall demonstrate a CO_2e emissions reduction in accordance with Section 602.2 and Equation 6-2 for CO_2e.

$$zEPI = 57 \times (EUIp/EUI) \quad \textbf{(Equation 6-1)}$$

where:

EUIp = the proposed energy use index in source kBtu/sf-y for the proposed design of the building and its site calculated in accordance with Section 602.1.2.

EUI = the base annual energy use index in source kBtu/sf-y for a baseline building and its site calculated in accordance with Section 602.1.2.

602.1.2 Base annual energy use index. The proposed energy use index (EUIp) of the building and building site shall be calculated in accordance with Equation 6-1 and Appendix G to ASHRAE 90.1, as modified by Sections 602.1.2.1 through 602.1.2.3. The annual energy use shall include all energy used for building functions and its anticipated occupancy.

602.1.2.1 Modifications to Appendix G of ASHRAE 90.1. The performance rating in Section G1.2 of ASHRAE 90.1 shall be based on energy use converted to consistent units in accordance with Sections 602.1.2.2 and 602.1.2.3, instead of energy cost.

602.1.2.2 Electric power. In calculating the annual energy use index, electric energy used shall be consistent units by converting the electric power use at the utility meter or measured point of delivery to Btus and multiplying by the conversion factor in Table 602.1.2.1 based on the geographical location of the building.

TABLE 602.1.2.1
ELECTRICITY GENERATION ENERGY CONVERSION FACTORS BY EPA eGRID SUB-REGION[a]

eGRID 2007 SUB-REGION ACRONYM	eGRID 2007 SUB-REGION NAME	ENERGY CONVERSION FACTOR
AKGD	ASCC Alaska Grid	2.97
AKMS	ASCC Miscellaneous	1.76
ERCT	ERCOT All	2.93
FRCC	FRCC All	2.97
HIMS	HICC Miscellaneous	3.82
HIOA	HICC Oahu	3.14
MORE	MRO East	3.40
MROW	MRO West	3.41
NYLI	NPCC Long Island	3.20
NEWE	NPCC New England	3.01
NYCW	NPCC NYC/Westchester	3.32
NYUP	NPCC Upstate NY	2.51
RFCE	RFC East	3.15
RFCM	RFC Michigan	3.05
RFCW	RFC West	3.14
SRMW	SERC Midwest	3.24
SRMV	SERC Mississippi Valley	3.00
SRSO	SERC South	3.08
SRTV	SERC Tennessee Valley	3.11
SRVC	SERC Virginia/Carolina	3.13
SPNO	SPP North	3.53
SPSO	SPP South	3.05
CAMX	WECC California	2.61
NWPP	WECC Northwest	2.26
RMPA	WECC Rockies	3.18
AZNM	WECC Southwest	2.95

a. Sources: EPA eGrid2007 version 1.1, 2005 data; EPA eGrid regional gross grid loss factors; EIA Table 8.4a (Sum tables 8.4b and 8.4c) and Table 8.2c (Breakout of Table 8.2b), 2005 data.

602.1.2.3 Nonrenewable energy. In calculating the annual energy use index for fuel other than electrical power, energy use shall be converted to consistent units by multiplying the nonrenewable energy fossil fuel use at the utility meter or measured point of delivery to Btu's and multiplying by the conversion factor in Table 602.1.2.2. The conversion factor for energy sources not included in Table 602.1.2.2 shall be 1.1. Conversion factors for purchased district heating shall be 1.35 for hot water and 1.45 for steam. The conversion factor for district cooling shall be 0.33 times the value in Table 602.1.2.1 based on the EPA eGRID Sub-region in which the building is located.

TABLE 602.1.2.2
U.S. AVERAGE BUILDING FUELS ENERGY CONVERSION FACTORS BY FUEL TYPE[a]

FUEL TYPE	ENERGY CONVERSION FACTOR
Natural Gas	1.09
Fuel Oil	1.13
LPG	1.12

a. Source: Gas Technology Institute Source Energy and Emissions Analysis Tool.

602.1.3 Registered design professional in responsible charge of building energy simulation. For purposes of this section, and where it is required that documents be prepared by a *registered design professional*, the *code official* is authorized to require the owner to engage and designate on the building permit application a *registered design professional* who shall act as the registered design professional in responsible charge of building energy simulation. Modelers engaged by the registered design professional in responsible charge of building energy simulation shall be certified by an *approved* accrediting entity. Where the circumstances require, the owner shall designate a substitute registered design professional in responsible charge of building energy simulation who shall perform the duties required of the original registered design professional in responsible charge of building energy simulation. The *code official* shall be notified in writing by the owner whenever the registered design professional in responsible charge of building energy simulation is changed or is unable to continue to perform the duties.

602.2 Annual direct and indirect CO_2e emissions. The CO_2e emissions calculations for the building and building site shall be determined in accordance with Sections 602.2.1 and 602.2.2. The emissions associated with the proposed design shall be less than or equal to the CO_2e emissions associated with the standard reference design in accordance with Equation 6-2.

$$CO_2e\ pd \geq (zEPI \times CO_2e\ srbd)/57 \qquad \textbf{(Equation 6-2)}$$

where:

zEPI = the minimum score in accordance with Section 602.1.1.

CO_2e pd = emissions associated with the proposed design.

CO_2e sr*bd* = emissions associated with the standard reference budget design in accordance with Section 602.1.2.

602.2.1 Onsite electricity. Emissions associated with use of electric power shall be based on electric power excluding any renewable or recovered waste energy covered under Section 602.2.1. Emissions shall be calculated by converting the electric power used by the building at the electric utility meter or measured point of delivery, to MWHs, and multiplying by the CO_2e conversion factor in Table 602.2.1 based on the EPA eGRID Sub-region in which the building is located.

TABLE 602.2.1
ELECTRICITY EMISSION RATE BY EPA eGRID SUB-REGION[a]

eGRID 2007 SUB-REGION ACRONYM	eGRID 2007 SUB-REGION NAME	2005 CO_2e RATE (lbs/MWh)
AKGD	ASCC Alaska Grid	1270
AKMS	ASCC Miscellaneous	515
ERCT	ERCOT All	1417
FRCC	FRCC All	1416
HIMS	HICC Miscellaneous	1595
HIOA	HICC Oahu	18591
MORE	MRO East	1971
MROW	MRO West	1957
NYLI	NPCC Long Island	1651
NEWE	NPCC New England	999
NYCW	NPCC NYC/Westchester	874
NYUP	NPCC Upstate NY	774
RFCE	RFC East	1224
RFCM	RFC Michigan	1680
RFCW	RFC West	1652
SRMW	SERC Midwest	1966
SRMV	SERC Mississippi Valley	1094
SRSO	SERC South	1601
SRTV	SERC Tennessee Valley	1623
SRVC	SERC Virginia/Carolina	1220
SPNO	SPP North	2106
SPSO	SPP South	1780
CAMX	WECC California	768
NWPP	WECC Northwest	958
RMPA	WECC Rockies	1999
AZNM	WECC Southwest	1391

a. Sources: EPA eGRID2007 Version 1.1, 2005 data; EPA eGrid regional gross grid loss factor.

602.2.2 Onsite nonrenewable energy. Emissions associated with the use of nonrenewable energy sources other than electrical power such as natural gas, fuel oil, and propane shall be calculated by multiplying the fossil fuel energy used by the building and its site at the utility meter by the national emission factors in Table 602.2.2 and the conversions required by this section. Emissions associated with fossil fuels not specified in Table 602.2.2 shall be calculated by multiplying the fossil fuel used by the building at the utility meter by 250. Emissions associated with purchased district energy shall be calculated by multiplying the energy used by the building at the utility meter by 150 for hot water and steam, and for district cooling, the factors from Table 602.2.2 based on the EPA eGRID Sub-region in which the building is located.

TABLE 602.2.2
FOSSIL FUEL EMISSION FACTORS

EMISSION RATE (lb/MMbtu HHV)	NATURAL GAS AS STATIONARY FUEL	FUEL OIL AS STATIONARY FUEL	PROPANE AS STATIONARY FUEL
CO_2e	137.35	200.63	162.85

For SI: MMBtu = 1,000,000 Btu = 10 terms: HHV = High-heating value.

602.2.3 Annual direct and indirect CO_2e emissions associated with onsite use of fossil fuels and purchased district energy. Emissions associated with the use of natural gas, fuel oil and, propane shall be calculated by multiplying the natural gas, fuel oil, and propane delivered to the building at the utility meter by the corresponding emission factors in Table 602.2.2. Emissions associated with fossil fuels not listed shall be calculated by multiplying the fossil fuel delivered to the building at the utility meter by 250. Emissions associated with purchased district heating shall be calculated by multiplying the heating energy delivered to the building at the utility meter by 150 for hot water and steam, and for district cooling, the factors from Table 602.2.1 based on the EPA eGRID Sub-region in which the building is located.

SECTION 603
ENERGY METERING, MONITORING AND REPORTING

603.1 Purpose. Buildings that consume energy shall comply with Section 603. The purpose of this section is to provide requirements that will ensure that buildings are constructed or altered in a way that will provide the capability for their energy use, production and reclamation to be measured, monitored and reported. This includes the design of energy distribution systems so as to isolate load types, the installation of or ability to install in the future meters, devices and a data acquisition system, and the installation of, or the ability to provide, public displays and other appropriate reporting mechanisms in the future.

All forms of energy delivered to the building and building site, produced on the building site or in the building and reclaimed at the building site or in the building shall be metered and all energy load types measured in accordance with this section.

603.1.1 Buildings with tenants. In buildings with tenants, the metering required by Section 603.3 shall be collected for the entire building and for each tenant individually. Tenants shall have access to all data collected for their space.

603.2 Energy distribution design requirements and load type isolation in buildings. Energy distribution systems within, on or adjacent to and serving a building shall be designed such that each primary circuit, panel, feeder, piping system or supply mechanism supplies only one energy use type as defined in Sections 603.2.1 through 603.2.5. The energy use type served by each distribution system shall be clearly designated on the energy distribution system with the use served, and adequate space shall be provided for installation of metering equipment or other data collection devices, temporary or permanent, to measure their energy use. The energy distribution system shall be designed to facilitate the collection of data for each of the building energy use categories in Section 603.4 and for each of the end use categories listed in Sections 603.2.1 through 603.2.5. Where there are multiple buildings on a building site, each building shall comply separately with the provisions of Section 603.

Exception: Buildings designed and constructed such that the total usage of each of the load types described in Sections 603.2.1 through 603.2.5 shall be permitted to be measured through the use of installed sub-meters or other equivalent methods as *approved.*

603.2.1 HVAC system total energy use. The HVAC system total energy use category shall include all energy used to heat, cool, and provide ventilation to the building including, but not limited to, fans, pumps, boiler energy, chiller energy and hot water.

603.2.2 Lighting system total energy use. The lighting system total energy use category shall include all interior and exterior lighting used in occupant spaces and common areas.

603.2.3 Plug loads. The plug loads energy use category shall include all energy use by devices, appliances and equipment connected to convenience receptacle outlets.

603.2.4 Process loads. The process loads energy use category shall include the energy used by any single load associated with activities within the building, such as, but not limited to, data centers, manufacturing equipment and commercial kitchens, that exceeds 5 percent of the peak connected load of the whole building.

603.2.5 Energy used for building operations loads and other miscellaneous loads. The category of energy used for building operations loads and other miscellaneous loads shall include all vertical transportation systems, automatic doors, motorized shading systems, ornamental fountains and fireplaces, swimming pools, inground spas, snow-melt systems, exterior lighting that is mounted on the building or used to illuminate building facades and the use of any miscellaneous loads in the building not specified in Sections 603.2.1 through 603.2.4.

603.3 Energy-type metering. Buildings shall be provided with the capability to determine energy use and peak demand as provided in this section for each of the energy types specified in Sections 603.3.1 through 603.3.7. Utility energy meters or supplemental sub-meters are permitted to be used to collect whole building data, and shall be equipped with a local data port connected to a data acquisition system in accordance with Section 603.5.

603.3.1 Gaseous fuels. Gaseous fuels including, but not limited to, natural gas, LP gas, coal gas, hydrogen, landfill gas, digester gas and biogas shall be capable of being metered at the building site to determine the gross consumption and peak demand of each different gaseous fuel by each building on a building site. The installation of gas meters and related piping shall be in accordance with the *International Fuel Gas Code*.

603.3.2 Liquid fuels. Liquid fuels including, but not limited, to fuel oil, petroleum-based diesel, kerosene, gasoline, bio diesel, methanol, ethanol and butane shall be capable of being metered at the building site to allow a determination of the gross consumption and peak demand of each liquid fuel use by each building on a building site. The installation of meters and related piping shall be in accordance with the *International Mechanical Code*.

603.3.3 Solid fuels. Solid fuels including, but not limited to, coal, charcoal, peat, wood products, grains, and municipal waste shall be capable of having their use determined at the building site to allow a determination of the gross consumption and peak demand of each solid fuel use by each building on a building site.

603.3.4 Electric power. Electric power shall be capable of being metered at the building site to allow a determination of the gross consumption and peak demand by each building on a building site. The installation of electric meters and related wiring shall be in accordance with NFPA 70.

603.3.5 District heating and cooling. Hot water, steam, chilled water, and brine shall be capable of being metered at the building site, or where produced on the building site, to allow a determination of the gross consumption of heating and cooling energy by each building on a building site. Energy use associated with the production of hot water, steam, chilled water or brine shall be determined based on the fuel used.

603.3.6 Combined heat and power. Equipment and systems with a connected load greater than 125,000 Btu/hr (36.63 kW) providing combined heat and power (CHP) shall be capable of being metered to allow a determination of the gross consumption of each form of delivered energy to the equipment. The output of CHP shall be metered in accordance with the applicable portions of Section 603 based on the forms of output from the CHP.

603.3.7 Renewable and waste energy. Equipment and systems providing energy from renewable or waste energy sources which is included in the determination of the building zEPI, shall be capable of being metered to allow a determination of the output of equipment and systems in accordance with Sections 603.3.7.1 through 603.3.7.5.

603.3.7.1 Solar electric. Equipment and systems providing electric power through conversion of solar energy directly to electric power shall be capable of being metered so that the peak electric power (kW) provided to the building and its systems or to off-site entities can be determined at 15-minute intervals and the amount of electric power (kWh) provided to the building and its systems can be determined at intervals of 1 hour or less.

603.3.7.2 Solar thermal. Equipment and systems providing heat to fluids or gases through the capture of solar energy shall be capable of being metered so that the peak thermal energy (Btu/h) provided to the building and its systems or to off-site entities can be determined at 15-minute intervals and the amount of heat captured (Btu) for delivery to the building and its systems can be determined intervals of 1 hour or less.

Exception: Systems with a rated output of less than 100 kBtu/hr shall not be required to have the capacity to be metered.

603.3.7.3 Waste heat. Equipment and systems providing energy through the capture of waste heat shall be capable of being metered so that the amount of heat captured and delivered to the building and its systems can be determined at intervals of 1 hour or less.

Exception: Systems with a rated output of less than 100 kBtu/hr shall not be required to have the capacity to be metered.

603.3.7.4 Wind power systems. Equipment and systems providing electric power through conversion of wind energy directly to electric power shall be capable of being metered so that the peak electric power (kW) provided to the building and its systems or to off-site entities can be determined at 15-minute intervals and the amount of electric power (kWh) provided to the building and its systems can be determined at intervals of 1 hour or less.

603.3.7.5 Other renewable energy electric production systems. Equipment and systems providing electric power through conversion of other forms of renewable energy directly to electric power shall be capable of being metered so that the peak electric power (kW) provided to the building and its systems or to off-site entities can be determined at 15-minute intervals and the amount of electric power (kWh) provided to the building and its systems can be determined at intervals of 1 hour or less.

603.4 Energy load type sub-metering. For buildings that are not less than 25,000 square feet (2323 m^2) in *total building floor area* the energy use of the categories specified in Section 603.2 shall be metered through the use of sub-meters or other *approved*, equivalent methods meeting the capability requirements of Section 603.3.

603.4.1 Buildings less than 25,000 square feet. For buildings that are less than 25,000 square feet (2323 m^2) in *total building floor area*, the energy distribution system shall be designed and constructed to accommodate the future installation of sub-meters and other *approved* devices in accordance with Section 603.4. This includes, but is not limited to, providing access to distribution lines and ensuring adequate space for the installation of sub-meters and other *approved* devices.

603.5 Minimum energy measurement and verification. Meters, sub-meters, and other *approved* devices installed in compliance with Sections 603.3 and 603.4 shall be connected to a data acquisition and management system capable of storing not less than 36-months worth of data collected by all meters and other *approved* devices and transferring the data in real time to a display as required in Section 603.6.

603.5.1 Annual emissions. The data acquisition and management system shall be capable of providing the data necessary to calculate the annual CO_2e emissions associated with the operation of the building and its systems using the results of annual energy use measured in accordance with Section 603.5. The calculation shall be based on energy measured for each form of energy delivered to the site on an annual basis. Where reporting of emissions is required, the determination of emissions shall be in accordance with Section 602.2.3.

603.6 Energy display. A permanent, readily accessible and visible display shall be provided adjacent to the main building entrance or on a publicly available Internet web site. The display shall be capable of providing all of the following:

1. The current energy demand for the whole building level measurements, updated for each fuel type at the intervals specified in Section 603.3.
2. The average and peak demands for the previous day and the same day the previous year.
3. The total energy usage for the previous 18 months.

SECTION 604 AUTOMATED DEMAND-RESPONSE (AUTO-DR) INFRASTRUCTURE

604.1 Establishing an open and interoperable automated demand-response (Auto-DR) infrastructure. Where this section is indicated to be applicable in Table 302.1, buildings that contain heating, ventilating, air-conditioning (HVAC) or lighting systems shall comply with Sections 604.1 through 604.4. A building energy management and control system (EMCS) shall be provided and integrated with building HVAC systems controls and lighting systems controls to receive an open and interoperable automated demand-response (Auto-DR) relay or Internet signal. Building HVAC and lighting systems and specific building energy-using components shall incorporate preprogrammed demand response strategies that are automated with a demand response automation Internet software client.

Exception: Auto-DR infrastructure is not required for the following:

1. Buildings located where the electric utility or regional Independent System Operator (ISO) or Regional Transmission Operator (RTO) does not offer a demand response program to buildings regulated by this code.
2. Buildings with a peak electric demand not greater than 0.75 times that of the standard reference design.
3. Buildings that have incorporated onsite renewable energy generation to provide 20 percent or more of the building's energy demand.

604.2 Software clients. Demand response automation software clients shall be capable of communicating with a demand response automation server via the Internet or other communication relay.

604.3 Heating, ventilating and air-conditioning (HVAC) systems. The Auto-DR strategy for HVAC systems shall be capable of reducing the building peak cooling or heating HVAC demand by not less than 10 percent when signaled from the electric utility, regional independent system operator (ISO) or regional transmission operator (RTO), through any combination of the strategies and systemic adjustments, including, but not limited to the following:

1. Space temperature setpoint reset.
2. Increasing chilled water supply temperatures or decreasing hot water supply temperatures.
3. Increasing or decreasing supply air temperatures for variable air volume (VAV) systems.
4. Limiting capacity of HVAC equipment that has variable or multiple-stage capacity control.
5. Cycling of HVAC equipment or turning off noncritical equipment.
6. Disabling HVAC in unoccupied areas.
7. Limiting the capacity of chilled water, hot water, and refrigerant control valves.
8. Limiting the capacity of supply and exhaust fans, without reducing the outdoor air supply below the minimum required by Chapter 4 of the *International Mechanical Code*, or the minimum required by ASHRAE 62.1.
9. Limiting the capacity of chilled water or hot water supply pumps.
10. Anticipatory control strategies to precool or preheat in anticipation of a peak event.

Exception: The Auto-DR strategy is not required to include the following buildings and systems:

1. Hospitals and critical emergency response facilities.
2. Life safety ventilation for hazardous materials storage.
3. Building smoke exhaust systems.
4. Manufacturing process systems.

604.3.1 Rebound avoidance. The Auto-DR strategy shall include logic to prevent a rebound peak. When the signal for Auto-DR is ended, a gradual return to normal heating, ventilation and air-conditioning (HVAC) equipment operations shall be part of the Auto-DR strategy, through any combination of the strategies and systemic adjustments, including, but not limited to the following:

1. Where close to the unoccupied period, the Auto-DR period shall be extended using a rebound avoidance, extended Auto-DR control strategy until the initiation of the unoccupied period.
2. Rebound avoidance, slow recovery control strategies, gradually increasing or decreasing space temperature setpoints or a variance in the timing by cooling or heating zone.
3. Rebound avoidance, slow recovery control strategies, gradually increasing or decreasing zone supply air temperatures.
4. Rebound avoidance, slow recovery control strategies, gradually increasing or decreasing chilled water temperatures or decreasing hot water temperatures.
5. Rebound avoidance, sequential equipment recovery strategies, gradually restoring demand limited equipment capacity.
6. Rebound avoidance, sequential equipment recovery strategies, gradually restoring equipment that was turned off during the Auto-DR period.
7. Rebound avoidance, slow recovery control strategies, gradually increasing capacity for air moving and pumping systems.
8. Rebound avoidance, sequential equipment recovery or rebound avoidance, slow recovery control where chilled water or hot water and other capacity control valves are sequentially or gradually allowed to return to normal operation, respectively.

604.4 Lighting. In Group B office spaces, the Auto-DR system shall be capable of reducing total connected power of lighting as determined in accordance with Section C405.5 of the *International Energy Conservation Code* by not less than 15 percent.

Exception: The following buildings and lighting systems need not be addressed by the Auto-DR system:

1. Buildings or portions associated with lifeline services.
2. Luminaires on emergency circuits.
3. Luminaires located in emergency and life safety areas of a building.
4. Lighting in buildings that are less than 5,000 square feet (465 m^2) in total area.
5. Luminaires located within a daylight zone that are dimmable and connected to automatic *daylight controls* complying with Section C405.2.2.3.2 of the *International Energy Conservation Code*.
6. Signage used for emergency, life safety or traffic control purposes.

SECTION 605
BUILDING ENVELOPE SYSTEMS

605.1 Prescriptive compliance. Where buildings are designed using the prescriptive-based compliance path in accordance with Section 601.3.2, *building thermal envelope* systems shall comply with the provisions of Section C402 of the *International Energy Conservation Code* and the provisions of this section.

605.1.1 Insulation and fenestration criteria. The *building thermal envelope* shall exceed the requirements of Tables C402.1.2 and C402.3 of the *International Energy Conservation Code* by not less than 10 percent. Specifically, for purposes of compliance with this code, each U-factor, C-factor, F-factor and SHGC in the specified tables shall be reduced by 10 percent to determine the prescriptive criteria for this code. In Sky Type "C" locations specified in Section 808.4, the skylights shall not exceed 5 percent of the building roof area.

605.1.1.1 Permanent shading devices for fenestration. Vertical fenestration within 45 degrees (785 rad) of the nearest west, south, and east cardinal ordinate shall be shaded by permanent horizontal exterior projections with a projection factor greater than or equal to 0.25. Where different windows or glass doors have different projection factor values, each shall be evaluated separately, or an area-weighted projection factor value shall be calculated and used for all windows and glass doors. Horizontal projections shall extend laterally beyond the edge of the glazing not less than one-half of the height of the glazing, except at building corners.

Exception: Shading devices are not required for the following buildings and fenestrations:

1. Buildings located in hurricane-prone regions in accordance with Section 1609.2 of the *International Building Code* or on any other building with a mean roof height exceeding the height limits specified in Table 1504.8 of the *International Building Code* based on the exposure category and basic wind speed at the building site.
2. Where fenestration is located in a building wall that is within 18 inches (457 mm) of the lot line.
3. Where equivalent shading of the fenestration is provided by buildings, structures, geological formations, or permanent exterior projections that are not horizontal, as determined by sun angle studies at the peak solar altitude on the spring equinox, and three hours before and after the peak solar altitude on the spring equinox.
4. Where fenestration contains dynamic glazing that has a lower labeled solar heat gain coefficient (SHGC) equal to or less than 0.12, and the ratio of the higher and lower labeled visible transmittance (VT) is greater than or equal to 5. Dynamic glazing shall be automatically controlled to modulate, in multiple steps, the amount of solar gain and light transmitted into the space in response to daylight levels or solar intensity. Functional testing of controls shall be conducted in accordance with Section C408.3.1 of the *International Energy Conservation Code.*

605.1.2 Air leakage. The *building thermal envelope* shall be durably sealed to limit air leakage in accordance with Section C402.4 of the *International Energy Conservation Code* and the provisions of this section.

605.1.2.1 Air barriers. A continuous air barrier shall be provided for buildings in climate zones 1 through 8 in accordance with Section C402.4.1 of the *International Energy Conservation Code.* The exception in Section C402.4.1 of the *International Energy Conservation Code* shall not apply.

605.1.2.2 Testing requirement. The *building thermal envelope* air tightness shall be considered to be acceptable where the tested air leakage of the total area of the *building thermal envelope* is less than 0.25 cfm/ft^2 under a pressure differential of 0.3 in water column (1.57 lb/ft^2) (1.25 L/s.m^2 under a pressure differential of 75 Pa). Testing shall occur after rough-in and after installation of penetrations of the building envelope, including penetrations for utilities, heating, ventilating and air-conditioning (HVAC) systems, plumbing, and electrical equipment and appliances. Testing shall be done in accordance with ASTM E 779.

605.1.2.3 Air curtains. Where a building entrance is required to be protected with a vestibule in accordance with the *International Energy Conservation Code*, an air curtain tested in accordance with ANSI/AMCA 220 is permitted to be used as an alternative to separate conditioned space from the exterior.

605.2. Roof replacement. Above-deck insulation for roof replacement on an existing building with insulation entirely above the deck and where the roof slope is less than two units vertical in 12 units horizontal (17-percent slope) shall be in accordance with Section 1003.2.7.

SECTION 606
BUILDING MECHANICAL SYSTEMS

606.1 Prescriptive compliance. Where buildings are designed using the prescriptive-based compliance path in accordance with Section 601.3.2, building mechanical systems shall comply with the provisions of the *International Energy Conservation Code* and the provisions of this section.

606.2 HVAC equipment performance requirements. Heating, ventilation and air-conditioning (HVAC) equipment shall comply with Sections 606.2.1 and 606.2.2.

606.2.1 Equipment covered by federal standards. Equipment covered by federal minimum efficiency standards shall comply with the minimum efficiency requirements of the *International Energy Conservation Code.*

606.2.2 Equipment not covered by federal standards. Equipment not covered by federal minimum efficiency standards shall comply with the minimum efficiency requirements of this section.

606.2.2.1 Ground source heat pumps. The efficiency of ground source heat pumps shall comply with the provisions of Table 606.2.2.1 based on the applicable referenced test procedure.

TABLE 606.2.2.1
ENERGY-EFFICIENCY CRITERIA FOR GROUND SOURCE HEAT PUMPS

PRODUCT TYPE	MINIMUM EER	MINIMUM COP	TEST PROCEDURE
Water-to-Air Closed loop	14.1	3.3	ISO 13256-1
Water-to-Air Open loop	16.2	3.6	ISO 13256-1
Water-to-Water Closed loop	15.1	3.0	ISO 13256-2
Water-to-Water Open loop	19.1	3.4	ISO 13256-2
Direct Expansion (DX) or Direct GeoExchange (DGX)	15.0	3.5	AHRI 870

EER = Energy efficiency ratio, COP = Coefficient of performance.

606.2.2.2 Multi-stage ground source heat pumps. The efficiency of multi-stage ground source heat pumps shall comply with the provisions of Table 606.2.2.1 based on the applicable referenced test procedure.

606.2.2.3 Minimum fan efficiency. Stand-alone supply, return and exhaust fans designed for operating with motors over 750 watts (1 hp) shall have an energy efficiency classification of not less than FEG71 as defined in AMCA 205. The total efficiency of the fan at the design point of operation shall be within 10 percentage points of either the maximum total efficiency of the fan or the static efficiency of the fan.

606.3 Duct and plenum insulation, sealing and testing. Supply and return air ducts and plenums, air handlers and filter boxes shall be insulated and sealed in accordance with Section C403.2.7.1.1 of the *International Energy Conservation Code*. The exception in Section C403.2.7.1.1 shall not apply.

606.3.1 Duct air leakage testing. Ductwork that is designed to operate at static pressures greater than 3 inches water column (747 Pa) and all ductwork located outdoors shall be leak-tested in accordance with the SMACNA *HVAC Air Duct Leakage Test Manual*. Representative sections totaling not less than 25 percent of the total installed duct area for the designated pressure class shall be tested. Positive pressure testing is acceptable for negative pressure ductwork. Duct systems with pressure ratings in excess of 3 inches water column (747 Pa) shall be identified on the construction documents. Duct leakage shall not exceed the rate determined in accordance with Equation 6-3.

$$F = C_L P^{0.65}$$ **(Equation 6-3)**

where:

F = maximum leakage in cfm/100 ft^2 duct surface area;

C_L = 4, duct leakage class, cfm/100 ft^2 at 1 inch water column.

P = test pressure, which shall be equal to the design duct pressure class rating inches of water column.

606.4 Heating, ventilating and air-conditioning (HVAC) piping insulation. Piping with a nominal diameter greater than $^1/_4$ inch (6.4 mm), including associated valves, fittings and piping system components, in heating, ventilating and air-conditioning (HVAC) systems shall be thermally insulated in accordance with Table 606.4. For insulation outside of the conductivity ranges specified in Table 606.4, the minimum thickness of the insulation shall be determined in accordance with Equation 6-4.

$$T = r\,[(1 + t/r)^{K/k} - 1]$$ **(Equation 6-4)**

where:

T = minimum insulation thickness (inches).

r = actual outside radius of pipe (inches).

t = insulation thickness specified in Table 606.4 for applicable fluid temperature and pipe size.

K = Conductivity of alternate material at mean rating temperature indicated for the applicable fluid temperature (Btu × in/h × ft^2 × °F).

k = the upper value of the conductivity range specified in Table 606.4 for the applicable fluid temperature.

Building cavities and interstitial framing spaces shall be large enough to accommodate the combined diameter of the pipe plus the insulation, plus the full thickness of the insulation plus any other objects in the cavity that the piping must cross.

Exception: Piping insulation is not required for the following:

1. Factory-installed piping within HVAC equipment tested and rated in accordance with Section 606.2.
2. Piping conveying fluids having a design operating temperature range between 60°F (15.6°C) and 105°F (40.6°C).
3. Piping conveying fluids not heated or cooled such as roof and condensate drains, cold water supply, and natural gas piping.
4. Where heat gain or heat loss will not increase energy usage such as liquid refrigerant piping.
5. Piping having an outside diameter of 1 inch (25 mm) or less, associated with strainers, control valves, and balancing valves.

TABLE 606.4
MINIMUM PIPE INSULATION THICKNESS

FLUID	CONDUCTIVITY Btu-in/(h × ft^2 × F)	RATIO OF WALL THICKNESS OF PIPE INSULATION TO NOMINAL PIPE DIAMETER[a, b]
Steam	0.27 – 0.34	≥ 2:1
Hot Water	0.22 – 0.29	≥ 1:1
Chilled Water	0.22 – 0.28	≥ 1:1

For SI: 1 inch = 25.4 mm, 1 Btu-in = W/m × K.

a. The proportions apply to all nominal pipe diameters greater than $^1/_4$ inch and less than or equal to 2 inches. For nominal pipe diameters larger than 2 inches, outside diameter, the minimum wall thickness of the insulation shall be equal to the wall thickness required for 2-inch pipe.

b. These thicknesses are based on energy-efficiency considerations only.

606.5 Economizers. Economizers shall comply with the requirements of the *International Energy Conservation Code*, except as noted herein.

606.5.1 Economizer systems. Each cooling system that has a fan shall include either an air economizer complying with Section 606.5.1.1 or a water economizer complying with Section 606.5.1.2.

Exception: Economizers are not required for the following.

1. Individual fan-cooling units with a supply capacity less than the minimum listed in Table 606.5.1(1).
2. In Group I-2 occupancies, hospitals, and Group B occupancies, ambulatory care facilities, where more than 75 percent of the air designed to be supplied by the system is to spaces that are required to be humidified above a 35°F (1.7°C) dew-point temperature to comply with applicable codes or accreditation standards. In other occupancies, where more than 25 percent of the air designed to be supplied by the system is to spaces that are designed to be humidified above a 35°F (1.7°C) dew-point temperature to satisfy process needs.
3. Systems that include a condenser heat recovery system that is designed to utilize 60 percent of the peak heat rejection load at design conditions and there is a documented need for that rejected heat for either service hot water or space heating during peak heat rejection design conditions.
4. Systems that serve spaces estimated as having a sensible cooling load at design conditions, excluding transmission and infiltration loads, of less than or equal to transmission and infiltration losses at the temperature and relative humidity design conditions in accordance with Section 6.1 of ASHRAE 55.
5. Where the use of outdoor air for cooling will affect supermarket open refrigerated casework systems.
6. Where the cooling efficiency is equal to, or greater than, the efficiency improvement requirements in Table 606.5.1(2).

606.5.1.1 Air economizers. Air economizers shall be designed in accordance with Sections 606.5.1.1.1 through 606.5.1.1.4.

TABLE 606.5.1(1)
ECONOMIZER REQUIREMENTS

CLIMATE ZONES	ECONOMIZER REQUIREMENT
1A, 1B	No requirement
2A, 2B, 3A, 3B, 3C, 4A, 4B, 4C, 5A, 5B, 5C, 6A, 6B, 7, 8	Economizers on all cooling systems having a capacity ≥ 33,000 Btu/h[a]

For SI: 1 British thermal unit per hour = 0.293 W.

a. The total capacity of all systems without economizers shall not exceed 480,000 Btu/h per building or 20 percent of the building's air economizer capacity, whichever is greater.

606.5.1.1.1 Design capacity. Air economizer systems shall be capable of modulating outdoor air and return air dampers to provide up to 100 percent of the design supply air quantity as outdoor air for cooling.

606.5.1.1.2 Control signal. Economizer dampers shall be capable of being sequenced with the mechanical cooling equipment and shall not be controlled by only mixed air temperature.

Exception: The use of mixed air temperature limit control shall be permitted for systems controlled from space temperature, such as single-zone systems.

606.5.1.1.3 High-limit shutoff. Air economizers shall be capable of automatically reducing outdoor air intake to the design minimum outdoor air quantity when the outdoor air intake will not reduce cooling energy usage. High-limit shutoff control types for specific climates shall be chosen from Table 606.5.1.1.3(1). High-limit shutoff control settings for the Table 606.5.1.1.3(1) control types shall be as specified in Table 606.5.1.1.3(2).

606.5.1.1.4 Relief of excess outdoor air. Systems shall provide a means to relieve excess outdoor air during air economizer operation to prevent overpressurizing of the building. The relief air outlets shall be located to avoid recirculation of the relief of air into the building.

TABLE 606.5.1(2)
EQUIPMENT EFFICIENCY PERFORMANCE EXCEPTION FOR ECONOMIZERS

CLIMATE ZONES	COOLING EQUIPMENT EFFICIENCY IMPROVEMENT (%)[a]
2A	17
2B	21
3A	27
3B	32
4A	42
4B	49

IPLV = Integrated part load value, IEER = Integrated energy-efficiency ratio, SEER = Seasonal energy-efficiency rating, EER = Energy-efficiency ratio, COP = Coefficient of performance

a. Where a unit is rated with an IPLV, IEER or SEER, the minimum values for these metrics shall be increased by the percentage listed in the table in order to eliminate the required air or water economizer. Where a unit is rated only with a full load metric such as EER or COP cooling, these metrics shall be increased by the percentage shown.

606.5.1.2 Water economizer systems for HVAC equipment. Economizer systems for heating, ventilating and air-conditioning (HVAC) equipment shall be designed in accordance with Sections 606.5.1.2.1 through 606.5.1.2.4.

606.5.1.2.1 Design capacity. Water economizer systems shall be capable of cooling supply air by indirect evaporation and providing up to 100 percent of the expected system cooling load at outdoor air

TABLE 606.5.1.1.3(1)
HIGH-LIMIT SHUTOFF CONTROL OPTIONS FOR AIR ECONOMIZERS

CLIMATE ZONES	ALLOWED CONTROL TYPES	PROHIBITED CONTROL TYPES
1B, 2B, 3B, 3C, 4B, 4C, 5B, 5C, 6B, 7, 8	Fixed dry bulb Differential dry bulb Electronic enthalpy[a] Differential enthalpy Dew-point and dry-bulb temperatures	Fixed enthalpy
1A, 2A, 3A, 4A	Fixed enthalpy Electronic enthalpy[a] Differential enthalpy Dew-point and dry-bulb temperatures	Fixed dry bulb Differential dry bulb
All other climates zones	Fixed dry bulb Differential dry bulb Fixed enthalpy Electronic enthalpy[a] Differential enthalpy Dew-point and dry-bulb temperatures	—

a. Electronic enthalpy controllers are devices that use a combination of humidity and dry-bulb temperature in their switching algorithm.

temperatures of 50°F (10°C) dry bulb/45°F (7.2°C) wet bulb and below.

Exception: Systems in which a water economizer is used and where dehumidification requirements cannot be met using outdoor air temperatures of 50°F (10°C) dry bulb/45°F (7.2°C) wet bulb, shall satisfy 100 percent of the expected system cooling load at 45°F (7.2°C) dry bulb/40°F (4.4°C) wet bulb.

606.5.1.2.2 Maximum pressure drop. Precooling coils and water-to-water heat exchangers used as part of a water economizer system shall have a water-side pressure drop of less than 15 feet of water column (44 835 Pa) including the control valve or a secondary loop shall be created so that the coil or heat exchanger pressure drop is not seen by the circulating pumps when the system is in the normal cooling noneconomizer mode.

606.5.1.2.3 Integrated economizer control. Economizer systems shall be integrated with the mechanical cooling system and shall be capable of providing partial cooling whether or not additional mechanical cooling is required to meet the remainder of the cooling load.

606.5.1.2.4 Economizer heating system impact. Heating, ventilating and air-conditioning (HVAC) system design and economizer controls shall be so that economizer operation does not increase the building heating energy use during normal operation.

Exception: Economizers on variable air volume (VAV) systems that cause zone level heating to increase because of reduction in supply air temperature.

606.6 Variable air volume (VAV) fan control. Individual fans with motors equal to or greater than 1.0 horsepower (0.746 kW) shall be one of the following:

1. Driven by a mechanical or electrical variable speed drive.
2. Driven by a vane-axial fan with variable-pitch blades.

TABLE 606.5.1.1.3(2)
HIGH-LIMIT SHUTOFF CONTROL SETTING FOR AIR ECONOMIZERS

DEVICE TYPE	CLIMATE ZONE	REQUIRED HIGH LIMIT (Economizer off when)	
		Equation	Description of equation
Fixed dry bulb	1B, 2B, 3B, 3C, 4B, 4C, 5B, 5C, 6B, 7, 8	$T_{OA} > 75°F$	Outdoor air temperature (T_{OA}) is greater than 75°F
	5A, 6A, 7A	$T_{OA} > 70°F$	Outdoor air temperature (T_{OA}) is greater than 70°F
	All other zones	$T_{OA} > 65°F$	Outdoor air temperature (T_{OA}) is greater than 65°F
Differential dry bulb	1B, 2B, 3B, 3C, 4B, 4C, 5A, 5B, 5C, 6A, 6B, 7, 8	$T_{OA} > T_{RA}$	Outdoor air temperature (T_{OA}) is greater than return air temperature (T_{RA})
Fixed enthalpy	All	$h_{OA} > 28$ Btu/lb[a]	Outdoor air enthalpy (h_{OA}) is greater than 28 Btu/lb of dry air[a]
Electronic enthalpy	All	$(T_{OA}/RH_{OA}) > A$	Outdoor air temperature (T_{OA}) divided by RH_{OA} is greater than the "A" setpoint curve[b]
Differential enthalpy	All	$h_{OA} > h_{RA}$	Outdoor air enthalpy (h_{OA}) is greater than return air enthalpy (h_{RA})
Dew-point and dry bulb temperatures	All	$DP_{OA} > 55°F$ or $T_{OA} > 75°F$	Outside dew point (DP_{OA}) is greater than 55°F or Outdoor air dry bulb (T_{OA}) is greater than 75°F

For SI: °C = [(°F) - 32]/1.8, 1 foot = 304.8 mm, 1 British thermal unit per pound = 2326 J/Kg.

a. At altitudes substantially different than sea level, the fixed enthalpy limit shall be set to the enthalpy value at 75°F and 50-percent relative humidity. As an example, at approximately 6000 feet elevation the fixed enthalpy limit is approximately 30.7 Btu/lb.

b. Setpoint "A" corresponds to a curve on the psychometric chart that goes through a point at approximately 75°F and 40-percent relative humidity and is nearly parallel to dry-bulb lines at low humidity levels and nearly parallel to enthalpy lines at high-humidity levels.

3. Provided with controls or devices that will result in fan motor demand of not more than 30 percent of its design wattage at 50 percent of design airflow when static pressure set point equals one-third of the total design static pressure, based on manufacturer's certified fan data.

Static pressure sensors used to control VAV fans shall be placed in a position so that the controller set point is not greater than one-third of the total design fan static pressure, except for systems with direct digital control. Where this results in the sensor being located downstream of major duct branching, multiple sensors shall be installed in each major branch to ensure that the static pressure can be maintained in each branch.

For systems with direct digital control of individual zone boxes reporting to the central control panel, the static pressure set point shall be reset based on the zone requiring the most pressure. The set point shall be reset lower until one zone damper is wide open.

Exception: Systems without zone dampers are exempt from the static pressure reset requirements.

606.7 Kitchen exhaust systems. Kitchen ventilation and exhaust systems shall be in accordance with the *International Mechanical Code* and this section. Kitchen ventilation systems that deliver conditioned supply air to any space containing a kitchen hood shall not be capable of exceeding the greater of the following:

1. The ventilation rate required to supply the space conditioning load; or
2. The hood exhaust flow minus the available transfer air from adjacent spaces. For the purposes of this section, available transfer air is considered to be that portion of outdoor ventilation air not required to satisfy other exhaust needs, such as restrooms, and not required to maintain pressurization of adjacent spaces.

Where the total hood exhaust airflow rate of kitchen hoods in the space is greater than 5,000 cfm (2360 L/s) each hood shall have an exhaust rate in not greater than 110 percent of the minimum exhaust rate required by the *International Mechanical Code* and the ventilation system shall comply with one of the following:

1. Not less than 50 percent of replacement air is transfer air that would otherwise be exhausted.
2. Demand ventilation systems that are capable of reducing exhaust and replacement air system airflow rates by not less than 50 percent for not less than 75 percent of the exhaust air. The demand ventilation system shall include controls necessary to modulate airflow in response to appliance operation and to maintain full capture and containment of smoke, effluent and combustion products during cooking and when idle.
3. Listed energy recovery devices with a sensible heat recovery effectiveness of not less than 40 percent shall provided for not less than 50 percent of the total exhaust air.

Where a single hood, or hood section, is installed over appliances with different duty ratings, the maximum allowable flow rate for the hood or hood section shall be based on the requirements for the appliance with the highest duty rating located under the hood or hood section.

Exception: Where not less than 75 percent of the replacement air provided by the kitchen ventilation and exhaust system is transfer air that would otherwise be exhausted, the provisions of this section shall not apply.

606.8 Laboratory exhaust systems. Laboratory exhaust systems shall comply with the provisions of the *International Energy Conservation Code* except as specified in Section 606.8.1.

606.8.1 Laboratory exhaust systems. Buildings with laboratory exhaust systems having a total exhaust rate greater than 5,000 cfm (2360 L/s) shall be provided with one or more of the following:

1. A variable air volume (VAV) laboratory exhaust and room supply system capable of reducing exhaust and makeup air flow rates to the minimum required in the *International Mechanical Code*.
2. A heat recovery system to precondition makeup air from laboratory exhaust so that the percentage that the exhaust and makeup air flow rates can be reduced from design conditions plus the sensible recovery effectiveness percentage totals not less than 50 percent.
3. Direct makeup auxiliary air supply equal to not less than 75 percent of the exhaust air flow rate capable of being heated and cooled to the design temperatures specified in Section C302.1 of the *International Energy Conservation Code*.

606.9 Control of HVAC in Group R-1 sleeping rooms. In Group R-1 occupancies, each sleeping room shall be provided with a dedicated system to control automatically the heating, ventilating and air-conditioning (HVAC) systems to control the energy consumption during unoccupied periods. The controls shall be designed to raise cooling and lower heating temperature set points by at least 4°F (-15.6°C) during periods when the sleeping room is unoccupied.

Exception: Automatic controls are not required in Group R-1 occupancies with fewer than 20 sleeping rooms.

SECTION 607
BUILDING SERVICE WATER HEATING SYSTEMS

607.1 Prescriptive compliance. Where buildings are designed using the prescriptive-based compliance path in accordance with Section 601.3.2, service water heating systems shall comply with the provisions of the *International Energy Conservation Code* and the provisions of this section.

607.2 Service water heating (SWH) equipment performance requirements. Service water heating equipment shall comply with Sections 607.2.1 and 607.2.2.

607.2.1 Equipment covered by federal standards. Equipment covered by federal minimum efficiency standards shall comply with the minimum efficiency requirements of the *International Energy Conservation Code*.

607.2.2 Water heater controls for dwelling units. Water heaters installed in dwelling units in buildings shall be equipped with external water temperature thermostat controls. The controls shall allow the occupant to set the water temperature at a setting that is below 100°F (38°C) and greater than or equal to 50°F (10°C).

607.3 Pools, hot tubs and spas. Pools, hot tubs and spas shall comply with the efficiency requirements of the *International Energy Conservation Code*.

607.3.1 Pools in conditioned space. For pools that are located within the conditioned space, not less than 25 percent of the annual energy consumption of pool operation and not less than 50 percent of the peak design space heating, ventilation, and cooling requirements for the space in which the pool is located shall be by one or both of the following:

1. An onsite renewable energy system.
2. A heat recovery system.

607.4 Snowmelt systems. Snow melt systems shall comply with the requirements of the *International Energy Conservation Code*. Hydronic systems shall supplement not less than 25 percent of the design snow melting total annual consumption measured in Btu/ft^2 (J/m^2), energy per unit area. Electric systems shall supplement not less than 50 percent of the design snow melt peak load demand. These requirements shall be supplied by one or both of the following:

1. An onsite renewable energy system.
2. A heat recovery system.

Exception: Emergency service ingress and egress are exempt from the requirements of Section 607.4.

607.5 Waste water heat recovery system. The following building types shall be provided with a waste water heat recovery system that will preheat the incoming water used for hot water functions by not less than 10°F (5.6°C):

1. Group A-2, restaurants and banquet halls;
2. Group F, laundries;
3. Group R-1, boarding houses (transient), hotels (transient), motels (transient);
4. Group R-2 buildings;
5. Group A-3, health clubs and spas; and
6. Group I-2, hospitals, psychiatric hospitals and nursing homes.

Exception: Waste water heat recovery systems are not required for single-story slab-on-grade and single-story on crawl-space buildings.

607.6 Service water heating piping insulation. Service water heating piping shall be thermally insulated in accordance with Table 606.4. Where hot water distribution piping is installed within attics and crawlspaces, the insulation shall continue to cover the pipe for a distance not less than 6 inches (152 mm) beyond the *building thermal envelope*. Where hot water distribution piping is installed within walls, the insulation shall completely surround the pipe with not less than 1 inch (25 mm) of insulation. Where hot water piping is installed in a wall cavity of insufficient size to accommodate the pipe and insulation levels of Table 606.4, the insulation thickness shall be permitted to have the maximum thickness that the wall cavity can accommodate, but not less than $^1/_2$-inch (12 mm) thick.

Exception: Insulation is not required for the following:

1. Factory-installed piping within service water heating equipment tested and rated in accordance with Section 606.4.
2. Piping conveying fluids that is neither heated nor cooled, including cold water supply and natural gas piping.
3. Hot water supply piping exposed under sinks, lavatories and similar fixtures.
4. Hot water distribution piping buried within blown-in or sprayed roof/ceiling insulation, such as fiberglass or cellulose, where the insulation completely and continuously surrounds the pipe.

607.6.1 Buried piping. Service hot water piping installed within a slab or below grade shall be insulated in accordance with Section 607.6 and shall be placed within a physically protective, waterproof channel or sleeve having internal dimensions large enough so that the piping and insulation can be removed and replaced, and maintain its dimensional integrity during and after construction.

Exception: For piping other than that located under building slabs, insulation is not required where the insulation manufacturer stipulates that the pipe insulation will maintain its insulating value in underground applications in damp soil where installed in accordance with the manufacturer's instructions.

607.7 Circulating hot water systems. Controls that allow continuous, timer, or water temperature-initiated (aquastat) operation of a circulating pump are prohibited. Gravity or thermosyphon circulation loops are prohibited. Pumps on circulating hot water systems shall be activated on demand by either a hard-wired or wireless activation control of one of the following types:

1. A normally open, momentary contact switch.
2. Motion sensors that make momentary contact when motion is sensed. After the signal is sent, the sensor shall go into a lock out mode for not less than 5 minutes to prevent sending a signal to the electronic controls while the circulation loop is still hot.
3. A flow switch.
4. A door switch.

The controls for the pump shall be electronic and operate on the principal of shutting off the pump with a rise in temperature. Electronic controls shall have a lock-out to prevent operation at temperatures greater than 105°F (41°C) in the event of failure of the device that senses temperature rise. The electronic controls shall have a lock out mode for not more than 5 minutes that prevents extended operation of the pump if the sensor fails or is damaged.

SECTION 608
BUILDING ELECTRICAL POWER AND LIGHTING SYSTEMS

608.1 General. Where buildings are designed using the prescriptive-based compliance path in accordance with Section 601.3.2, building electrical power and lighting systems shall comply with the provisions of the *International Energy Conservation Code* and the provisions of Section 608.

608.1.1 Occupant sensor controls. Occupant sensor controls shall comply with Section C405.2 of the *International Energy Conservation Code*.

608.1.2 Time switch controls. Time switch controls shall comply with Section C405.2 of the *International Energy Conservation Code*.

608.1.3 Automatic daylight controls. Automatic *daylight controls* shall comply with Section C405.2 of the *International Energy Conservation Code*.

608.2 Sleeping unit controls. Sleeping units in Group R-1 and R-2 occupancies shall have an automatic control system or device that shuts off permanently wired luminaires and switched receptacles, except those in bathrooms, within 30 minutes of the unit being vacated.

Exception: Sleeping unit controls are not required in sleeping units where permanently wired luminaires and switched receptacles, except those in bathrooms, are connected to a captive key control.

608.2.1 Sleeping unit bathroom controls. Permanently wired luminaires located in bathrooms within sleeping units in Group R-1 and R-2 occupancies shall be equipped with occupant sensors that require manual intervention to energize circuits.

Exception: Not more than 5 watts of lighting in each bathroom shall be permitted to be connected to the captive key control at the main room entry instead of being connected to the occupant sensor control.

608.3 Interior light reduction controls. Occupant sensor controls shall be provided to automatically reduce connected lighting power by not less than 45 percent during periods when occupants are not present in the following locations:

1. Corridors and enclosed stairwells;
2. Storage and stack areas not open to the public; and
3. Parking garages.

Exception: Automatic power reduction is not required for the following:

1. Where occupant sensor controls are overridden by time switch controls that keep lights on continuously during peak occupancy periods.
2. Means of egress lighting required by the *International Building Code* or the *International Fire Code*.

608.4 Exterior lighting controls. Exterior lighting shall comply with the requirements of Sections 608.4.1 and 608.4.2.

608.4.1 Exterior light reduction. Exterior lighting shall be controlled by a time switch and configured so that the total exterior lighting power is automatically reduced by not less than 30 percent within 2 hours after facility operations conclude.

Exception: Exterior lighting need not be controlled for the following occupancies and conditions:

1. Group H occupancies.
2. Group I-3 occupancies
3. Lighting that is connected to occupant sensor controls.
4. Means of egress lighting required by the *International Building Code* or the *International Fire Code*.
5. Solar powered luminaires that are not connected to a centralized power source.

608.4.2 Exterior lighting and signage shutoff. The lighting of building facades, signage, and landscape features shall be controlled by a time switch control configured so that the lighting automatically shuts off from within 1 hour after facility operations conclude until within 1 hour before facility operations begin or as established by the jurisdiction. Where facility operations are continuous, decorative lighting of building facades and landscape features shall automatically shut off from midnight until 6:00 a.m.

608.5 Automatic daylight controls. Automatic *daylight controls* shall be provided in daylit areas complying with Section 808.3.1 or Section 808.3.2 to control the lights serving those areas. General lighting in a sidelighting daylit area that is within one window head height shall be separately controlled by automatic *daylight controls*.

Exception: Automatic *daylight controls* are not required for the following spaces and equipment:

1. Toplighting daylit areas where the skylight is located in a portion of the roof that is shaded during the peak sun angle on the summer solstice by permanent features of the building or by permanent features of adjacent buildings.
2. Sidelighting daylit areas where the fenestration is located in an obstructed exterior wall that does not face a public way or a yard or court complying with Section 1206 of the *International Building Code* or where the distance to any buildings, structures, or geological formations in front of the wall is less than two times the height of the buildings, structures, or geological formations.
3. Daylit areas served by less than 90 watts of lighting.
4. Spaces where medical care is directly provided.
5. Spaces within dwelling units or sleeping units.
6. Lighting required to comply with Section C405.2.3 of the *International Energy Conservation Code*.

608.6 Plug load controls. Receptacles and electrical outlets in the following spaces shall be controlled by an occupant sensor or time switch as follows:

1. In Group B office spaces without furniture systems incorporating wired receptacles, not less than one con-

trolled receptacle shall be provided for each 50 square feet (4.65 m^2).

2. In Group B office spaces with furniture systems incorporating wired receptacles, not less than one controlled circuit shall be provided at each electrical outlet used for powering furniture systems.
3. In classrooms in Group B and Group E occupancies, not less than four controlled receptacles shall be provided in each classroom.
4. In copy rooms, print shops, and computer labs, not less than one controlled receptacle shall be provided for each data jack.
5. In spaces with an overhead cabinet above a counter or work surface, not less than one controlled receptacle shall be provided for each work surface.

608.6.1 Distribution and marking. Controlled receptacles and electrical outlets shall be distributed in a reasonably uniform pattern throughout each space. Controlled receptacles shall be marked to differentiate them from uncontrolled receptacles.

608.6.2 Furniture systems. Furniture systems incorporating wired receptacles shall include not less than two receptacles at each workstation that are connected to a controlled circuit.

608.6.3 Computer office equipment. Computer monitors, plug in space heaters, air purifiers, radios, computer speakers, coffee makers, fans, and task lights located in spaces with controlled receptacles shall be plugged into controlled receptacles.

608.6.4 Audio and visual systems. Displays, projectors, and audio amplifiers in Group B and Group E classrooms, conference and meeting rooms, and multipurpose rooms shall be controlled by an occupant sensor.

608.6.5 Water dispensers. Water dispensers that utilize energy to cool or heat drinking water shall be controlled by time switch controls.

608.6.6 Refrigerator and freezer cases. Lighting integral to vending machines and refrigerator and freezer cases shall be controlled by an occupant sensor or a time switch.

608.7 Fuel gas lighting systems. Fixtures that generate illumination by combustion of fuel gas shall be included in lighting power calculations required under Sections C405.5 and C405.6 of the *International Energy Conservation Code* by converting the maximum rated Btu/h of the luminaire into watts using Equation 6-5.

Wattage Equivalent = Maximum btu/h rating of the fuel gas lighting system/3.413. **Equation 6-5**

Exception: Fuel gas lighting at historic buildings in accordance with Section C101.4.2 of the *International Energy Conservation Code* is not included in the calculation.

608.7.1 Continuously burning pilot lights. Fixtures that generate illumination by combustion of fuel gas shall not contain continuously burning pilot lights.

608.8 Electrical system efficiency. Electrical systems shall comply with Section 608.8.1.

608.8.1 Prescriptive compliance. Prescriptive compliance for electrical systems shall be in accordance with Sections 608.8.1.1 through 608.8.1.3.

608.8.1.1 Transformer efficiency. Distribution transformers installed on the load side of the service disconnecting means shall comply with the provisions of Tables 608.8.1.1(1), 608.8.1.1(2) and 608.8.1.1(3), and the Energy Policy Act of 2005 as applicable.

Exception: The following transformers are exempt from the efficiency standards of Section 608.8.1.1:

1. Transformers not covered by the Energy Policy Act of 2005.
2. Transformers for special purpose applications, and not used in general purpose applications.
3. Transformers with multiple voltage taps where the highest tap is not less than 20 percent more than the lowest tap.
4. Drive transformers, rectifier transformers, auto-transformers, uninterruptible power supply transformers, impedance transformers, regulating transformers, sealed and non-ventilating transformers, machine tool transformers, welding transformers, grounding transformers, and testing transformers.

608.8.1.2 Voltage drop in feeders. The voltage drop in feeder conductors shall not exceed 1.5 percent at design load.

608.8.1.3 Voltage drop in branch circuits. The voltage drop in branch circuit conductors shall not exceed 1.5 percent at design load.

TABLE 608.8.1.1(1)
LOW-VOLTAGE DRY-TYPE DISTRIBUTION TRANSFORMERS
(Maximum 600 Volt Primary)[a]

SINGLE PHASE		THREE PHASE	
kVA Rating	Minimum Efficiency (%)	kVA Rating	Minimum Efficiency (%)
15	97.7	15	97.0
25	98.0	30	97.5
37.5	98.2	45	97.7
50	98.3	75	98.0
75	98.5	112.5	98.2
100	98.6	150	98.3
167	98.7	225	98.5
250	98.8	300	98.6
333	98.9	500	98.7
—	—	750	98.8
—	—	1000	98.9

a. All efficiency values for low-voltage transformers are at 35 percent of nameplate-rated load, determined in accordance with the DOE test procedure. 10 CFR Part 431, Sub-part K, Appendix A.

TABLE 608.8.1.1(2)
MEDIUM-VOLTAGE DRY-TYPE DISTRIBUTION TRANSFORMERS
(Maximum 34,500 Volt Primary, Maximum 600 Volt Secondary)[a]

SINGLE PHASE				THREE PHASE			
kVA Rating	20-45 kV BIL Minimum Efficiency (%)	46-95 kV BIL Minimum Efficiency (%)	>96 kV BIL Minimum Efficiency (%)	kVA Rating	20-45 kV BIL Minimum Efficiency (%)	46-95 kV BIL Minimum Efficiency (%)	>96 kV BIL Minimum Efficiency (%)
15	98.10	97.86	—	15	97.50	97.18	—
25	98.33	98.12	—	30	97.90	97.63	—
37.5	98.49	98.30	—	45	98.10	97.86	—
50	98.60	98.42	—	75	98.33	98.12	—
75	98.73	98.57	98.53	112.5	98.49	98.30	—
100	98.82	98.67	98.63	150	98.60	98.42	—
167	98.96	98.83	98.80	225	98.73	98.57	98.53
250	99.07	98.95	98.91	300	98.82	98.67	98.63
333	99.14	99.03	98.99	500	98.96	98.83	98.80
500	99.22	99.12	99.09	750	99.07	98.95	98.91
667	99.27	99.18	99.15	1000	99.14	99.03	98.99
833	99.31	99.23	99.20	1500	99.22	99.12	99.09
—	—	—	—	2000	99.27	99.18	99.15
—	—	—	—	2500	99.31	99.23	99.20

BIL = Basic impulse insulation level.

a. All efficiency values for medium-voltage transformers are at 50 percent of nameplate-rated load, determined in accordance with the DOE test procedure. 10 CFR Part 431, Sub-part K, Appendix A.

TABLE 608.8.1.1(3)
MEDIUM-VOLTAGE LIQUID-IMMERSED DISTRIBUTION TRANSFORMERS
(Maximum 34,500 Volt Primary, Maximum 600 Volt Secondary)[a]

SINGLE PHASE		THREE PHASE	
kVA Rating	Minimum Efficiency (%)	kVA Rating	Minimum Efficiency (%)
10	98.62	15	98.36
15	98.76	30	98.62
25	98.91	45	98.76
37.5	99.01	75	98.91
50	99.08	112.5	99.01
75	99.17	150	99.08
100	99.23	225	99.17
167	99.25	300	99.23
250	99.32	500	99.25
333	99.36	750	99.32
500	99.42	1000	99.36
667	99.46	1500	99.42
883	99.49	2000	99.46
—	—	2500	99.49

a. All efficiency values for medium-voltage transformers are at 50 percent of nameplate-rated load, determined in accordance with the DOE test procedure. 10 CFR Part 431, Sub-part K, Appendix A.

608.9 Exterior lighting. Exterior lighting shall comply with Sections C405.6.1 and C405.6.2 of the *International Energy Conservation Code* regardless of how the power for that lighting is supplied.

Exception: Lighting for the following purposes is exempt:

1. Where *approved* because of historical, safety, signage, or emergency lighting considerations.
2. Roadway lighting required by governmental authorities.

608.10 Verification of lamps and ballasts. Prior to issuance of a certificate of occupancy, the field inspector shall confirm the installation of luminaires, type and quantity; lamps, type, wattage and quantity, and ballasts, type and performance for not less than one representative luminaire of each type, for consistency with the *approved* construction documents. Where a discrepancy is found, energy calculations shall be revised and resubmitted.

608.11 Verification of lighting controls. Prior to issuance of a certificate of occupancy, the field inspector shall confirm the installation of lighting controls shown on the *approved* construction documents. Where a discrepancy is found, the installation shall be reviewed for conformance to the *International Energy Conservation Code* and Sections 608.2, 608.3, 608.4, 608.5, and 608.6.

608.12 Main electrical panel rating. The main electrical service entrance panel for the building shall be listed and labeled as a suitable connection to an onsite renewable energy source.

SECTION 609
SPECIFIC APPLIANCES AND EQUIPMENT

609.1 General. This section provides requirements for appliances and equipment installed in the building or on the building site. Permanent appliances and equipment shall comply with the provisions of Section 609.2, and portable appliances and equipment shall comply with the provisions of Section 609.3.

Exception: Section 609 does not apply to appliances and equipment in compliance with Sections 605 through 608 and those specified in Table 609.1.

TABLE 609.1
APPLIANCES AND EQUIPMENT COVERED BY FEDERAL EFFICIENCY STANDARDS

RESIDENTIAL PRODUCTS	COMMERCIAL PRODUCTS
Battery chargers[a]	Automatic ice makers
Ceiling fans and ceiling fanlight kits	Commercial clothes washers
Clothes dryers	Distribution transformers
Clothes washers	Electric motors[a]
Dehumidifiers	HD lamps[a]
Dishwashers	Metal halide lamp fixtures
Fluorescent and incandescent lamps	Refrigerated beverage vending machines[a]
Fluorescent lamp ballasts[a]	Walk-in coolers and walk-in freezers
Microwave ovens[a]	
Ranges and ovens	
Refrigerators, refrigerator-freezers, and freezers	
Room air conditioners	
Torchieres	

a. These products currently have no federal standards. NOTE: U.S. Department of Energy rulemakings are underway or scheduled.

609.2 Permanent appliances and equipment. Appliances and equipment that are permanently connected to the building energy supply systems shall comply with the provisions of Sections 609.2.1 through 609.2.4 as applicable. Such appliances and equipment shall be listed and labeled and installed in accordance with the manufacturer's installation instructions and the provisions and terms of their listing, the *International Energy Conservation Code*, *International Fuel Gas Code*, *International Mechanical Code*, *International Plumbing Code* and *International Building Code*, and shall be provided with controls and energy monitoring systems as required by this code.

609.2.1 Elevators. Elevator systems shall comply with Sections 609.2.1.1 through 609.2.1.2.3.

609.2.1.1 Lighting. The total lighting in each elevator cab shall be not less than 35 lumens per watt, based on the total lumens from lamps divided by the total wattage of the luminaires in the cab, but not including luminaires of signals and displays.

609.2.1.2 Power conversion system. Power conversion systems for traction elevators shall comply with Sections 609.2.1.2.1 through 609.2.1.2.3.

609.2.1.2.1 Motor. Induction motors with a Class IE2 efficiency rating, as defined by IEC EN 60034-30, or alternative technologies, such as permanent magnet synchronous motors that have equal or better efficiency, shall be used.

609.2.1.2.2 Transmission. Transmissions shall not reduce the efficiency of the combined motor/transmission below that shown for the Class IE2 motor. Gearless machines shall be assumed to have a 100-percent transmission efficiency.

609.2.1.2.3 Drive. Potential energy released during motion shall be recovered.

609.2.1.3 Ventilation. Cab ventilation fans shall have an efficacy greater than or equal to 3.0 cfm per watt (0.085 m^3/min./watt).

609.2.1.4 Standby mode. When the elevator is stopped, not occupied, and with doors closed, lighting, ventilation, and cab displays shall be capable of being de-energized within 5 minutes of stopping, and re-energized prior to opening the doors. Power shall cease to be applied to the door motor after the elevator is stopped, lighting is de-energized, and no one is in the cab, and re-energized upon the next passenger arrival. In buildings with multiple elevators serving the same floors, not less than half of the elevators shall be capable of switching to sleep, low-power mode, during periods of low traffic.

609.2.1.5 Guides. Elevator car guides shall be of the roller type, in order to reduce frictional energy losses. Counterweights with sliding guides shall be balanced in order to minimize frictional losses associated with the counterweight guides.

609.2.2 Escalators and moving walkways. Escalators and moving walkways shall comply with Sections 609.2.2.1 through 609.2.2.5.

609.2.2.1 Lighting. Light sources, including, but not limited to, balustrade lighting, comb-plate lighting and step demarcation lighting, shall have an efficacy of not less than 35 lm/W, based on the total lumens from lamps divided by the total wattage of the luminaires provided on the escalator or moving walk.

609.2.2.2 Drive system. Induction motors with a class IE3 efficiency rating, as defined by IEC EN 60034-30, or permanent magnet synchronous motors shall be used.

609.2.2.3 Energy recovery. Down-running escalators equipped with direct variable frequency drives shall use regenerative drives and return recovered energy to the building electrical power system.

609.2.2.4 Handrails. Handrails shall use friction-reducing measures, such as, but not limited to, rollers in newels.

609.2.2.5 Standby mode. During standby mode, escalators and moving walkways shall be capable of being automatically slowed to not greater than 50 percent of nominal speed. Escalators and moving walkways shall be capable of being automatically turned off when the building is unoccupied or outside of facility operations. In locations where multiple escalators serve the same passenger load, not less than 50 percent of the escalators shall have the capability of being turned off in response to reduced occupant traffic.

609.2.3 Commercial food service equipment. Not less than 50 percent of the commercial food service equipment installed shall comply with energy efficiency and water use as identified on Table 609.2.3, based on aggregate energy input rating.

609.2.4 Conveyors. Motors associated with conveyors shall be sized to meet the expected load and designed to run within 90 percent of capacity at all times the conveyor is expected to operate. Conveyor motors shall be provided with sleep mode controls. Two-speed motors and adjustable-speed drives shall be provided where load weights are expected to vary. Readily accessible controls shall be provided to allow for manual shut off of the conveyor when the conveyor is not needed. Conveyor systems shall be designed to use gravity feed where conditions allow and arranged so that long straight runs are provided with as few drives as possible.

SECTION 610
BUILDING RENEWABLE ENERGY SYSTEMS

610.1 Renewable energy systems requirements. Buildings that consume energy shall comply with this section. Each building or surrounding lot or building site where there are multiple buildings on the building site shall be equipped with one or more renewable energy systems in accordance with this section.

Renewable energy systems shall comply with the requirements of Section 610.2 for solar photovoltaic systems, Section 610.3 for wind systems, or Section 610.4 for solar water heating systems, and Section 610.5 for performance monitoring and metering of these systems as *approved* by the *code official*. These systems shall be commissioned in accordance with the requirements of Section 611.

Exception: Renewable energy systems are not required for the following:

1. Buildings or building sites where there are multiple buildings on the building site providing not less than 2 percent of the total estimated annual energy use of the building, or collective buildings on the site, with onsite renewable energy using a combination of renewable energy generation systems complying with the requirements of Section 610.2, 610.3, or 610.4.
2. Where not less than 4 percent of the total annual building energy consumption from renewable generation takes the form of a 10-year commitment to *renewable energy credit* ownership, confirmed by the *code official*.
3. Where the combined application of onsite generated renewable energy and a commitment to *renewable energy credit* ownership as confirmed by the *code official*, totals not less than 4 percent of the total annual building energy consumption from renewable generation.

610.1.1 Building performance-based compliance. Buildings and surrounding property or building sites where there are multiple buildings on the building site, that are designed and constructed in accordance with Section 601.3.1, performance-based compliance, shall be equipped with one or more renewable energy systems that have the capacity to provide not less than 2 percent of the total calculated annual energy use of the building, or collective buildings on the site.

610.1.2 Building prescriptive compliance. Buildings and surrounding property or building sites where there are multiple buildings on the building site, that are designed and constructed in accordance with Section 601.3.2, prescriptive compliance, shall be equipped with one or more renewable energy systems that have the capacity to provide not less than 2 percent of the total estimated annual energy use of the building, or collective buildings on the building site, with onsite renewable energy by calculation demonstrating that onsite renewable energy production has a rating of not less than 1.75 Btu/h (0.5 W) or not less than 0.50 watts per square foot of conditioned floor area, and using any single or combination of renewable energy generation systems meeting the requirements of Sections 610.2, 610.3, or 610.4.

610.2 Solar photovoltaic systems. Solar photovoltaic systems shall be sized to provide not less than 2 percent of the total estimated annual electric energy consumption of the building, or collective buildings on the building site in accordance with Section 610.1.1 or 610.1.2.

610.2.1 Limitation. Solar photovoltaic systems shall not be used to comply with Section 610.1 where building sites have total global insolation levels lower than 2.00 kWh/m^2/day as determined in accordance with NREL SERI TR-642-761.

610.2.2 Requirements. The installation, inspection, maintenance, repair and replacement of solar photovoltaic systems and system components shall comply with the manufacturer's instructions, Section 610.2.2.1, the *International Fire Code*, the *International Building Code* and NFPA 70.

610.2.2.1 Performance verification. Solar photovoltaic systems shall be tested on installation to verify that the installed performance meets the design specifications. A report of the tested performance shall be provided to the building owner.

TABLE 609.2.3
COMMERCIAL FOOD SERVICE EQUIPMENT—ENERGY-EFFICIENCY AND WATER USE REQUIREMENTS

APPLIANCE TYPE	ENERGY-EFFICIENCY REQUIREMENTS	MAXIMUM WATER USE
Combination Oven/Steamer		
Electric[a]	N/A	3.5 gal/hr/pan
Gas[a]	N/A	3.5 gal/hr/pan
Dishwashers		
Door type, high temp[b]	idle rate ≤ 0.7 kW	0.95 gal/ rack
Door type, low temp[b]	idle rate ≤ 0.6 kW	1.18 gal/ rack
Multiple tank conveyor, high temp[c]	idle rate ≤ 2.0 kW	0.54 gal/ rack
Multiple tank conveyor, low temp[c]	idle rate ≤ 2.0 kW	0.54 gal/ rack
Pot pan and utensil[d]	N/A	2.2 gal/ rack
Rackless conveyor[d]	N/A	2.2 gallons/minute
Single tank conveyor, high temp[c]	idle rate ≤ 1.5 kW	0.7 gal/ rack
Single tank conveyor, low temp[c]	idle rate ≤ 1.5 kW	0.79 gal/ rack
Under counter, high temp[b]	idle rate ≤ 0.5 kW	1.0 gal/ rack
Under counter, low temp[b]	idle rate ≤ 0.5 kW	1.7 gal/ rack
Freezers		
Chest[e]	daily energy ≤ 0.270V + 0.130 kWh/day	N/A
Reach-in, solid door, 0 ≤ V < 15 ft^3 [e]	daily energy ≤ 0.250V + 1.250 kWh/day	N/A
Reach-in, solid door, 15 ≤ V < 30 ft^3 [e]	daily energy ≤ 0.4V – 1.000 kWh/day	N/A
Reach-in, solid door, 30 ≤ V < 50 ft^3 [e]	daily energy ≤ 0.163V + 6.125 kWh/day	N/A
Reach-in, solid door, 50 ≤ V ft^3 [e]	daily energy ≤ 0.158V + 6.333 kWh/day	N/A
Reach-in, transparent door, 0 ≤ V < 15 ft^3 [e]	daily energy ≤ 0.607V + 0.893 kWh/day	N/A
Reach-in, transparent door, 15 ≤ V < 30 ft^3 [e]	daily energy ≤ 0.733V – 1.000 kWh/day	N/A
Reach-in, transparent door, 30 ≤ V < 50 ft^3 [e]	daily energy ≤ 0.250V + 13.50 kWh/day	N/A
Reach-in, transparent door, 50 ≤ V ft^3 [e]	daily energy ≤ 0.450V + 3.50 kWh/day	N/A
Fryers		
Deep fat, electric[f]	efficiency ≥ 50% and idle rate ≤ 9000 Btu/h	N/A
Deep fat, gas[f]	efficiency ≥ 80% and idle rate ≤ 1.0 kW	N/A
Large vat, electric[g]	efficiency ≥ 80% and idle rate ≤ 1.1 kW	N/A
Large vat, gas[g]	efficiency ≥ 50% and idle rate ≤ 12000 Btu/h	N/A
Griddles		
Double-sided, electric[h]	efficiency ≥ 70% and idle rate ≤ 355 W/sq. ft.	N/A
Double-sided, gas[h]	efficiency ≥ 38% and idle rate ≤ 2650 Btu/h/sq. ft.	N/A
Single-sided, electric[i]	efficiency ≥ 70% and idle rate ≤ 355 W/sq. ft.	N/A
Single-sided, gas[i]	efficiency ≥ 38% and idle rate ≤ 2650 Btu/h/sq. ft.	N/A
Hot Food Holding Cabinets		
13 ≤ V ≥ 28 ft^3 [j]	Idle Rate ≤ 2V + 254 Watts	N/A
V < 13 ft^3 [j]	Idle Rate ≤ 21.5V Watts	N/A
V > 28 ft^3 [j]	Idle Rate ≤ 3.8V + 203.5 Watts	N/A

(continued)

TABLE 609.2.3—continued
COMMERCIAL FOOD SERVICE EQUIPMENT—ENERGY-EFFICIENCY AND WATER USE REQUIREMENTS

APPLIANCE TYPE	ENERGY-EFFICIENCY REQUIREMENTS	MAXIMUM WATER USE
	Ice Machines	
Ice making head, H > 450 lb/day[k]	energy ≤ 6.20 – 0.0010H kWh/100 lb ice	25 gal/100 lb ice
Ice making head, H < 450 lb/day[k]	energy ≤ 9.23 – 0.0077H kWh/100 lb ice	25 gal/100 lb ice
Remote condensing unit w/o remote compressor, H < 1000 lb/day[k]	energy ≤ 8.05 – 0.0035H kWh/100 lb ice	25 gal/100 lb ice
Remote condensing unit w/o remote compressor, H > 1000 lb/day[k]	energy ≤ 4.64 kWh/100 lb ice	25 gal/100 lb ice
Remote condensing unit with remote compressor, H < 934 lb/day[k]	energy ≤ 8.05 – 0.0035H kWh/100 lb ice	25 gal/100 lb ice
Remote condensing unit with remote compressor, H > 934 lb/day[k]	energy ≤ 4.82 kWh/100 lb ice	25 gal/100 lb ice
Self-contained unit, H < 175 lb/day[k]	energy ≤ 16.7 – 0.0436H kWh/100 lb ice	35 gal/100 lb ice
Self-contained unit, H > 175 lb/day[k]	energy ≤ 9.11 kWh/100 lb ice	35 gal/100 lb ice
	Convection Ovens	
Full-size electric[l]	efficiency ≥ 70% and idle rate ≤ 1.6 kW	0.25 gals/hr
Full-size gas[l]	efficiency ≥ 44% and idle rate ≤ 13000 Btu/h	0.25 gals/hr
Half-size electric[l]	efficiency ≥ 70% and idle rate ≤ 1.0 kW	0.25 gals/hr
	Refrigerators	
Chest[e]	daily energy ≤ 0.125V + 0.475 kWh/day	N/A
Reach-in, solid door, 0 ≤ V < 15 ft^3 [e]	daily energy ≤ 0.089V + 1.411 kWh/day	N/A
Reach-in, solid door, 15 ≤ V < 30 ft^3 [e]	daily energy ≤ 0.037V + 2.200 kWh/day	N/A
Reach-in, solid door, 30 ≤ V < 50 ft^3 [e]	daily energy ≤ 0.056V + 1.635 kWh/day	N/A
Reach-in, solid door, 50 ≤ V ft^3 [e]	daily energy ≤ 0.06V + 1.416 kWh/day	N/A
Reach-in, transparent door, 0 ≤ V < 15 ft^3 [e]	daily energy ≤ 0.118V + 1.382 kWh/day	N/A
Reach-in, transparent door, 15 ≤ V < 30 ft^3 [e]	daily energy ≤ 0.140V + 1.050 kWh/day	N/A
Reach-in, transparent door, 30 ≤ V < 50 ft^3 [e]	daily energy ≤ 0.088V + 2.625 kWh/day	N/A
Reach-in, transparent door, 50 ≤ V ft^3 [e]	daily energy ≤ 0.110V + 1.500 kWh/day	N/A
	Steam Cookers	
With drain connection, electric[m]	N/A	5 gal/hour/pan
With drain connection, gas[m]	N/A	5 gal/hour/pan
No drain connection, electric[m]	efficiency ≥ 50% and idle rate ≤ 135W/pan	2 gal/hour/pan
No drain connection, gas[m]	efficiency ≥ 38% and idle rate ≤ 2100 Btu/h/pan	2 gal/hour/pan
Water-cooled refrigeration equipment	Not allowed unless on a closed-loop system or cooling tower	—

For SI: °C = [°F – 32]/1.8, 1 Btu/h = 0.29 W.

a. Maximum water use as determined by ASTM F 2861.

b. Idle rate as determined by ASTM F 1696 and water use as determined by ANSI/NSF 3.

c. Idle rate as determined by ASTM F 1920 and water use as determined by ANSI/NSF 3.

d. Water use as determined by ANSI/NSF 3.

e. Daily energy use as determined by ANSI/ASHRAE Standard 72 with temperature set points at 38°F for medium temp refrigerators, 0°F for low temp freezers, and -15°F for ice cream freezers.

f. Heavy-load cooking-energy efficiency and idle rate as determined by ASTM F 1361.

g. Heavy-load (French fry) cooking-energy efficiency and idle rate as determined by ASTM F 2144.

h. Heavy-load cooking-energy efficiency and idle rate as determined by ASTM F 1605.

i. Heavy-load cooking-energy efficiency and idle rate as determined by ASTM F 1275.

j. Idle rate as determined by ASTM F 2140.

k. Energy and water use as determined by ARI 810.

l. Heavy-load (potato) cooking-energy efficiency and idle rate as determined by ASTM F 1496.

m. Heavy-load (potato) cooking-energy efficiency and idle rate as determined by ASTM F 1484.

610.3 Wind energy systems. Wind energy systems shall be designed, constructed and sized to provide not less than 2 percent of the total estimated annual electric energy consumption of the building, or collective buildings on the building site in accordance with NFPA 70 and Section 610.1.1 or 610.1.2.

610.3.1 Installation, location and structural requirements. Wind energy systems shall be located on the building, adjacent to the building, or on the building site.

610.4 Solar water heating equipment. Not less than 10 percent of the building's annual estimated hot water energy usage shall be supplied by onsite solar water heating equipment.

610.5 Renewable energy system performance monitoring and metering. Renewable energy systems shall be metered and monitored in accordance with Sections 610.5.1 and 610.5.2.

610.5.1 Metering. Renewable energy systems shall be metered separately from the building's electrical and fossil fuel meters. Renewable energy systems shall be metered to measure the amount of renewable electric or thermal energy generated on the building site in accordance with Section 603.

610.5.2 Monitoring. Renewable energy systems shall be monitored to measure the peak electric or thermal energy generated by the renewable energy systems during the building's anticipated peak electric or fossil fuel consumption period in accordance with Section 603.

SECTION 611
ENERGY SYSTEMS COMMISSIONING AND COMPLETION

611.1 Mechanical systems commissioning and completion requirements. Within 60 days from approval conducting the final mechanical inspection, the *registered design professional* shall provide evidence of mechanical systems commissioning and completion of the mechanical system installation to the *code official*, in accordance with the *International Energy Conservation Code.*

Drawing notes shall clearly indicate provisions for commissioning and completion requirements in accordance with this section and are permitted to refer to specifications for further requirements. Copies of all documentation shall be given to the owner and made available to the *code official* upon request.

611.1.1 Commissioning plan. A commissioning plan shall be developed by a *registered design professional* or approved agency and shall include as a minimum all of the following items:

1. A narrative describing the activities that will be accomplished during each phase of commissioning, including guidance on who accomplishes the activities and how they are completed.
2. Equipment and systems to be tested including, but not limited to, the specific equipment, appliances or systems to be tested and the number and extent of tests.
3. Functions to be tested including, but not limited to, calibrations and economizer controls.
4. Conditions under which the test shall be performed including, but not limited to, affirmation of winter and summer design conditions and full outside air.
5. Measurable criteria for performance.

611.1.2 Systems adjusting and balancing. HVAC systems shall be balanced in accordance with generally accepted engineering standards. Air and water flow rates shall be measured and adjusted to deliver final flow rates within the tolerances provided in the product specifications. Test and balance activities shall include, at a minimum, the provisions of Sections 611.1.2.1 and 611.1.2.2.

611.1.2.1 Air systems balancing. Each supply air outlet and zone terminal device shall be equipped with a means for air balancing in accordance with the *International Mechanical Code*. Discharge dampers are prohibited on constant volume fans and variable volume fans with motors of 10 hp (7.35 kW) and larger. Air systems shall be balanced in a manner to first minimize throttling losses then, for fans with system power of greater than 1 hp (735 W), fan speed shall be adjusted to meet design flow conditions.

Exception: Fans with fan motor horsepower of 1 hp (735 W) or less.

611.1.2.2 Hydronic systems balancing. Individual hydronic heating and cooling coils shall be equipped with means for balancing and measuring flow. Hydronic systems shall be proportionately balanced in a manner to first minimize throttling losses, then the pump impeller shall be trimmed or pump speed shall be adjusted to meet design flow conditions. Each hydronic system shall have either the capability to measure pressure across the pump, or shall have test ports at each side of each pump.

Exceptions:

1. Pumps with pump motors of 5 hp (3677 W) or less.
2. Where throttling results in not greater than 5 percent of the nameplate horsepower draw above that required if the impeller were trimmed.

611.1.3 Functional performance testing. Functional performance testing shall be in accordance with the requirements of Sections 611.1.3.1, 611.1.3.2 and 611.1.3.3.

611.1.3.1 Equipment. Equipment functional performance testing shall demonstrate the installation and operation of components, systems, and system-to-system interfacing relationships in accordance with *approved* plans and specifications so that operation, function, and maintenance serviceability for each of the commissioned systems is confirmed. Testing shall include all specified modes of control and sequence of operation, including under full-load, part-load and all of the following emergency conditions:

1. Each mode as described in the sequence of operation.
2. Redundant or automatic backup mode.
3. Performance of alarms.
4. Mode of operation upon a loss of power and restoration of power.

611.1.3.2 Controls. HVAC control systems shall be tested to document that control devices, components, equipment, and systems are calibrated, adjusted and operated in accordance with the *approved* plans and specifications. Sequences of operation shall be functionally tested to document that they operate in accordance with the *approved* plans and specifications.

611.1.3.3 Economizers. Air economizers shall undergo a functional test to determine that they operate in accordance with the manufacturer's specifications.

611.1.4 Preliminary commissioning report. A preliminary report of commissioning test procedures and results shall be completed and certified by the *registered design professional* or approved agency and provided to the building owner. The report shall be identified as "Preliminary Commissioning Report" and shall identify all of the following:

1. Itemization of deficiencies found during testing required by this section that have not been corrected at the time of report preparation.
2. Deferred tests that cannot be performed at the time of report preparation because of climatic conditions.
3. Climatic conditions required for performance of the deferred tests.

611.1.4.1 Acceptance. Buildings, or portions thereof, shall not pass the final mechanical inspection until such time as the *code official* has received a letter of transmittal from the building owner acknowledging that the building owner has received the Preliminary Commissioning Report.

611.1.4.2 Copy. At the request of the *code official*, a copy of the Preliminary Commissioning Report shall be made available for review.

611.1.4.3 Certification. A certification, signed and sealed by the *registered design professional*, documenting that the mechanical and service water heating systems comply with Sections C403 and C404 of the *International Energy Conservation Code*, shall be provided to the *code official*.

611.1.5 Completion requirements. The construction documents shall specify that the requirements described in this section be provided to the building owner within 90 days of the date of receipt of the certificate of occupancy.

611.1.5.1 Drawings. Construction documents shall include the location of and performance data pertaining to each piece of equipment.

611.1.5.2 Manuals. An operating and maintenance manual in accordance with industry-accepted standards shall be provided and shall include all of the following:

1. Submittal data stating equipment size and selected options for each piece of equipment requiring maintenance.
2. Manufacturer's operation manuals and maintenance manuals for each piece of equipment requiring maintenance, except equipment not furnished as part of the building project. Required routine maintenance shall be clearly identified.
3. Names and addresses of not less than one service agency.

A systems manual shall be provided and shall include all of the following:

1. HVAC controls system maintenance and calibration information, including wiring diagrams, schematics, and control sequence descriptions. Desired or field-determined setpoints shall be permanently recorded on control drawings at control devices or, for digital control systems, in programming comments.
2. A complete narrative of how each system is intended to operate, including recommended setpoints, seasonal changeover information and emergency shutdown operation.
3. Control sequence descriptions for lighting, domestic hot water heating and all renewable energy systems complete with a description of how these systems connect to, and are controlled in conjunction with, the overall building system.

611.1.5.3 System balancing report. A written report describing the activities and measurements completed in accordance with Section 611.1.2 shall be provided.

611.1.5.4 Final commissioning report. A complete report of test procedures and results identified as "Final Commissioning Report" shall be completed and provided to the building owner. The report shall include all of the following:

1. Results of all functional performance tests.
2. Disposition of all deficiencies found during testing, including details of corrective measures used or proposed.
3. All functional performance test procedures used during the commissioning process including measurable criteria for test acceptance, provided herein for repeatability.

Exception: Deferred tests that were not performed at the time of report preparation because of climatic conditions.

611.1.5.5 Post-occupancy recommissioning. The commissioning activities specified in Sections 611.1.2 through 611.1.5 shall be repeated 18 to 24 months after certificate of occupancy. Systems and control devices that are not functioning properly shall be repaired or replaced. Adjustments to calibration settings shall be documented. This documentation shall be provided to the building owner.

611.2 Sequence of operation. A sequence of operation shall be developed and finalized upon commissioning, when the operational details are initialized and validated. A sequence of operation shall be the final record of system operation, and shall be included on the control diagram "as-builts," or as part of the education and operation and maintenance document that is provided to the owner.

611.3 Lighting and electrical systems commissioning and completion requirements. Prior to issuance of a certificate of occupancy, the *registered design professional* shall provide evidence of lighting and electrical systems commissioning and completion in accordance with the *International Energy Conservation Code* and the provisions of this section.

Drawing notes shall specify the provisions for commissioning and completion requirements in accordance with this section and are permitted to refer to specifications for further requirements. Copies of all documentation shall be given to the owner and made available to the *code official* upon request in accordance with Sections 611.2.4 and 611.2.5

611.3.1 Preconstruction documentation, lighting. Construction and owner education documents shall include floor plans, diagrams and notations of sufficient clarity describing the types of, location and operational requirements of all lighting controls including a sequence of operation and preliminary intended setpoints for all dimming systems and automatic *daylight controls*, demonstrating conformance to the provisions of this code, relevant laws, ordinances, rules and regulations, as *approved* by the *code official*.

611.3.2 Verification. The approved agency conducting commissioning shall verify that controls have been installed in accordance with the *approved* construction documents. Any discrepancies shall be reviewed for compliance with Section 608 and the requirements of Section C405.2 of the *International Energy Conservation Code*.

611.3.3 Commissioning. Lighting controls shall be commissioned in accordance with this section.

611.3.3.1 Occupant sensors. It shall be verified that the functional testing in accordance with Section C405.2 of the *International Energy Conservation Code* has been performed.

611.3.3.2 Automatic daylight controls. Automatic *daylight controls* shall be commissioned in accordance with all of the following:

1. It shall be verified that the placement and orientation of each sensor is consistent with the manufacturer's instructions. If not, the sensor shall be relocated or replaced.
2. Control systems shall be initially calibrated to meet settings and design intent established in the construction documents.
3. Prior to calibration of systems controlling dimmable luminaires, all lamps shall be seasoned in accordance with the recommendations of the lamp manufacturer.
4. Where located inside buildings, calibration of open-loop *daylight controls*, which receive illumination from natural light only, shall not occur until fenestration shading devices such as blinds or shades have been installed and commissioned.
5. Calibration of closed-loop *daylight controls*, that receive illumination from both natural and artificial light, shall not occur until furniture systems and interior finishes have been installed, and any fenestration shading devices such as blinds or shades have been installed and commissioned.
6. Calibration procedures shall be in accordance with the manufacturer's instructions.

611.3.3.3 Time switch and programmable schedule controls. Lighting controls installed in accordance with Section 608 shall be programmed. Scheduling shall incorporate weekday, weekend and holiday operating times, including leap year and daylight savings time corrections. It shall be verified that system overrides work and are located in compliance with Section C405.2 of the *International Energy Conservation Code*.

611.3.3.4 Dimming systems with preset scenes. For programmable dimming systems, it shall be verified that automatic shutoff and manual overrides are working and that programming is complete. Prior to programming, all lamps shall be seasoned in accordance with NEMA LSD 23.

611.3.4 Post-commissioning documentation. The following documentation shall be provided to the building owner in accordance with Section 903.

1. Settings determined during commissioning activities outlined in Section 611.3.3.
2. A narrative describing the intent and functionality of all controls including any capability for users to override a schedule or master command.
3. Specification sheets for all lighting equipment and controls.
4. Operation manuals for each lighting control device. Required maintenance and maintenance schedules shall be clearly identified. Documentation and instructions necessary for building maintenance personnel to maintain and recalibrate lighting systems and controls.
5. An annual inspection schedule for lighting controls.
6. Troubleshooting information for fluorescent dimming systems and the remediation of switching issues such as false-ons and false-offs.

611.3.5 Post-occupancy recommissioning. The commissioning activities in Section 611.3.3 shall be repeated 18 to 24 months after issuance of the certificate of occupancy. Control devices that are not functioning properly shall be repaired or replaced. Adjustments to calibration settings shall be documented. This documentation shall be provided to the building owner.

611.4 Building envelope systems commissioning and completion requirements. Prior to issuance of a certificate of occupancy, the *registered design professional* shall provide evidence of *building thermal envelope* systems commissioning and completion to the building owner in accordance with the *International Energy Conservation Code* and the provisions of this section.

Construction documents shall specify the provisions for commissioning and completion requirements in accordance with this section and are permitted to refer to specifications for further requirements. Copies of all documentation shall be given to the building owner and made available to the *code official* upon request in accordance with Sections 611.4.1 and 611.4.2.

611.4.1 Preconstruction documentation, building thermal envelope. Construction and owner education documents shall indicate the location, nature and extent of the work proposed and show the functional requirements and operation of all *building thermal envelope* systems demonstrating conformance to the provisions of this code, relevant laws, ordinances, rules and regulations, as *approved* by the *code official*.

611.4.2 Verification. The approved agency conducting commissioning shall verify that *building thermal envelope* systems have been installed in accordance with the *approved* construction documents. Any discrepancies shall be reviewed for compliance with requirements of the *International Energy Conservation Code* and this code.

CHAPTER 7

WATER RESOURCE CONSERVATION, QUALITY AND EFFICIENCY

SECTION 701
GENERAL

701.1 Scope. The provisions of this chapter shall establish the means of conserving water, protecting water quality and providing for safe water consumption.

SECTION 702
FIXTURES, FITTINGS, EQUIPMENT AND APPLIANCES

702.1 Fitting and fixture consumption. Fixtures shall comply with Table 702.1 and the following:

1. For dwelling unit and guestroom shower compartments with a floor area of not greater than 2600 in^2 (1.7 m^2), the combined flow rate from shower water outlets that are capable of operating simultaneously including rain systems, waterfalls, body sprays and jets shall not exceed 2.0 gallons per minute (gpm) (7.6 L/min). Where the floor area of such shower compartments is greater than 2600 in^2 (1.7 m^2), the combined flow rate from simultaneously operating shower water outlets shall not exceed 2.0 gpm (7.6 L/min) for each additional 2600 in^2 (1.7 m^2) of floor area or portion thereof.
2. In gang shower rooms, the combined flow rate from shower water outlets that are capable of operating simultaneously including rain systems, waterfalls, body sprays and jets shall not exceed 2.0 gpm (7.6 L/min) for every 1600 in^2 (1.01 m^2) or portion thereof of room floor area.
3. In shower compartments required to comply with the requirements of Chapter 11 of the *International Building Code*, the combined flow rate from shower water outlets that are capable of operating simultaneously including rain systems, waterfalls, body sprays and jets shall not exceed 4.0 gpm (15.1 L/min) for every 2600 in^2 (1.7 m^2) or portion thereof of room floor area.

702.2 Combination tub and shower valves. Tub spout leakage from combination tub and shower valves that occurs when the outlet flow is diverted to the shower shall not exceed 0.1 gpm, measured in accordance with the requirements of ASME A112.18.1/CSA B125.1.

702.3 Food establishment prerinse spray valves. Food establishment prerinse spray valves shall have a maximum flow rate in accordance with Table 702.1 and shall shut off automatically when released.

702.4 Drinking fountain controls. Drinking fountains equipped with manually controlled valves shall shut off automatically upon the release of the valve. Metered drinking fountains shall comply with the flow volume specified in Table 702.1.

TABLE 702.1
MAXIMUM FIXTURE AND FITTING FLOW RATES FOR REDUCED WATER CONSUMPTION

FIXTURE OR FIXTURE FITTING TYPE	MAXIMUM FLOW RATE
Showerhead[a]	2.0 gpm and WaterSense labeled
Lavatory faucet and bar sink—private	1.5 gpm
Lavatory faucet—public (metered)	0.25 gpc[b]
Lavatory faucet—public (nonmetered)	0.5 gpm
Kitchen faucet—private	2.2 gpm
Kitchen and bar sink faucets in other than dwelling units and guestrooms	2.2 gpm
Urinal	0.5 gpf and WaterSense labeled or nonwater urinal
Water closet—public and remote[c]	1.6 gpf
Water closet—public and nonremote	1.28 gpf average[d, e]
Water closet-tank type, private	1.28 gpf and WaterSense labeled[d]
Water closet—flushometer type, private	1.28 gpf[e]
Prerinse spray valves	1.3 gpm
Drinking fountains (manual)	0.7 gpm
Drinking fountains (metered)	0.25 gpc[b]

For SI: 1 foot = 304.8 mm, 1 gallon per cycle (gpc) = 3.8 Lpc, 1 gallon per flush (gpf) = 3.8 Lpf, 1 gallon per minute (gpm) = 3.8 Lpm.

a. Includes hand showers, body sprays, rainfall panels and jets. Showerheads shall be supplied by automatic compensating valves that comply with ASSE 1016 or ASME A112.18.1/CSA B125.1 and that are specifically designed to function at the flow rate of the showerheads being used.

b. Gallons per cycle of water volume discharged from each activation of a metered faucet.

c. A remote water closet is a water closet located not less than 30 feet upstream of other drain line connections or fixtures and is located where less than 1.5 drainage fixture units are upstream of the drain line connection.

d. The effective flush volume for a dual-flush water closet is defined as the composite, average flush volume of two reduced flushes and one full flush.

e. In public settings, the maximum water use of a dual flush water closet is based solely on its full flush operation; not an average of full and reduced volume flushes.

702.5 Nonwater urinal connection. The fixture drain for nonwater urinals shall connect to a branch drain that serves one or more lavatories, water closets or water-using urinals that discharge upstream of such urinals.

702.6 Appliances. Sections 702.6.1 through 702.6.4 shall regulate appliances that are not related to space conditioning.

702.6.1 Clothes washers. Clothes washers of the type in the ENERGY STAR program as defined in "ENERGY STAR® Program Requirements, Product Specification for Clothes Washers, Eligibility Criteria," shall have a water factor (WF) not exceeding 6.0 and a *modified energy factor* (MEF) of not less than 2.0.

702.6.2 Ice makers. Ice makers shall not be water cooled. Ice makers producing cubed-type ice shall be ENERGY STAR qualified as commercial ice machines. Ice makers of a type not currently ENERGY STAR qualified, such as flake, nugget or continuous-type ice makers, shall not exceed the total water use of 25 gallons per 100 pounds (208 L per 100 kg) of ice produced.

702.6.3 Steam cookers. Steam cookers shall consume not more than the amounts indicated in Table 610.2.3.

702.6.4 Dishwashers. Dishwashers shall be ENERGY STAR qualified where an ENERGY STAR category exists for the specific dishwasher type. Where an ENERGY STAR category does not exist, the dishwasher shall be in accordance with Table 702.6.4.

**TABLE 702.6.4
MAXIMUM WATER CONSUMPTION
FOR COMMERCIAL DISHWASHERS**

DISHWASHER TYPE	MAXIMUM WATER CONSUMPTION
Rackless conveyor	2.2 gallons per minute
Utensil washer	2.2 gallons per rack

For SI: 1 gallon per minute = 3.785 Lpm.

702.7 Municipal reclaimed water. Where required by Table 302.1 and where municipal reclaimed water is accessible and allowed for such use by the laws, rules and ordinances applicable in the jurisdiction, it shall be supplied to water closets, water-supplied urinals, water-supplied trap primers and applicable industrial uses. A municipal reclaimed water supply shall be deemed accessible where the supply is not greater that 150 percent of the distance that the potable water supply is from the lot boundary or the supply is within 100 feet (30.5 m) of a potable water supply that serves the lot.

702.8 Efficient hot and tempered water distribution. Hot and tempered water distribution shall comply with either the maximum pipe length or maximum pipe volume limits in this section. Hot and tempered water shall be delivered to the outlets of individual showers, combination tub-showers, sinks, lavatories, dishwashers, washing machines and hot water hose bibbs in accordance with Section 702.8.1 or Section 702.8.2. For purposes of this section, references to pipe shall include tubing. For purposes of this section, the source of hot or tempered water shall be considered to be a water heater, boiler, circulation loop piping or electrically heat-traced piping.

702.8.1 Maximum allowable pipe length method. The maximum allowable pipe length from the source of hot or tempered water to the termination of the fixture supply pipe shall be in accordance with the maximum pipe length columns in Table 702.8.2. Where the length contains more than one size of pipe, the largest size shall be used for determining the maximum allowable length of the pipe in Table 702.8.2.

702.8.2 Maximum allowable pipe volume method. The water volume in the piping shall be calculated in accordance with Section 702.8.2.1. The maximum volume of hot or tempered water in the piping to public lavatory faucets, metering or nonmetering, shall be 2 ounces (0.06 L). For fixtures other than public lavatory faucets, the maximum volume shall be 64 ounces (1.89 L) for hot or tempered water from a water heater or boiler; and 24 ounces (0.7 L) for hot or tempered water from a circulation loop pipe or an electrically heat-traced pipe.

702.8.2.1 Water volume determination. The volume shall be the sum of the internal volumes of pipe, fittings, valves, meters and manifolds between the source of hot water and the termination of the fixture supply pipe. The volume shall be determined from the liquid ounces per foot column of Table 702.8.2. The volume contained within fixture shutoff valves, flexible water supply connectors to a fixture fitting, or within a fixture fitting shall not be included in the water volume determination. Where hot or tempered water is supplied by a circulation loop pipe or an electrically heat-traced pipe, the volume shall include the portion of the fitting on the source pipe that supplies water to the fixture.

**TABLE 702.8.2
MAXIMUM LENGTH OF PIPE OR TUBE**

NOMINAL PIPE OR TUBE SIZE (inch)	LIQUID OUNCES PER FOOT OF LENGTH	MAXIMUM PIPE OR TUBE LENGTH		
		System without a circulation loop or heat-traced line (feet)	System with a circulation loop or heat-traced line (feet)	Lavatory faucets – public (metering and nonmetering) (feet)
$^1/_4$ [a]	0.33	50	16	6
$^5/_{16}$ [a]	0.5	50	16	4
$^3/_8$ [a]	0.75	50	16	3
$^1/_2$	1.5	43	16	2
$^5/_8$	2	32	12	1
$^3/_4$	3	21	8	0.5
$^7/_8$	4	16	6	0.5
1	5	13	5	0.5
$1^1/_4$	8	8	3	0.5
$1^1/_2$	11	6	2	0.5
2 or larger	18	4	1	0.5

For SI: 1 inch = 25.4 mm, 1 foot = 304.8 mm, 1 gallon per minute = 3.785 L/m, 1 ounce = 29.6 ml.

a. The flow rate for $^1/_4$-inch size pipe or tube is limited to 0.5 gallons per minute; for $^5/_{16}$-inch size, it is limited to 1 gpm; for $^3/_8$-inch size, it is limited to 1.5 gpm.

702.9 Trap priming water. Potable water shall not be used for trap priming purposes where an alternate nonpotable onsite water distribution system, a reclaimed water distribution system or a gray water distribution system is available.

702.9.1 Continuous operation prohibited. Trap primers that allow continuous water flow shall be prohibited.

702.9.2 Volume limitation. Trap primers shall be of the type that use not more than 30 gallons (114 L) per year per trap.

702.9.3 Water criteria. Where nonpotable water is available and is already being used to supply plumbing fixtures, such water shall be used to supply trap primers.

702.10 Water-powered pumps. Water-powered pumps shall not be used as the primary means of removing ground water from sumps. Where used as an emergency backup pump for the primary pump, the primary pump shall be an electrically powered pump and the water-powered pump shall be equipped with an auditory alarm that indicates when the water-powered pump is operating. The alarm shall have a minimum sound pressure level rating of 85 dB measured at a distance of 10 feet (3048 mm). Where water-powered pumps are used, they shall have a water-efficiency factor of pumping not less than 2 gallons (7.6 L) of water to a height of 8 feet (2438 mm) for every 1 gallon (3.8 L) of water used to operate the pump, measured at a water pressure of 60 psi (413.7 kPa). Pumps shall be clearly marked as to the gallons (liters) of water pumped per gallon (liters) of potable water consumed.

702.11 Food service handwashing faucets. Faucets for handwashing sinks in food service preparation and serving areas shall be of the self-closing type.

702.12 Dipper wells. The water supply to a dipper well shall have a shutoff valve and flow control valve. Water flow into a dipper well shall not exceed 1 gpm (3.78 Lpm) at a supply pressure of 60 psi (413.7 kPa).

702.13 Automated vehicle wash facilities. Not less than 50 percent of the water used for the rinsing phase of the wash cycle at automated vehicle wash facilities shall be collected to be reused for the washing phase. Towel and chamois washing machines shall have high-level water cutoffs. Except for water recirculated within the facility, potable and nonpotable water use for automobile washing shall not exceed 40 gallons (151 L) per vehicle for in-bay automatic washing and 35 gallons (132.5 L) per vehicle for conveyor and express-type car washing.

Exception: Bus and large commercial vehicle washing facilities.

702.14 Self-service vehicle wash facilities. Spray wand nozzles used at self-service vehicle wash facilities shall discharge not more than 3 gpm (11.4 Lpm). Faucets for chamois wringer sinks shall be of the self-closing type.

702.15 Vehicle washing facilities. Waste water from reverse osmosis water treatment systems installed in vehicle washing facilities shall discharge to the washing phase water holding tank.

702.16 Food waste disposers. The water flow into a commercial food waste disposer in a food establishment shall be controlled by a load-sensing device such that the water flow does not exceed 1 gpm (3.78 Lpm) under no-load operating conditions and 8 gpm (30.2 Lpm) under full-load operating conditions.

702.17 Combination ovens. Combination ovens shall consume not more than 3.5 gallons (13.25 L) per hour per steamer pan in any operational mode. Water consumption shall be tested in accordance with the requirements of ASTM F 1639.

702.18 Autoclaves and sterilizers. Autoclaves and sterilizers requiring condensate tempering systems shall be of the type that does not require potable water to be blended with the discharge water to reduce the temperature of discharge.

702.18.1 Vacuum autoclaves and sterilizers. Vacuum sterilizers shall be prohibited from utilizing venturi-type vacuum mechanisms using water.

702.19 Liquid ring vacuum pumps. Except where the discharge is contaminated with hazardous materials or pathogens, the discharge water from liquid ring vacuum pumps shall be recovered for reuse within the pump or for other onsite applications.

702.20 Film processors. The cooling water discharge from water-cooled film processors shall be recovered and reused within the processor or for other onsite applications.

SECTION 703
HVAC SYSTEMS AND EQUIPMENT

703.1 Hydronic closed systems. Closed loop hydronic heating and cooling systems, and ground-source heat pump systems shall not be connected to a potable makeup water supply.

703.2 Humidification systems. Except where greater humidity is required for medical, agricultural, archival or scientific research purposes, humidification systems shall be disabled and locked-out when the relative humidity in the space served is greater than 55 percent.

703.3 Condensate coolers and tempering. Potable water shall not be used as tempering water for sanitary discharge where the tempering water volume requirement for the application exceeds 200 gallons per day (757 liters per day). Where the tempering water volume required for the application is 200 gallons per day (757 liters per day) or less and potable water is used for tempering, water flow control devices shall be installed. Such control devices shall limit the flow rate of tempering water to that which is necessary to limit the temperature of the waste discharge to a maximum of 140°F (60°C). Such devices shall have a maximum flow rate of 200 gallons per day (757 liters per day).

703.4 Condensate drainage recovery. Condensate shall be collected and reused onsite for applications such as, but not limited to, water features, fountains, gray water collection systems and rainwater collection systems. Where onsite applications for condensate reuse are not available and the community sanitary sewer authority provides return credit for sanitary sewage or recycles sewage into a nonpotable water supply, condensate shall be discharged to the sanitary sewer system except where prohibited by the authority having jurisdiction.

703.5 Heat exchangers. Once-through cooling shall be prohibited. Heat exchangers shall be connected to a recirculating water system such as a chilled water loop, cooling tower loop or similar recirculating system.

703.6 Humidifier discharge. Water discharge from flow-through-type humidifiers and from the draining and flushing operations of other types of humidifiers shall be collected for reuse where a collection and reuse system exists.

703.7 Cooling towers, evaporative condensers and fluid coolers. Cooling towers, evaporative condensers, and fluid coolers shall be installed in accordance with the requirements of Section 908 of the *International Mechanical Code*.

703.7.1 Location. Cooling towers, evaporative condensers and fluid coolers shall be located on the property as required for buildings in accordance with the *International Building Code* and shall be located so as to prevent the discharge vapor plumes from entering occupied spaces. Plume discharges shall be not less than 5 feet (1524 mm) above and 20 feet (6096 mm) away from any ventilation inlet to a building.

703.7.2 Once-through cooling. The use of potable water for once-through or single-pass cooling operations is prohibited.

703.7.3 Metering. The metering of mechanical systems, system components, equipment and appliances shall be conducted in accordance with Section 705.2.

703.7.4 Controllers and alarms. Cooling towers, evaporative condensers, and fluid coolers shall be equipped with conductivity controllers and overflow alarms.

703.7.5 Drift. Cooling towers, evaporative condensers and fluid coolers shall produce drift losses of not greater than 0.002 percent of the recirculated water volume for counter-flow systems, and not greater than 0.005 percent of the recirculated water for cross-flow systems.

703.7.6 Water quality. Where nonpotable water is used within cooling towers, evaporative condensers and fluid coolers, it shall conform to the water quality and treatment requirements of the jurisdiction having authority and the water chemistry guidelines recommended by the equipment manufacturers.

703.7.7 Discharge. The discharge water from cooling towers used for air-conditioning systems shall be in compliance with Table 703.7.7. Where the discharge water is not captured for reuse, it shall be discharged and treated in accordance with jurisdictional requirements, if applicable.

Exception: Discharge water with total dissolved solids in excess of 1,500 ppm (1,500 mg/L), or silica in excess of 120 ppm (120 mg/L) measured as silicon dioxide shall not be required to meet the minimum parameters specified in Table 703.7.7.

TABLE 703.7.7
MINIMUM CYCLES OF CONCENTRATION FOR DISCHARGE WATER

MAKEUP WATER TOTAL HARDNESS (mg/L)[a]	MAXIMUM CYCLES OF CONCENTRATION
< 200	5
≥ 200	3.5

a. Total hardness concentration expressed as calcium carbonate.

703.8 Wet-hood exhaust scrubber systems. Where wet-hood exhaust scrubber systems are used, they shall incorporate a water recirculation system. The makeup water supplies for such systems shall be metered in accordance with Section 705.1.

703.8.1 Washdown systems. Hoods incorporating washdown or rinsing systems for perchloric acid and similar chemicals shall utilize self-closing valves. Such systems shall be designed to drain automatically after each washdown process has been completed.

703.8.2 Water sources. Where suitable alternate onsite nonpotable water or municipal reclaimed water is available, makeup water supplies to the recirculation system of wet-hood exhaust scrubbers shall utilize alternate onsite nonpotable water or municipal reclaimed water of a water quality appropriate for the application.

703.9 Evaporative cooling. Evaporative cooling systems shall use less than 4 gallons of water per ton-hour (4.2 L per kWh) of cooling capacity when system controls are set to the maximum water use. The amount of water use shall be expressed in maximum water use per ton-hour (kWh) of cooling capacity and shall be marked on the equipment, included in product user manuals, included in product information literature and included in manufacturer's instructions. Water use information shall be readily available at the time of code compliance inspection.

703.9.1 Overflow alarm. Cooling systems shall be equipped with an overflow alarm to alert building owners, tenants or maintenance personnel when the water refill valve continues to allow water to flow into the reservoir when the reservoir is full. The alarm shall have a minimum sound pressure level rating of 85 dB measured at a distance of 10 feet (3048 mm).

703.9.2 Automatic pump shutoff. Cooling systems shall automatically cease pumping water to the evaporation pads when sensible heat reduction is not needed.

703.9.3 Cooler reservoir discharge. A water quality management system such as a timer or water quality sensor shall be required. Where timers are used, the time interval between the discharge events of the water reservoir shall be set to 6 hours or greater of cooler operation. Continuous discharge or continuous bleed systems shall be prohibited.

703.9.4 Discharge water reuse. Discharge water shall be reused where appropriate applications exist on site. Where a nonpotable water source system exists on site, evaporative cooler discharge water shall be collected and discharged to such collection system.

Exception: Where the reservoir water will adversely affect the quality of the nonpotable water supply making the nonpotable water unusable for its intended purposes.

703.9.5 Discharge water to drain. Where discharge water is not required to be recovered for reuse, the sump overflow pipe shall not directly connect to a drain. Where the discharge water is discharged into a sanitary drain, an air gap of not less than 6 inches (150 mm) shall be required between the termination of the discharge pipe and the drain opening. The discharge pipe shall terminate in a location that is readily visible to the building owners, tenants or maintenance personnel.

SECTION 704 WATER TREATMENT DEVICES AND EQUIPMENT

704.1 Water softeners. Water softeners shall comply with Sections 704.1.1 through 704.1.4.

704.1.1 Demand-initiated regeneration. Water softeners shall be equipped with demand-initiated regeneration control systems. Such control systems shall automatically initiate the regeneration cycle after determining the depletion, or impending depletion of softening capacity.

704.1.2 Water consumption. Water softeners shall have a maximum water consumption during regeneration of 5 gallons (18.9 L) per 1000 grains (17.1 g/L) of hardness removed as measured in accordance with NSF 44.

704.1.3 Waste connections. Waste water from water softener regeneration shall not discharge to reclaimed water collection systems and shall discharge in accordance with the *International Plumbing Code*.

704.1.4 Efficiency and listing. Water softeners that regenerate in place, that are connected to the water system they serve by piping not exceeding $1^1/_4$ inches (31.8 mm) in diameter, or that have a volume of 3 cubic feet (0.085 m^3) or more of cation exchange media shall have a rated salt efficiency of not less than 4,000 grains of total hardness exchange per pound of salt (477 g of total hardness exchange per kg of salt), based on sodium chloride equivalency and shall be listed and labeled in accordance with NSF 44. All other water softeners shall have a rated salt efficiency of not less than 3,500 grains of total hardness exchange per pound of salt (477 g of total hardness exchange per kg of salt), based on sodium chloride equivalency.

704.2 Reverse osmosis water treatment systems. Point-of-use reverse osmosis treatment systems shall be listed and labeled in accordance with NSF 58. The discharge pipe from a reverse osmosis drinking water treatment unit shall connect to the building drainage system in accordance with Section 611.2 of the *International Plumbing Code*. Point-of-use reverse osmosis systems shall be equipped with an automatic shutoff valve that prevents the production of reject water when there is no demand for treated water.

704.3 Onsite reclaimed water treatment systems. Onsite reclaimed water treatment systems, including gray water reuse treatment systems and waste water treatment systems, used to produce nonpotable water for use in water closet and urinal flushing, surface irrigation and similar applications shall listed and labeled to NSF 350.

SECTION 705 METERING

705.1 Metering. Water consumed from any source associated with the building or building site shall be metered. Each potable and reclaimed source of water, and each onsite nonpotable water source, shall be metered separately. Meters shall be installed in accordance with the requirements of the *International Plumbing Code*. For the purposes of Section 705.1.1, Each meter identified in Table 705.1.1 shall be capable of communicating water consumption data remotely and at a minimum, be capable of providing daily data with electronic data storage and reporting capability that can produce reports that show daily, monthly, and annual water consumption.

705.1.1 Metering. All potable and nonpotable water supplied to the applications listed in Table 705.1.1 shall be individually metered in accordance with the requirements indicated in Table 705.1.1. Similar appliances and equipment shall be permitted to be grouped and supplied from piping connected to a single meter.

SECTION 706 NONPOTABLE WATER REQUIREMENTS

706.1 Scope. The provisions of this section shall govern the use of nonpotable water and the construction, installation, and design of systems utilizing nonpotable water. The use and application of nonpotable water shall comply with laws, rules and ordinances applicable in the jurisdiction.

706.2 Signage required. Where nonpotable water is used for a water use application, signage shall be provided that reads as follows: "Nonpotable water is utilized for [APPLICATION NAME]. Caution: nonpotable water. DO NOT DRINK." The words shall be legibly and indelibly printed on a sign constructed of corrosion-resistant waterproof material. The letters of the words shall be not less than 0.5 inches (13 mm) in height and of a color in contrast to the background on which they are applied. In addition to the required wordage, the pictograph shown in Figure 706.2 shall appear on the signage required by this section. The required location of the signage and pictograph shall be in accordance with the applicable section of this code that requires the use of nonpotable water.

706.3 Water quality. Nonpotable water for each end use application shall meet the minimum water quality requirements as established for the application by the laws, rules and ordinances applicable in the jurisdiction.

TABLE 705.1.1
METERING REQUIREMENTS

APPLICATION	REQUIREMENTS
Irrigation	Irrigation systems that are automatically controlled shall be metered.
Tenant spaces	Tenant spaces that are estimated to consume over 1000 gallons of water per day shall be metered individually.
Onsite water collection systems	The makeup water lines supplying onsite water collection systems shall be metered.
Ornamental water features	Ornamental water features with a permanently installed water supply shall be required to utilize a meter on makeup water supply lines.
Pools and in-ground spas	Indoor and outdoor pools and in-ground spas shall be required to utilize a meter on makeup water supply lines.
Cooling towers	Cooling towers of 100 tons capacity or greater or groups of towers shall be required to utilize a meter on makeup water and blow-down water supply lines.
Steam boilers	The makeup water supply line to steam boilers anticipated to draw more than 100,000 gallons annually or having a rating of 500,000 Btu/h or greater shall be metered.
Industrial processes	Industrial processes consuming more than 1,000 gallons per day on average shall be metered individually.
Evaporative coolers	Evaporative coolers supplying in excess of 0.6 gpm, on average, makeup water shall be.
Fluid coolers and chillers	Water-cooled fluid coolers and chillers that do not utilize closed-loop recirculation shall be metered.
Makeup water for closed loop systems such as chilled water and hydronic systems	Makeup water supplying systems of 50 tons of cooling capacity or 500,000 Btu/h of heating capacity shall be metered.
Roof spray systems	Roof spray systems for irrigating vegetated roofs or thermal conditioning shall be metered.

For SI: 1 gallon = 3.8 L, 1 gallon per minute = 3.8 Lpm, 1 ton = 12,000 Btu, 1 British thermal unit per hour = 0.00029 kWh.

FIGURE 706.2
PICTOGRAPH—DO NOT DRINK

SECTION 707
RAINWATER COLLECTION AND DISTRIBUTION SYSTEMS

707.1 Scope. The provisions of this section shall govern the construction, installation, alteration, and repair of rainwater collection and conveyance systems.

707.2 Potable water connections. Where a potable system is connected to a rainwater collection and conveyance system, the potable water supply shall be protected against backflow in accordance with Section 608 of the *International Plumbing Code*.

707.3 Nonpotable water connections. Where nonpotable water from different sources is combined in a system, the system shall comply with the most stringent of the requirements of this code that are applicable to such sources.

707.4 Installation. Except as provided for in this section, all systems shall be installed in compliance with the provisions of the *International Plumbing Code* and the manufacturer's instructions.

707.5 Rainwater collected for landscape irrigation. Rainwater collected on the surface of the building site, or from the roof surfaces of the building, and used for landscape irrigation purposes shall not be limited regarding the method of application. Rainwater collected from elevated building locations that is to be used in building site irrigation, shall comply with the provisions of Section 707 with the exception of Sections 707.11.1, 707.11.1.1 and 707.11.7.3.

707.6 Approved components and materials. Piping, plumbing components, and materials used in the collection and conveyance systems shall be manufactured of material *approved* for the intended application and compatible with any disinfection and treatment systems used.

707.7 Insect and vermin control. Inlets and vents to the system shall be protected to prevent the entrance of insects and vermin into storage tanks and piping systems. Screens installed on vent pipes, inlets, and overflow pipes shall have an aperture of not greater than $^1/_{16}$ inch (1.6 mm) and shall be close fitting. Screen materials shall be compatible with contacting system components and shall not accelerate corrosion of system components.

707.8 Drainage. Water drained from the roof washer or debris excluder shall not be drained to the sanitary sewer. Such water shall be diverted from the storage tank and discharge in a location that will not cause erosion or damage to property. Roof washers and debris excluders shall be provided with an automatic means of self-draining between rain events, and shall not drain onto roof surfaces.

707.9 Freeze protection. Where sustained freezing temperatures occur, provisions shall be made to keep storage tanks and the related piping from freezing.

707.10 Trenching requirements. All water service piping, including piping containing rainwater, shall be separated from the building sewer by 5 feet (1524 m) of undisturbed or compacted earth. Water service pipes, potable and nonpotable, shall not be located in, under or above cesspools, septic tanks, septic tank drainage fields or seepage pits. Buried rainwater collection and distribution piping shall comply with the requirements of Section 306 of the *International Plumbing Code* for support, trenching, bedding, backfilling and tunneling.

Exceptions:

1. The required separation distance shall not apply where the bottom of the water service pipe within 5 feet (1524 mm) of the sewer is a minimum of 12 inches (305 mm) above the top of the highest point of the sewer and the pipe materials shall comply with the *International Plumbing Code* for such applications.
2. Water service pipe is permitted to be located in the same trench with a building sewer, provided such sewer is constructed of materials that comply with the *International Plumbing Code* for such installations.
3. The required separation distance shall not apply where a potable or nonpotable water service pipe crosses a sewer pipe provided the water service pipe is sleeved to not less than 5 feet (1524 mm) horizontally from the sewer pipe centerline on both sides of such crossing with pipe materials that comply with the *International Plumbing Code* for such applications.
4. Irrigation piping located outside of a building and downstream of the backflow preventer is not required to meet the trenching requirements where rainwater is used for outdoor applications.

707.11 Rainwater catchment and collection systems. The design of rainwater collection and conveyance systems shall conform to accepted engineering practice.

707.11.1 Collection surface. Rainwater shall be collected only from above-ground impervious roofing surfaces constructed from *approved* materials. Collection of water from vehicular parking or pedestrian surfaces shall be prohibited except where the water is used exclusively for landscape irrigation. Overflow and bleed-off pipes from roof-mounted appliances including but not limited to evaporative coolers, water heaters, and solar water heaters shall not discharge onto rainwater collection surfaces.

707.11.1.1 Potable water applications. Where collected water is to be treated to potable water standards, wood or cedar shake roofing materials, roofing materials treated with biocides, and lead flashing are prohibited on collection surfaces. Painted surfaces are acceptable only where paint has been certified to ensure that the toxicity level of the paint is acceptable for drinking water contact. Lead, chromium or zinc-based paints are not permitted on rainwater collection surfaces. Flat roofing products shall be certified to NSF P151. Rainwater shall not be collected from vegetated roof systems.

707.11.2 Debris excluders. Downspouts and leaders shall be connected to a roof washer and shall be equipped with a debris excluder or equivalent device to prevent the contamination of collected rainwater with leaves, sticks, pine needles and similar material. Debris excluders and equivalent devices shall be self-cleaning.

707.11.3 Roof gutters and downspouts. Gutters and downspouts shall be constructed of materials that are compatible with the collection surface and the rainwater quality for the desired end use. Joints shall be water tight. Where the collected rainwater is to be used for potable applications, gutters, downspouts, flashing and joints shall be constructed of materials *approved* for drinking water applications.

707.11.3.1 Slope. Roof gutters, leaders, and rainwater collection piping shall slope continuously toward collection inlets. Gutters and downspouts shall have a slope of not less than 1 unit in 96 units along their entire length, and shall not permit the collection or pooling of water at any point.

Exception: Siphonic drainage systems installed in accordance with the manufacturer's installation instructions shall not be required to have slope.

707.11.3.2 Size. Gutters and downspouts shall be installed and sized in accordance with Section 1106.6 of the *International Plumbing Code*.

707.11.3.3 Cleanouts. Cleanouts shall be provided in the water conveyance system so as to allow access to all filters, flushes, pipes and downspouts.

707.11.4 Collection pipe materials. In buildings where rainwater collection and conveyance systems are installed, drainage piping *approved* for use within plumbing drainage systems shall be utilized to collect rainwater and convey it to the storage tank. Vent piping *approved* for use within plumbing venting systems shall be utilized for all vents within the rainwater system. Drains to a storm water discharge shall use *approved* waste piping.

707.11.4.1 Joints. Collection piping conveying rainwater shall utilize joints *approved* for use with the distribution piping and appropriate for the intended applications as specified in the *International Plumbing Code*.

707.11.4.2 Size. Collection piping conveying rainwater from collection surfaces shall be sized in accordance with Chapter 11 of the *International Plumbing Code* and local rainfall rates.

707.11.4.3 Marking. Additional marking of rainwater collection piping shall not be required beyond that required for sanitary drainage, waste, and vent piping by the *International Plumbing Code*.

707.11.5 Filtration. Collected rainwater shall be filtered to the level required for the intended end use. Filters shall be accessible for inspection and maintenance.

707.11.6 Disinfection. Where the intended application and initial quality of the collected rainwater requires disinfection or other treatment or both, the collected rainwater shall be treated as needed to ensure that the required water quality is delivered at the point of use. Where chlorine is used for disinfection or treatment, water shall be tested for residual chlorine in accordance with ASTM D 1253. The levels of residual chlorine shall not exceed the levels allowed for the intended use in accordance with the requirements of the jurisdiction.

707.11.7 Storage tank. The design of the storage tank shall be in accordance with Sections 707.11.7.1 through 707.11.7.10.

707.11.7.1 Location. Storage tanks shall be installed either above or below grade. Above-grade storage tanks shall be protected from direct sunlight and shall be constructed using opaque, UV-resistant materials including, but not limited to, heavily tinted plastic, fiberglass, lined metal, concrete, wood, or painted to prevent algae growth, or shall have specially constructed sun barriers including, but not limited to, installation in garages, crawlspaces, or sheds. Storage tanks and their manholes shall not be located directly under any soil or waste piping or any source of contamination. Rainwater storage tanks shall be located with a minimum horizontal distance between various elements as indicated in Table 707.11.7.1.

TABLE 707.11.7.1
LOCATION OF RAINWATER STORAGE TANKS

ELEMENT	MINIMUM HORIZONTAL DISTANCE FROM STORAGE TANK (feet)
Critical root zone (CRZ) of protected trees	2
Lot line adjoining private lots	5
Seepage pits	5
Septic tanks	5

For SI: 1 foot = 304.8 mm.

707.11.7.2 Materials. Where water is collected onsite, it shall be collected in an *approved* tank constructed of durable, nonabsorbent and corrosion-resistant materials. Storage vessels shall be compatible with the material being stored.Where collected water is to be treated to potable water standards, tanks shall be constructed of materials in accordance with NSF 61. Storage tanks shall be constructed of materials compatible with the type of disinfection system used to treat water upstream of the tank and used to maintain water quality within the tank.

707.11.7.2.1 Wooden tanks. Wooden storage tanks shall not be required to have a liner. Where a tank is lined and used for potable water, the liner shall be in accordance with NSF standards. Where unlined tanks are used, the species of wood shall be decay resistant and untreated.

707.11.7.3 Makeup water. Where an uninterrupted supply is required for the intended application, potable or municipally supplied reclaimed or recycled water shall be provided as a source of makeup water for the storage tank. The potable or reclaimed or recycled water supply shall be protected against backflow in accordance with the *International Plumbing Code*.

707.11.7.4 Overflow. The storage tank shall be equipped with an overflow pipe having the same or larger area as the sum of the areas of all tank inlet pipes. The overflow pipe shall be protected from insects or vermin and the discharge from such pipe shall be disposed of in a manner consistent with storm water runoff requirements of the jurisdiction. The overflow pipe shall discharge at a sufficient distance from the tank to avoid damaging the tank foundation or the adjacent property. The overflow drain shall not be equipped with a shutoff valve. A minimum of one cleanout shall be provided on each overflow pipe in accordance with Section 708 of the *International Plumbing Code*.

707.11.7.5 Access. A minimum of one access opening shall be provided to allow inspection and cleaning of the tank interior. Access openings to storage tanks and other vessels shall have an *approved* locking device or shall otherwise be protected from unauthorized access. Below-grade storage tanks, located outside of the building, shall be provided with either a manhole not less than 24 inches (610 mm) square or a manhole with an inside diameter of not less than 24 inches (610 mm). Manholes shall extend not less than 4 inches (102 mm) above ground or shall be designed so as to prevent water infiltration. Finish grade shall be sloped away from the manhole to divert surface water from the manhole. Each manhole cover shall be secured to prevent unauthorized access. Service ports in manhole covers shall be not less than 8 inches (203 mm) in diameter and shall be not less than 4 inches (102 mm) above the finished grade level. The service port shall be secured to prevent unauthorized access.

Exception: Storage tanks having a volume of less than 800 gallons (3028 L) and installed below grade shall not be required to be equipped with a manhole where provided with a service port that is not less than 8 inches (203 mm) in diameter.

707.11.7.6 Venting. Tanks shall be provided with a vent sized in accordance with the *International Plumbing Code* and based on the diameter of the tank influent pipe. Tank vents shall not be connected to sanitary drainage system vents.

707.11.7.7 Inlets. Storage tank inlets shall be designed to introduce water into the tank with minimum turbulence, and shall be located and designed to avoid agitating the contents of the storage tank.

707.11.7.8 Outlets. Outlets shall be located not less than 4 inches (102 mm) above the bottom of the storage tanks and shall not skim water from the surface.

707.11.7.9 Draining of tanks. Where tanks require draining for service or cleaning, tanks shall be drained by using a pump or by a drain located at the lowest point in the tank. The discharge from draining the tank shall be disposed of in a manner consistent with the storm water runoff requirements of the jurisdiction and at a sufficient distance from the tank to avoid damaging the tank foundation.

707.11.7.10 Marking and signage. Each storage tank shall be marked with its rated capacity. Storage tanks shall bear signage that reads as follows: "CAUTION: NONPOTABLE WATER – DO NOT DRINK." Where an opening is provided that could allow the entry of personnel, the opening shall bear signage that reads as follows: "DANGER – CONFINED SPACE." Markings shall be indelibly printed on a tag or sign constructed of corrosion-resistant waterproof material mounted on the tank or shall be indelibly printed on the tank. The letters of words shall be not less than 0.5 inches (13 mm) in height and shall be of a color that contrasts with the background on which they are applied.

707.11.8 Valves. Valves shall be supplied in accordance with Section 707.11.8.1.

707.11.8.1 Backwater valve. Backwater valves shall be installed on each overflow and tank drain pipe. Backwater valves shall be installed so that access is provided to the working parts for service and repair.

707.11.9 Roof washer. A sufficient amount of rainwater shall be diverted at the beginning of each rain event, and not allowed to enter the storage tank, to wash accumulated debris from the collection surface. The amount of rainfall to be diverted shall be field adjustable as necessary to minimize storage tank water contamination. The roof washer shall not rely on manually operated valves or devices, and shall operate automatically. Diverted rainwater shall not be drained to the roof surface, and shall be discharged in a manner consistent with the storm water runoff requirements of the jurisdiction. Roof washers shall be accessible for maintenance and service.

707.11.10 Vent piping. Storage tanks shall be provided with a vent in accordance with the requirements of Section 707.11.7.6. Vents shall be sized in accordance with the *International Plumbing Code*, based on the aggregate diameter of storage tank influent pipe(s). Vents shall be protected from contamination by means of a U-bend installed with the opening directed downward or an *approved* cap. Vent outlets shall extend a minimum of 4 inches (102 mm) above grade, or as necessary to prevent surface water from entering the storage tank. Vent openings shall be protected against the entrance of vermin and insects in accordance with the requirements of Section 707.7.

707.11.11 Pumping and control system. Mechanical equipment including pumps, valves and filters shall be easily accessible and removable in order to perform repair, maintenance and cleaning. Where collected rainwater is to be treated to potable water standards, the pump and all other pump components shall be listed, labeled and *approved* for use with potable water systems. Pressurized water shall be supplied at a pressure appropriate for the application and within the range specified by the *International Plumbing Code*. Where water could be supplied at an excessive pressure, a pressure-reducing valve shall be installed in accordance with the requirements of the *International Plumbing Code*.

707.11.11.1 Water-pressure-reducing valve or regulator. Where the rainwater pressure supplied by the pumping system exceeds 80 psi (552 kPa) static, a pressure-reducing valve shall be installed to reduce the pressure in the rainwater distribution system piping to 80 psi (552 kPa) static or less. Pressure-reducing valves shall be specified and installed in accordance with Section 604.8 of the *International Plumbing Code*.

707.11.12 Distribution pipe. Distribution piping shall comply with Sections 707.11.12.1 through 707.11.12.4.

707.11.12.1 Materials. Distribution piping conveying rainwater shall conform to the standards and requirements specified by the *International Plumbing Code* for nonpotable or potable water, as applicable.

707.11.12.2 Joints. Distribution piping conveying rainwater shall utilize joints *approved* for use with the distribution piping and appropriate for the intended applications as specified in the *International Plumbing Code*.

707.11.12.3 Size. Distribution piping conveying rainwater shall be sized in accordance with the *International Plumbing Code* for the intended application.

707.11.12.4 Marking. Nonpotable rainwater distribution piping shall be of the color purple and shall be embossed or integrally stamped or marked with the words: "CAUTION: NONPOTABLE WATER – DO NOT DRINK" or shall be installed with a purple identification tape or wrap. Identification tape shall be not less than 3 inches (76 mm) wide and have white or black lettering on purple field stating "CAUTION: NONPOTABLE WATER – DO NOT DRINK." Identification tape shall be installed on top of nonpotable rainwater distribution pipes, fastened not greater than every 10 feet (3048 mm) to each pipe length and run continuously the entire length of the pipe. Lettering shall be readily observable within the room or space where the piping is located.

Exception: Piping located outside of the building and downstream of the backflow preventer is not required to be purple where rainwater is used for outdoor applications.

707.12 Tests and inspections. Tests and inspection shall be performed in accordance with Sections 707.12.1 through 707.12.10.

707.12.1 Drainage and vent tests. The testing of rainwater collection piping, overflow piping, vent piping and storage tank drains shall be conducted in accordance with Section 312 of the *International Plumbing Code*.

707.12.2 Drainage and vent final test. A final test shall be applied to the rainwater collection piping, overflow piping, storage tank, and tank vent piping in accordance with Section 312.4 of the *International Plumbing Code*.

707.12.3 Water supply system test. The testing of makeup water supply piping and rainwater distribution piping shall be conducted in accordance with Section 312.5 of the *International Plumbing Code*.

707.12.4 Inspection and testing of backflow prevention assemblies. The testing of backflow preventers and backwater valves shall be conducted in accordance with Section 312.10 of the *International Plumbing Code*.

707.12.5 Inspection vermin and insect protection. Inlets and vents to the system shall be inspected to ensure that each is protected to prevent the entrance of insects or vermin into storage tank and piping systems in accordance with Section 707.8.

707.12.6 Roof gutter inspection and test. Roof gutters shall be inspected to verify that the installation and slope is in accordance with Section 707.11.3. Gutters shall be tested by pouring not less than 1 gallon (3.8 L) of water into the end of the gutter opposite the collection point. The gutter being tested shall not leak and shall not retain standing water.

707.12.7 Roofwasher test. Roofwashers shall be tested by introducing water into the gutters. Proper diversion of the first quantity of water in accordance with the requirements of Section 707.11.9 shall be verified.

707.12.8 Storage tank tests. Storage tanks shall be tested in accordance with the following:

1. Storage tanks shall be filled with water to the overflow line prior to and during inspection. Seams and joints shall be left exposed and the tank shall remain water tight without leakage for a period of 24 hours.
2. After 24 hours, supplemental water shall be introduced for a period of 15 minutes to verify proper drainage of the overflow system and verify that there are no leaks.
3. The makeup water system shall be observed for proper operation and successful automatic shutoff of the system at the refill threshold shall be verified.

707.12.9 Supply pressure test. The static water pressure at the point of use furthest from the supply shall be verified to be within the range required for the application, in accordance with Section 707.11.11.

707.12.10 Water quality test. The quality of the water for the intended application shall be verified at the point of use in accordance with the requirements of the jurisdiction. Except where site conditions as specified in ASTM E 2727 affect the rainwater, collected rainwater shall be considered to have the parameters indicated in Table 707.12.10.

TABLE 707.12.10
RAINWATER QUALITY

PARAMETER	VALUE
pH	6.0 – 7.0
Biological oxygen demand	Not greater than 10 mg/L
Nephelometric turbidity unit	Not greater than 2
Fecal coliform	No detectable fecal coli in 100 mL
Sodium	No detectable sodium in 100 mL
Chlorine	No detectable chlorine in 100 mL
Enteroviruses	No detectable enteroviruses in 100 mL

707.13 Operations and maintenance manuals. Operations and maintenance materials shall be supplied in accordance with 707.13.1 through 707.13.4.

707.13.1 Manual. A detailed operations and maintenance manual shall be supplied in hardcopy form with all rainwater collection systems.

707.13.2 Schematics. The manual shall include a detailed system schematic, the locations of all system components, and a list of all system components including manufacturer and model number.

707.13.3 Maintenance procedures. The manual shall provide a maintenance schedule and procedures for all system components requiring periodic maintenance. Consumable parts including filters shall be noted along with part numbers.

707.13.4 Operations procedures. The manual shall include system startup and shutdown procedures. The manual shall include detailed operating procedures for the system.

707.14 System abandonment. If the owner of a rainwater collection and conveyance system elects to cease use of, or fails to properly maintain such system, the system shall be abandoned and shall comply with the following:

1. System piping connecting to a utility-provided water system shall be removed or disabled.
2. The rainwater distribution piping system shall be replaced with an *approved* potable water supply piping system. Where an existing potable pipe system is already in place, the fixtures shall be connected to the existing system.
3. The storage tank shall be secured from accidental access by sealing or locking tank inlets and access points, or filling with sand or equivalent.

707.15 Potable water applications. Where collected rainwater is to be used for potable water applications, all materials contacting the water shall comply with NSF 61.

707.15.1 Water quality testing. Collected rainwater shall be tested. Accumulated water to be tested shall be the result of not less than two rainfall events. Testing shall be in accordance with Sections 707.15.1.1 and 707.15.1.2.

707.15.1.1 Test methods. Water quality testing shall be performed in accordance with the latest edition of

APHA–Standard Methods for the Examination of Water and Wastewater and in accordance with Sections 707.15.1.1.1 and 707.15.1.1.2.

707.15.1.1.1 Annual tests required. Accumulated rainwater shall be tested prior to initial use and annually thereafter for Escherichia coli, total coliform, heterotrophic bacteria and cryptosporidium.

707.15.1.1.2 Quarterly tests required. Accumulated rainwater shall be tested prior to initial use and quarterly thereafter for pH, filterable solids, residual chlorine if disinfection is used, and turbidity. The pH shall be tested in accordance with ASTM D 5464; filterable solids shall be tested in accordance with ASTM D 5907; residual chlorine shall be tested in accordance with ASTM D 1253 and turbidity shall be tested in accordance with ASTM D 6698.

707.15.1.2 Test records. Test records shall be retained for not less than two years.

SECTION 708
GRAY WATER SYSTEMS

708.1 Scope. The provisions of this section shall govern the construction, installation, alteration, and repair of gray water reuse systems.

708.2 Permits. Permits shall be required for the construction, installation, alteration, and repair of gray water systems. Construction documents, engineering calculations, diagrams, and other such data pertaining to the gray water system shall be submitted with each application for permit in accordance with the laws, rules and ordinances applicable in the jurisdiction.

708.3 Potable water connections. Where a potable water system is connected to a gray water system, the potable water supply shall be protected against backflow in accordance with Section 608 of the *International Plumbing Code*.

708.4 Nonpotable water connections. Where nonpotable water from different sources is combined in a system, the system shall comply with the most stringent of the requirements of this code that are applicable to such sources.

708.5 Installation. Except as provided for in this section, all systems shall be installed in compliance with the provisions of the *International Plumbing Code* and the manufacturer's instructions, as applicable.

708.5.1 Gray water systems for landscape irrigation. Gray water systems used for landscape irrigation purposes shall be limited to subsurface and surface irrigation applications. Gray water shall not be retained longer than 24 hours before being used for surface irrigation. Gray water to be used in gray water irrigation shall comply with the provisions of Section 708 with the exception of Sections 708.6 and 708.12.6.5. Subsurface gray water systems shall be in accordance with Section 708.14. Gray water shall be filtered by a 0.004-inch (100 micron) or finer filter. The control panel for the gray water irrigation system shall be provided with signage in accordance with Section 706.2.

708.6 Applications. Untreated gray water shall be utilized in accordance with Section 702 and local codes. Treated gray water shall be utilized in accordance with Section 706 and as permitted by local codes.

708.7 Approved components and materials. The piping, plumbing components, and materials used in gray water systems shall be manufactured of material *approved* for the intended application and compatible with any disinfection and treatment systems used.

708.8 Insect and vermin control. The inlets and vents to the system shall be protected to prevent insects and vermin from entering storage tanks and piping systems. Screens installed on vent pipes and overflow pipes shall have an aperture not greater than $^{1}/_{16}$ inch (1.6 mm) and shall be close-fitting. Screen materials shall be compatible with contacting system components and shall not accelerate corrosion of system components

708.9 Freeze protection. Where sustained freezing temperatures occur, provisions shall be made to keep storage tanks and the related piping from freezing.

708.10 Trenching requirements. Water service piping, including piping containing gray water, shall be separated from the building sewer by 5 feet (1524 m) of undisturbed or compacted earth. Gray water piping shall be separated from potable water piping underground by 5 feet (1524 m) of undisturbed or compacted earth. Nonpotable water service pipes shall not be located in, under or above cesspools, septic tanks, septic tank drainage fields or seepage pits. Buried gray water piping shall comply with the requirements of Section 306 of the *International Plumbing Code* for support, trenching, bedding, backfilling, and tunneling.

Exceptions:

1. The required separation distance shall not apply where the bottom of the gray water service pipe within 5 feet (1524 mm) of the sewer is not less than 12 inches (305 mm) above the top of the highest point of the sewer and the pipe materials comply with the requirements of the *International Plumbing Code* for such applications.
2. The required separation distance shall not apply where the bottom of the potable water service pipe within 5 feet (1524 mm) of the gray water pipe is not less than 12 inches (305 mm) above the top of the highest point of the gray water pipe and the pipe materials comply with the requirements of the *International Plumbing Code* for such applications.
3. Water service pipe is permitted to be located in the same trench with a building sewer, provided that such sewer is constructed of materials that comply with the requirements of the *International Plumbing Code* for such applications.
4. The required separation distance shall not apply where a potable or nonpotable water service pipe crosses a sewer pipe provided that the water service pipe is sleeved to not less than 5 feet (1524 mm) horizontally from the sewer pipe centerline on both sides of such crossing with pipe materials that com-

ply with the requirements of the *International Plumbing Code* for such applications.

5. The required separation distance shall not apply where a potable water service pipe crosses a gray water pipe provided that the potable water service pipe is sleeved for a distance of not less than 5 feet (1524 mm) horizontally from the centerline of the gray water pipe on both sides of such crossing with pipe materials that comply with the requirements of the *International Plumbing Code* for such applications.
6. Irrigation piping located outside of a building and downstream of the backflow preventer is not required to meet the trenching requirements where gray water is used for outdoor applications.

708.11 System abandonment. If the owner of a gray water system elects to cease use of, or fails to properly maintain such system, the system shall be abandoned and shall comply with the following:

1. System piping connecting to a utility-provided water system shall be removed or disabled.
2. Storage tanks shall be secured against accidental access by sealing or locking tank inlets and access points, or filling with sand or equivalent.

708.12 Gray water systems. The design of the gray water system shall conform to accepted engineering practice.

708.12.1 Gray water sources. Gray water reuse systems shall collect waste discharge from only the following sources: bathtubs, showers, lavatories, clothes washers, and laundry trays. Water from other *approved* nonpotable sources including swimming pool backwash operations, air conditioner condensate, rainwater, cooling tower blowdown water, foundation drain water, steam system condensate, fluid cooler discharge water, food steamer discharge water, combination oven discharge water, industrial process water, and fire pump test water shall also be permitted to be collected for reuse by gray water systems, as *approved* by the *code official* and as appropriate for the intended application.

708.12.1.1 Prohibited gray water sources. Waste water containing urine or fecal matter shall not be diverted to gray water systems and shall discharge to the sanitary drainage system of the building or premises in accordance with the *International Plumbing Code*. Water from reverse osmosis system reject water, water softener discharge water, kitchen sink waste water, dishwasher waste water, and waste water discharged from wet-hood scrubbers shall not be collected for reuse within a gray water system.

708.12.2 Traps. Traps serving fixtures and devices discharging waste water to gray water reuse systems shall have a liquid seal of not less than 2 inches (51 mm) and not more than 4 inches (102 mm). Where a trap seal is subject to loss by evaporation, a trap seal primer valve shall be installed in accordance with the *International Plumbing Code*.

708.12.3 Collection pipe. Gray water reuse systems shall utilize drainage piping *approved* for use within plumbing drainage systems to collect and convey untreated gray water. Vent piping *approved* for use within plumbing venting systems shall be utilized for vents within the gray water system. Drains to the sanitary sewer shall use *approved* waste piping.

708.12.3.1 Joints. Collection piping conveying untreated gray water shall utilize joints *approved* for use with the distribution piping and appropriate for the intended applications as specified in the *International Plumbing Code*.

708.12.3.2 Size. Collection piping conveying rainwater from collection surfaces shall be sized in accordance with storm drainage sizing requirements specified in the *International Plumbing Code*.

708.12.3.3 Marking. Additional marking of untreated gray water collection piping shall not be required beyond that required for sanitary drainage, waste, and vent piping by the *International Plumbing Code*.

708.12.4 Filtration. Collected gray water shall be filtered as required for the intended end use. Filters shall be accessible for inspection and maintenance. Filters shall utilize a pressure gage or other *approved* method to provide indication when a filter requires servicing or replacement. Filters shall be installed with shutoff valves installed immediately upstream and downstream to allow for isolation during maintenance.

708.12.5 Disinfection. Where the intended application for collected gray water requires disinfection or other treatment or both, collected gray water shall be disinfected as needed to ensure that the required water quality is delivered at the point of use. Where chlorine is used for disinfection or treatment, water shall be tested for residual chlorine in accordance with ASTM D 1253. The levels of residual chlorine shall not exceed the levels allowed for the intended use in accordance with the requirements of the jurisdiction. Untreated gray water shall be retained in collection reservoirs for a maximum of 24 hours in accordance with Section 708.12.6.1.

708.12.6 Storage tank. The design of the storage tank shall be in accordance with Sections 708.12.6.1 through 708.12.6.10.

708.12.6.1 Sizing. The holding capacity of the storage tank shall be sized in accordance with the anticipated demand. Where gray water is to be used in untreated form for groundwater recharge or subsurface irrigation, the storage tank shall be sized to limit the retention time of gray water to a maximum of 24 hours.

708.12.6.2 Location. Storage tanks shall be installed above or below grade. Above-grade storage tanks shall be protected from direct sunlight and shall be constructed using opaque, UV-resistant materials such as, but not limited to, heavily tinted plastic, fiberglass, lined metal, concrete, wood, or painted to prevent algae growth, or shall have specially constructed sun barriers including, but not limited to, installation in garages, crawlspaces, or sheds. Storage tanks and their manholes shall not be located directly under any soil or waste piping or any source of contamination. Gray water storage tanks shall be located with a minimum horizontal distance between various elements as indicated in Table 708.12.6.2. Storage tanks containing untreated gray water shall be located a minimum horizontal distance of 5 feet (1524 mm) from buildings, in addition to the requirements in Table 708.12.6.2.

TABLE 708.12.6.2
LOCATION OF GRAY WATER STORAGE TANKS

ELEMENT	MINIMUM HORIZONTAL DISTANCE FROM STORAGE TANK (feet)
Critical root zone (CRZ) of protected trees	2
Lot line adjoining private lots	5
Seepage pits	5
Septic tanks	5
Water wells	50
Streams, lakes, wetlands and other bodies of water	50
Water service	5
Public water main	10

For SI: 1 foot = 304.8 mm.

708.12.6.3 Materials. Where collected onsite, water shall be collected in an *approved* tank constructed of durable, nonabsorbent and corrosion-resistant materials. The storage tank shall be constructed of materials compatible with any disinfection systems used to treat water upstream of the tank and with any systems used to maintain water quality within the tank.

708.12.6.3.1 Wood tanks. Wooden storage tanks that are not equipped with a makeup water source shall be provided with a flexible liner.

708.12.6.4 Makeup water. Where an uninterrupted supply of makeup water is required for the intended application, potable or municipally supplied reclaimed/recycled water shall be provided as a source of makeup water for the storage tank. The potable, reclaimed or recycled water supply shall be protected against backflow by means of an air gap not less than 4 inches (102 mm) above the overflow or an *approved* backflow device in accordance with the *International Plumbing Code*. There shall be a full-open valve located on the makeup water supply line to the storage tank. Inlets to storage tank shall be controlled by fill valves or other automatic supply valves installed so as to prevent the tank from overflowing and to prevent the water level from dropping below a predetermined point. Where makeup water is provided, the water level shall not be permitted to drop below the gray water inlet or the intake of any attached pump.

708.12.6.5 Overflow. The storage tank shall be equipped with an overflow pipe having the same or larger area as the sum of the areas of all reservoir inlet pipes. The overflow pipe shall be trapped and shall be indirectly connected to the sanitary drainage system. The overflow drain shall not be equipped with a shutoff valve. A minimum of one cleanout shall be provided on each overflow pipe in accordance with Section 708 of the *International Plumbing Code*.

708.12.6.6 Access. A minimum of one access opening shall be provided to allow inspection and cleaning of the tank interior. Access openings shall have an *approved* locking device or other *approved* method of securing access. Below-grade storage tanks, located outside of the building, shall be provided with either a manhole not less than 24 inches (610 mm) square or a manhole with an inside diameter not less than 24 inches (610 mm) and extending not less than 4 inches (102 mm) above ground. Finished grade shall be sloped away from the manhole to divert surface water from the manhole. Each manhole cover shall have a locking device. Service ports in manhole covers shall be not less than 8 inches (203 mm) in diameter and shall be not less than 4 inches (102 mm) above the finished grade level. The service port shall have a locking cover or a brass cleanout plug.

Exception: Storage tanks under 800 gallons (3024 L) in volume installed below grade shall not be required to be equipped with a manhole, but shall have a service port not less than 8 inches (203 mm) in diameter.

708.12.6.7 Venting. The tank shall be provided with a vent sized in accordance with the *International Plumbing Code* and based on the diameter of the tank influent pipe. The reservoir vent shall not be connected to sanitary drainage vent system.

708.12.6.8 Outlets. Outlets shall be located not less than 4 inches (102 mm) above the bottom of the storage tank, and shall not skim water from the surface.

708.12.6.9 Drain. A drain shall be located at the lowest point of the storage tank and shall be indirectly connected to the sanitary drainage system. The total area of all drains shall not be smaller than the total area of all overflow pipes. Not less than one cleanout shall be provided on each drain pipe in accordance with Section 708 of the *International Plumbing Code*.

708.12.6.10 Signage. Each storage tank shall be marked with its rated capacity and the location of the upstream bypass valve. The contents of storage tanks shall be identified with the words "CAUTION: NON-POTABLE WATER – DO NOT DRINK." Where an opening is provided that could allow the entry of personnel, the opening shall be marked with the words, "DANGER – CONFINED SPACE." Markings shall be indelibly printed on a tag or sign constructed of corrosion-resistant waterproof material mounted on the tank or shall be indelibly printed on the tank. The letters of the words shall be not less than 0.5 inches (13 mm) in height and shall be of a color in contrast with the background on which they are applied.

708.12.7 Valves. Valves shall be supplied in accordance with Sections 708.12.7.1 and 708.12.7.2.

708.12.7.1 Bypass valve. One three-way diverter valve listed and labeled to NSF 50 or other *approved* device shall be installed on gray water collection piping upstream of each storage tank, or drainfield, as applicable, to divert untreated gray water sources to the sanitary sewer to allow servicing and inspection of the system. Bypass valves shall be installed downstream of fixture traps and vent connections Bypass valves shall be marked to indicate the direction of flow, connection and storage tank or drainfield connection. Bypass valves shall be installed in accessible locations. Two shutoff valves shall not be installed to serve as a bypass valve.

708.12.7.2 Backwater valve. Overflow and tank drain piping shall be protected against backwater conditions by the installation of one or more backwater valves. Backwater valves shall be installed so that access is provided to the working parts for service and repair.

708.12.8 Vent piping. Storage tanks shall be provided with a vent in accordance with the requirements of Section 708.12.6.8. Vents shall be sized in accordance with the *International Plumbing Code*, based on the aggregate diameter of storage tank influent pipes. Open vents shall be protected from contamination by means of a U-bend installed with the opening directed downward or an *approved* cap. Vent outlets shall extend not less than 4 inches (102 mm) above grade, or as necessary to prevent surface water from entering the storage tank. Vent openings shall be protected against the entrance of vermin and insects in accordance with the requirements of Section 708.8.

708.12.9 Pumping and control system. Mechanical equipment including pumps, valves and filters shall be accessible and removable in order to perform repair, maintenance and cleaning. Pressurized water shall be supplied at a pressure appropriate for the application and within the range specified by the *International Plumbing Code*. Where water could be supplied at an excessive pressure, a pressure-reducing valve shall be installed in accordance with the requirements of the *International Plumbing Code*.

708.12.9.1 Standby power. Where required for the intended application, automatically activated standby power, capable of powering all essential treatment and pumping systems under design conditions shall be provided.

708.12.9.2 Inlet control valve alarm. Makeup water systems shall be provided with a warning mechanism that alerts the user to a failure of the inlet control valve to close correctly. The alarm shall activate before the water within the collection reservoir storage tank begins to discharge into the overflow system.

708.12.9.3 Water-pressure-reducing valve or regulator. Where the gray water pressure supplied by the pumping system exceeds 80 psi (552 kPa) static, a pressure-reducing valve shall be installed to reduce the pressure in the gray water distribution system piping to 80 psi (552 kPa) static or less. Pressure-reducing valves shall be specified and installed in accordance with Section 604.8 of the *International Plumbing Code*.

708.12.10 Distribution pipe. Distribution piping shall comply with Sections 708.12.10.1 through 708.12.10.4.

708.12.10.1 Materials. Distribution piping conveying gray water shall conform to standards and requirements specified by the *International Plumbing Code*.

708.12.10.2 Joints. Distribution piping conveying gray water shall utilize joints *approved* for use with the distribution piping and appropriate for the intended applications as specified in the *International Plumbing Code*.

708.12.10.3 Size. Distribution piping conveying gray water shall be sized in accordance with the *International Plumbing Code* for the intended application or applications.

708.12.10.4 Marking. All gray water distribution piping shall be either the color purple and embossed or integrally stamped or marked "CAUTION: NONPOTABLE WATER – DO NOT DRINK" or shall be installed with a purple identification tape or wrap. Identification tape shall be not less than 3 inches (76 mm) wide and have white or black lettering on purple field stating "CAUTION: NONPOTABLE WATER – DO NOT DRINK." Identification tape shall be installed on top of gray water distribution pipes, fastened not greater than every 10 feet (3048 mm) to each pipe length and run continuously the entire length of the pipe. Lettering shall be readily observable within the room or space where the piping is located.

Exception: Outside of the building, purple piping is not required downstream of the backflow preventer where gray water is used for outdoor applications.

708.13 Tests and inspections. Tests and inspections shall be performed in accordance with Sections 708.13.1 through 708.13.8.

708.13.1 Drainage and vent test. A pressure test shall be applied to the gray water collection piping, overflow piping, storage tank drainage piping and tank vent piping in accordance with Section 312 of the *International Plumbing Code*.

708.13.2 Drainage and vent final test. A final test shall be applied to the gray water collection piping, overflow piping, and tank vent piping in accordance with Section 312.4 of the *International Plumbing Code*.

708.13.3 Water supply system test. The testing of makeup water supply piping and rainwater distribution piping shall be conducted in accordance with Section 312.5 of the *International Plumbing Code*.

708.13.4 Inspection and testing of backflow prevention assemblies. The testing of backflow preventers and backwater valves shall be conducted in accordance with Section 312.10 of the *International Plumbing Code*.

708.13.5 Inspection vermin and insect protection. Inlets and vents to the system shall be inspected to verify that each is protected to prevent the entrance of insects and vermin into the storage tank and piping systems in accordance with Section 708.8.

708.13.6 Storage tank tests. Storage tanks shall be tested in accordance with all of the following:

1. Storage tanks shall be filled with water to the overflow line prior to and during inspection. All seams and joints shall be left exposed and the tank shall remain water tight without leakage for a period of 24 hours.
2. After 24 hours, supplemental water shall be introduced for a period of 15 minutes to verify proper drainage of the overflow system and verify that there are no leaks.
3. Following the successful test of the overflow, the water level in the tank shall be reduced to a point that is 2 inches (51 mm) below the makeup water trigger point using the tank drain. The tank drain shall be observed for proper operation. The makeup water system shall be observed to verify proper operation, and successful automatic shutoff of the system at the refill threshold. Water shall not be drained from the overflow at any time during the refill test.

708.13.7 Supply pressure test. The static water pressure at the point of use furthest from the supply shall be verified to be within the range required for the application, in accordance with Section 707.12.9.

708.13.8 Water quality test. The quality of the water for the intended application shall be verified at the point of use in accordance with the requirements of the jurisdiction.

708.14 Subsurface gray water irrigation systems. Gravity subsurface gray water irrigation systems, where provided in accordance with Section 404.1.1, shall be designed and installed in accordance with Sections 708.14.1 through 708.14.6. Gray water collection and storage systems shall comply with this section and the provisions of Section 708 except for Sections 708.6 and 708.12.6.5.

708.14.1 Estimating gray water discharge. The irrigation system shall be sized in accordance with the gallons-per-day-per-occupant number based on the type of fixtures connected to the gray water system. The discharge shall be calculated by the following equation:

$$C = (A \times B) - D \qquad \textbf{(Equation 7-1)}$$

where:

A = Number of occupants:

Residential—For dwelling units regulated by this code in accordance with Section 101.3, the number of occupants shall be determined by the actual number of occupants, but not less than two occupants for one bedroom and one occupant for each additional bedroom.

Commercial—Number of occupants for buildings without dwelling units shall be determined by the *International Building Code*.

B = Estimated flow demands for each occupant:

Residential— For dwelling units regulated by this code in accordance with Section 101.2, 25 gallons per day (94.6 Lpd) per occupant for showers, bathtubs and lavatories and 15 gallons per day (56.7 Lpd) per occupant for clothes washers or laundry trays.

Commercial—For buildings, without dwelling units, based on type of fixture or water use records minus the discharge of fixtures other than those discharging gray water.

C = Estimated gallons (L) of gray water discharge based on the total number of occupants.

D = Estimated gallons (L) of gray water to be used within the interior of the building.

708.14.2 Percolation tests. The permeability of the soil in the proposed absorption system shall be determined by percolation tests or permeability evaluation.

708.14.2.1 Percolation tests and procedures. Not less than three percolation tests in each system area shall be conducted. The holes shall be spaced uniformly in relation to the bottom depth of the proposed absorption system. Additional percolation tests shall be made where necessary, depending on system design.

708.14.2.1.1 Percolation test hole. The test hole shall be dug or bored. The test hole shall have vertical sides and a horizontal dimension of 4 inches to 8 inches (102 mm to 203 mm). The bottom and sides of the hole shall be scratched with a sharp-pointed instrument to expose the natural soil. All loose material shall be removed from the hole and the bottom shall be covered with 2 inches (51 mm) of gravel or coarse sand.

708.14.2.1.2 Test procedure, sandy soils. The hole shall be filled with clearwater to a depth of not less than 12 inches (305 mm) above the bottom of the hole for tests in sandy soils. The time for this amount of water to seep away shall be determined, and this procedure shall be repeated if the water from the second filling of the hole seeps away in 10 minutes or less. The test shall proceed as follows:

1. Water shall be added to a point not more than 6 inches (152 mm) above the gravel or coarse sand.
2. Thereupon, from a fixed reference point, water levels shall be measured at 10-minute intervals for a period of 1 hour.
3. Where 6 inches (152 mm) of water seeps away in less than 10 minutes, a shorter interval between measurements shall be used, but in no case shall the water depth exceed 6 inches (152 mm). Where 6 inches (152 mm) of water seeps away in less than 2 minutes, the test shall be stopped and a rate of less than 3 minutes per inch (7.2 s/mm) shall be reported.

4. The final water level drop shall be used to calculate the percolation rate.

Soils not meeting the above requirements shall be tested in accordance with Section 708.14.2.1.3.

708.14.2.1.3 Test procedure, other soils. The hole shall be filled with clear water, and a water depth of not less than 12 inches (305 mm) shall be maintained above the bottom of the hole for a 4-hour period by refilling whenever necessary or by use of an automatic siphon. Water remaining in the hole after 4 hours shall not be removed. Thereafter, the soil shall be allowed to swell not less than 16 hours or more than 30 hours. Immediately after the soil swelling period, the measurements for determining the percolation rate shall be made as follows:

1. Any soil sloughed into the hole shall be removed and the water level shall be adjusted to 6 inches (152 mm) above the gravel or coarse sand.
2. From a fixed reference point, the water level shall be measured at 30-minute intervals for a period of 4 hours, unless two successive water level drops do not vary by more than $^{1}/_{16}$ inch (1.59 mm). Not less than three water level drops shall be observed and recorded.
3. The hole shall be filled with clear water to a point not more than 6 inches (152 mm) above the gravel or coarse sand whenever it becomes nearly empty. Adjustments of the water level shall not be made during the three measurement periods except to the limits of the last measured water level drop.
4. When the first 6 inches (152 mm) of water seeps away in less than 30 minutes, the time interval between measurements shall be 10 minutes and the test run for 1 hour. The water depth shall not exceed 5 inches (127 mm) at any time during the measurement period.
5. The drop that occurs during the final measurement period shall be used in calculating the percolation rate.

708.14.2.1.4 Mechanical test equipment. Mechanical percolation test equipment shall be of an *approved* type.

708.14.3 Permeability evaluation. Soil shall be evaluated for estimated percolation based on soil structure and texture in accordance with accepted soil evaluation practices. Borings shall be made in accordance with Section 708.14.2.1 for evaluating the soil.

708.14.4 Subsurface landscape irrigation site location. The surface grade of all soil absorption systems shall be located at a point lower than the surface grade of any water well or reservoir on the same or adjoining lots. Where this is not possible, the irrigation system shall be located so that surface water drainage from the building site is not directed toward a well or reservoir. The soil absorption system shall be located with a minimum horizontal distance between various elements as indicated in Table 708.14.4 and as provided in Section 708.12.6.2. Surface water shall be diverted away from any soil absorption site on the same or adjoining lots.

TABLE 708.14.4
LOCATION OF GRAY WATER SYSTEM

ELEMENT	MINIMUM HORIZONTAL DISTANCE (feet) TO IRRIGATION DISPOSAL FIELD
Buildings	2
Lot lines other than lot lines adjoining public ways	5
Water wells	100
Streams, lakes, wetlands other bodies of water	100
Critical root zone (CRZ) of protected trees	2
Seepage pits	5
Septic tanks	5
Water service	5
Public water main	10

For SI: 1 foot = 304.8 mm.

708.14.5 Installation. Absorption systems shall be installed in accordance with Sections 708.14.5.1 through 708.14.5.5 to provide landscape irrigation without surfacing of gray water. Excavations shall not encroach upon the critical root zone (CRZ) of protected trees.

708.14.5.1 Absorption area. The total absorption area required shall be computed from the estimated daily gray water discharge and the design-loading rate based on the percolation rate for the site. The required absorption area equals the estimated gray water discharge divided by the design-loading rate from Table 708.14.5.1.

TABLE 708.14.5.1
DESIGN LOADING RATE

PERCOLATION RATE (minutes per inch)	DESIGN LOAD FACTOR (gallons per square foot per day)
Less than 10	1.2
10 to less than 30	0.8
30 to less than 45	0.72
45 and greater	0.4

For SI: 1 minute per inch = min/25.4 mm, 1 gallon per square foot = 40.7 L/m^2.

708.14.5.2 Seepage trench excavations. Seepage trench excavations shall be not less than 1 foot (304 mm) and not greater than 5 feet (1524 mm) wide. Trench excavations shall be spaced not less than 2 feet (610 mm) apart. The soil absorption area of a seepage trench shall be computed by using the bottom width of the trench multiplied by the length of pipe. Individual seepage trenches shall not exceed 100 feet (30 480 mm) in developed length.

708.14.5.3 Seepage bed excavations. Seepage bed excavations shall be not less than 5 feet (1524 mm) wide and shall have more than one distribution pipe.

The absorption area of a seepage bed shall be computed by using the bottom of the trench area. Distribution piping in a seepage bed shall be uniformly spaced a not greater than 5 feet (1524 mm) and not less than 3 feet (914 mm) apart, and not greater than 3 feet (914 mm) and not less than 1 foot (305 mm) from the sidewall or headwall.

708.14.5.4 Excavation and construction. The bottom of a trench or bed excavation shall be level. Seepage trenches or beds shall not be excavated where the soil is so wet that such material rolled between the hands forms a soil wire. All smeared or compacted soil surfaces in the sidewalls or bottom of seepage trench or bed excavations shall be scarified to the depth of smearing or compaction and the loose material removed. Where rain falls on an open excavation, the soil shall be left until sufficiently dry so a soil wire will not form when soil from the excavation bottom is rolled between the hands. The bottom area shall then be scarified and loose material removed.

708.14.5.5 Aggregate and backfill. Not less than a 6-inch-thick (152 mm) layer of aggregate ranging in size from $^1/_2$ to $2^1/_2$ inches (12.7 mm to 64 mm) shall be laid into the trench below the distribution piping elevation. The aggregate shall be evenly distributed in a layer not less than 2 inches (51 mm) thick over the top of the distribution pipe. The aggregate shall be covered with *approved* synthetic materials or 9 inches (229 mm) of uncompacted marsh hay or straw. Building paper shall not be used to cover the aggregate. Not less than 9 inches (229 mm) of soil backfill shall be placed on top of the synthetic material or marsh hay or straw.

708.14.6 Distribution piping. Distribution piping shall be not less than 3 inches (76 mm) in diameter. The top of the distribution pipe shall be not less than 8 inches (203 mm) below the original surface. The slope of the distribution pipes shall be not less than 2 inches (51 mm) and not greater than 4 inches (102 mm) per 100 feet (30 480 mm).

708.15 Operation and maintenance manuals. Operations and maintenance materials shall be supplied with gray water systems in accordance with Sections 708.15.1 through 708.15.4.

708.15.1 Manual. A detailed operations and maintenance manual shall be supplied in hardcopy form with all gray water systems.

708.15.2 Schematics. The manual shall include a detailed system schematic, locations of all system components, and a list of all system components including manufacturer and model number.

708.15.3 Maintenance procedures. The manual shall provide a maintenance schedule and procedures for all system components requiring periodic maintenance. Consumable parts including filters shall be noted along with part numbers.

708.15.4 Operations procedures. The manual shall include system startup and shutdown procedures. The manual shall include detailed operating procedures for the system.

SECTION 709
RECLAIMED WATER SYSTEMS

709.1 Scope. The provisions of this section shall govern the construction, installation, alteration, and repair of systems supplying nonpotable reclaimed water.

709.2 Permits. Permits shall be required for the construction, installation, alteration, and repair of reclaimed water systems. Construction documents, engineering calculations, diagrams, and other such data pertaining to the reclaimed system shall be submitted with each application for permit.

709.3 Potable water connections. Connections between a reclaimed water system and a potable water system shall be protected against backflow in accordance with Section 608 of the *International Plumbing Code*.

709.4 Installation. Except as provided for in this section, systems shall be installed in compliance with the provisions of the *International Plumbing Code* and the manufacturer's instructions, as applicable.

709.5 Applications. Reclaimed water shall be utilized in accordance with Section 706 and local codes.

709.5.1 Reclaimed water for landscape irrigation. Reclaimed water used for landscape irrigation purposes shall be limited to subsurface applications. Reclaimed water used in irrigation systems shall comply with the provisions of Section 709 except for Section 709.5. Reclaimed water shall be filtered by a 0.004-inch (100 micron) or finer filter. The control panel for the reclaimed water irrigation system shall be provided with signage in accordance with Section 706.2.

Exception: Subject to the approval of the *code official* based on the extent of purification occurring in reclamation process, reclaimed water shall be permitted in sprinkler irrigation applications.

709.6 Approved components and materials. Piping, plumbing components, and material used in the reclaimed water systems shall be manufactured of material *approved* for the intended application.

709.7 Water-pressure-reducing valve or regulator. Where the reclaimed water pressure supplied to the building exceeds 80 psi (552 kPa) static, a pressure-reducing valve shall be installed to reduce the pressure in the reclaimed water distribution system piping to 80 psi (552 kPa) static or less. Pressure-reducing valves shall be specified and installed in accordance with Section 604.8 of the *International Plumbing Code*.

709.8 Trenching requirements. Water service piping, including piping containing reclaimed water, shall be separated from the building sewer by 5 feet (1524 m) of undisturbed or compacted earth. Reclaimed water piping shall be separated from potable water piping underground by 5 feet (1524 m) of undisturbed or compacted earth. Reclaimed water service pipes shall not be located in, under or above cesspools, septic tanks, septic tank drainage fields or seepage pits. Buried reclaimed water piping shall comply with the requirements of Section 306 of the *International Plumbing Code* for support, trenching, bedding, backfilling and tunneling.

Exceptions:

1. The required separation distance shall not apply where the bottom of the reclaimed water service pipe within 5 feet (1524 mm) of the sewer is not less than 12 inches (305 mm) above the top of the highest point of the sewer and the pipe materials comply with the requirements of the *International Plumbing Code* for the application.
2. The required separation distance shall not apply where the bottom of the potable water service pipe within 5 feet (1524 mm) of the reclaimed water pipe is not less than 12 inches (305 mm) above the top of the highest point of the reclaimed water pipe and the pipe materials comply with the requirements of the *International Plumbing Code* for the application.
3. Water service pipe is permitted to be located in the same trench with a building sewer, provided such sewer is constructed of materials that comply with the requirements of the *International Plumbing Code* for the application.
4. The required separation distance shall not apply where a potable or nonpotable water service pipe crosses a sewer pipe provided the water service pipe is sleeved to not less than 5 feet (1524 mm) horizontally from the sewer pipe centerline on both sides of such crossing with pipe materials that comply with the requirements of the *International Plumbing Code* for the application.
5. The required separation distance shall not apply where a potable water service pipe crosses a reclaimed water pipe provided the potable water service pipe is sleeved to not less than 5 feet (1524 mm) horizontally from the reclaimed water pipe centerline on both sides of such crossing with pipe materials that comply with the requirements of the *International Plumbing Code* for the application.

709.9 Reclaimed water systems. The design of the reclaimed water systems shall conform to ASTM E 2635 and accepted engineering practice.

709.9.1 Distribution pipe. Distribution piping shall comply with Sections 709.9.1.1 through 709.9.1.4.

709.9.1.1 Materials. Distribution piping conveying reclaimed water shall conform to standards and requirements specified by the *International Plumbing Code.*

709.9.1.2 Joints. Distribution piping conveying reclaimed water shall utilize joints *approved* for use with the distribution piping and appropriate for the intended applications as specified in the *International Plumbing Code.*

709.9.1.3 Size. Distribution piping conveying reclaimed water shall be sized in accordance with the *International Plumbing Code* for the intended application.

709.9.1.4 Marking. Reclaimed water distribution piping shall be either the color purple and embossed or integrally stamped or marked "CAUTION: NONPOTABLE WATER – DO NOT DRINK" or be installed with a purple identification tape or wrap. Identification tape shall be not less than 3 inches (76 mm) wide and have white or black lettering on purple field stating "CAUTION: NONPOTABLE WATER – DO NOT DRINK." Identification tape shall be installed on top of reclaimed water distribution pipes, fastened not greater than every 10 feet (3048 mm) to each pipe length and run continuously the entire length of the pipe. Lettering shall be readily observable within the room or space where the piping is located.

Exception: Outside of the building, purple piping is not required downstream of the backflow preventer where reclaimed water is used for outdoor applications.

709.10 Tests and inspections. Tests and inspections shall be performed in accordance with Sections 709.10.1 and 709.10.2.

709.10.1 Water supply system test. The testing of makeup water supply piping and reclaimed water distribution piping shall be conducted in accordance with Section 312.5 of the *International Plumbing Code.*

709.10.2 Inspection and testing of backflow prevention assemblies. The testing of backflow preventers shall be conducted in accordance with Section 312.10 of the *International Plumbing Code.*

SECTION 710 ALTERNATE ONSITE NONPOTABLE WATER SOURCES

710.1 Alternate nonpotable sources of water. Other onsite sources of nonpotable water including, but not limited to, stormwater, reverse osmosis reject water, foundation drain water and swimming pool backwash water, shall be permitted to be used for nonpotable uses provided that they have been treated to the quality level necessary for their intended use and in accordance with requirements of the jurisdiction having authority.

CHAPTER 8

INDOOR ENVIRONMENTAL QUALITY AND COMFORT

SECTION 801 GENERAL

801.1 Scope and intent. The provisions of this chapter are intended to provide an interior environment that is conducive to the health of building occupants.

801.2 Indoor air quality management plan required. An indoor air quality management plan shall be developed. Such plan shall address the methods and procedures to be used during design and construction to obtain compliance with Sections 802 through 805.

SECTION 802 BUILDING CONSTRUCTION FEATURES, OPERATIONS AND MAINTENANCE FACILITATION

802.1 Scope. To facilitate the operation and maintenance of the completed building, the building and its systems shall comply with the requirements of Sections 802.2 and 802.3.

802.2 Air-handling system access. The arrangement and location of air-handling system components including, but not limited to, ducts, air handler units, fans, coils and condensate pans, shall allow access for cleaning and repair of the air-handling surfaces of such components. Access ports shall be installed in the air-handling system to permit such cleaning and repairs. Piping, conduits, and other building components shall not be located so as to obstruct the required access ports.

802.3 Air-handling system filters. Filter racks shall be designed to prevent airflow from bypassing filters. Access doors and panels provided for filter replacement shall be fitted with flexible seals to provide an effective seal between the doors and panels and the mating filter rack surfaces. Special tools shall not be required for opening access doors and panels. Filter access panels and doors shall not be obstructed.

SECTION 803 HVAC SYSTEMS

803.1 Construction phase requirements. The ventilation of buildings during the construction phase shall be in accordance with Sections 803.1.1 through 803.1.3.

803.1.1 Duct openings. Duct and other related air distribution component openings shall be covered with tape, plastic, sheet metal or shall be closed by an *approved* method to reduce the amount of dust and debris that collects in the system from the time of rough-in installation and until startup of the heating and cooling equipment. Dust and debris shall be cleaned from duct openings prior to system flush out and building occupancy.

803.1.2 Indoor air quality during construction. Temporary ventilation during construction shall be provided in accordance with Sections 803.1.2.1 through 803.1.2.3.

803.1.2.1 Ventilation. Ventilation during construction shall be achieved through openings in the building envelope using one or more of the following methods:

1. Natural ventilation in accordance with the provisions of the *International Building Code* or the *International Mechanical Code.*
2. Fans that produce a minimum of three air changes per hour.
3. Exhaust in the work area at a rate of not less than 0.05 cfm/ft^2 (0.24 L/s/in^2) and not less than 10 percent greater than the supply air rate so as to maintain negative pressurization of the space.

803.1.2.2 Protection of HVAC system openings. HVAC supply and return duct and equipment openings shall be protected during dust-producing operations.

803.1.2.3 Return air filters. Where a forced air HVAC system is used during construction, new return air filters shall be installed prior to system flush out and building occupancy.

803.1.3 Construction phase ductless system or filter. Where spaces are conditioned during the construction phase, space conditioning systems shall be of the ductless variety, or filters for ducted systems shall be rated at MERV 8 or higher in accordance with ASHRAE 52.2, and system equipment shall be designed to be compatible. Duct system design shall account for pressure drop across the filter.

803.2 Thermal environmental conditions for human occupancy. Buildings shall be designed in compliance with ASHRAE 55, Sections 6.1, "Design," and 6.2, "Documentation."

Exception: Spaces with special requirements for processes, activities, or contents that require a thermal environment outside of that which humans find thermally acceptable, such as food storage, natatoriums, shower rooms, saunas and drying rooms.

803.3 Environmental tobacco smoke control. Smoking shall not be allowed inside of buildings. Any exterior designated smoking areas shall be located not less than 25 ft (7.5 m) away from building entrances, outdoor air intakes, and operable windows.

803.4 Isolation of pollutant sources. The isolation of pollutant sources related to print, copy and janitorial rooms, garages and hangars shall be in accordance with Section 803.4.1.

803.4.1 Printer, copier and janitorial rooms. Enclosed rooms or spaces that are over 100 square feet (9.3 m^2) in area and that are used primarily as a print or copy facility containing five or more printers, copy machines, scanners, facsimile machines or similar machines in any combination, and rooms used primarily as janitorial rooms or closets where the use or storage of chemicals occurs, shall comply with all of the following:

1. The enclosing walls shall extend from the floor surface to the underside of the floor, roof deck or solid ceiling above and shall be constructed to resist the passage of airborne chemical pollutants and shall be constructed and sealed as required for 1-hour fire-resistance-rated construction assemblies. Alternatively, for janitorial rooms and closets, all chemicals shall be stored in *approved* chemical safety storage cabinets.
2. Doors in the enclosing walls shall be automatic or self-closing.
3. An HVAC system shall be provided that: provides separate exhaust airflow to the outdoors at a rate of not less than 0.50 cfm per square foot (2.4 L/s/m^2); that maintains a negative pressure of not less than 7 Pa within the room; and that prohibits the recirculation of air from the room to other portions of the building.

803.5 Filters. Filters for air-conditioning systems that serve occupied spaces shall be rated at MERV 11 or higher, in accordance with ASHRAE Standard 52.2, and system equipment shall be designed to be compatible. The air-handling system design shall account for pressure drop across the filter. The pressure drop across clean MERV 11 filters shall be not greater than 0.45 in. w.c. at 500 FPM (412 Pa at 2.54 m/s) filter face velocity. Filter performance shall be shown on the filter manufacturer's data sheet.

SECTION 804 SPECIFIC INDOOR AIR QUALITY AND POLLUTANT CONTROL MEASURES

804.1 Fireplaces and appliances. Where located within buildings, fireplaces, solid fuel-burning appliances, vented decorative gas appliances, vented gas fireplace heaters and decorative gas appliances for installation in fireplaces shall comply with Sections 804.1.1 through 804.1.3. Unvented room heaters and unvented decorative appliances, including alcohol burning, shall be prohibited.

804.1.1 Venting and combustion air. Fireplaces and fuel-burning appliances shall be vented to the outdoors and shall be provided with combustion air provided from the outdoors in accordance with the *International Mechanical Code* and the *International Fuel Gas Code*. Solid-fuel-burning fireplaces shall be provided with a means to tightly close off the chimney flue and combustion air openings when the fireplace is not in use.

804.1.2 Wood-fired appliances. Wood stoves and wood-burning fireplace inserts shall be listed and, additionally, shall be labeled in accordance with the requirements of the EPA Standards of Performance for New Residential Wood Heaters, 40 CFR Part 60, subpart AAA.

804.1.3 Biomass appliances. Biomass fireplaces, stoves and inserts shall be listed and labeled in accordance with ASTM E 1509 or UL 1482. Biomass furnaces shall be listed and labeled in accordance with CSA B366.1 or UL 391. Biomass boilers shall be listed and labeled in accordance with CSA B366.1 or UL 2523.

804.2 Post-construction, pre-occupancy baseline IAQ testing. Where this section is indicated to be applicable in Table 302.1, and after all interior finishes are installed, the building shall be tested for indoor air quality and the testing results shall indicate that the levels of VOCs meet the levels detailed in Table 804.2 using testing protocols in accordance with ASTM D 6196, ASTM D 5466, ASTM D 5197, ASTM D 6345, and ISO 7708. Test samples shall be taken in not less than one location in each 25,000 square feet (1860 m^2) of floor area or in each contiguous floor area.

Exceptions:

1. Group F, H, S and U occupancies shall not be required to comply with this section.
2. A building shall not be required to be tested where a similarly designed and constructed building as determined by the *code official*, for the same owner or tenant, has been tested for indoor air quality and the testing results indicate that the level of VOCs meet the levels detailed in Table 804.2.
3. Where the building indoor environment does not meet the concentration limits in Table 804.2 and the tenant does not address the air quality issue by mitigation and retesting, the building shall be flushed-out by supplying continuous ventilation with all air-handling units at their maximum outdoor air rate for at least 14 days while maintaining an internal temperature of at least 60°F (15.6°C), and relative humidity not higher than 60 percent. Occupancy shall be permitted to start 7 days after start of the flush-out, provided that the flush-out continues for the full 14 days.

SECTION 805 PROHIBITED MATERIALS

805.1 Scope. The use of the following materials shall be prohibited:

1. Asbestos-containing materials.
2. Urea-formaldehyde foam insulation.

TABLE 804.2
MAXIMUM CONCENTRATION OF AIR POLLUTANTS

MAXIMUM CONCENTRATION OF AIR POLLUTANTS RELEVANT TO IAQ	MAXIMUM CONCENTRATION, ug/m^3 (unless otherwise noted)
1-Methyl-2-pyrrolidinone[a]	160
1,1,1-Trichloroethane	1000
1,3-Butadiene	20
1,4-Dichlorobenzene	800
1,4-Dioxane	3000
2-Ethylhexanoic acid[a]	25
2-Propanol	7000
4-Phenylcyclohexene (4-PCH)[a]	2.5
Acetaldehyde	140
Acrylonitrile	5
Benzene	60
t-Butyl methyl ether	8000
Caprolactam[a]	100
Carbon disulfide	800
Carbon monoxide	9 ppm and no greater than 2 ppm above outdoor levels
Carbon tetrachloride	40
Chlorobenzene	1000
Chloroform	300
Dichloromethane	400
Ethylbenzene	2000
Ethylene glycol	400
Formaldehyde	27
n-Hexane	7000
Naphthalene	9
Nonanal[a]	13
Octanal[a]	7.2
Particulates (PM 2.5)	35 (24-hr)
Particulates (PM 10)	150 (24-hr)
Phenol	200
Styrene	900
Tetrachloroethene	35
Toluene	300
Total volatile organic compounds (TVOC)	500
Trichloroethene	600
Xylene isomers	700

a. This chemical has a limit only where carpets and fabrics with styrene butadiene rubber (SBR) latex backing material are installed as part of the base building systems.

SECTION 806 MATERIAL EMISSIONS AND POLLUTANT CONTROL

806.1 Emissions from composite wood products. Composite wood products used interior to the *approved* weather covering of the building shall comply with the emission limits or be manufactured in accordance with the standards cited in Table 806.1. Compliance with emission limits shall be demonstrated following the requirements of Section 93120 of Title 17, California Code of Regulations, *Airborne Toxic Control Measure to Reduce Formaldehyde Emissions from Composite Wood Products.*

Exceptions:

1. Composite wood products that are made using adhesives that do not contain urea-formaldehyde (UF) resins.
2. Composite wood products that are sealed with an impermeable material on all sides and edges.
3. Composite wood products that are used to make elements considered to be furniture, fixtures and equipment (FF&E) that are not permanently installed.

TABLE 806.1
COMPOSITE PRODUCTS EMISSIONS

PRODUCT	FORMALDEHYDE LIMIT[b] (ppm)	STANDARD
Hardwood plywood	0.05	—
Particle board	0.09	—
Medium-density fiberboard	0.11	—
Thin medium-density fiberboard[a]	0.13	—

a. Maximum thickness of $^5/_{16}$ inch (8 mm).
b. Phase 2 Formaldehyde Emissions Standards, Table 1, Section 93120, Title 17, California Code of Regulations; compliance shall be demonstrated in accordance with ASTM E 1333 or ASTM D 6007.

806.2 Adhesives and sealants. A minimum of 85 percent by weight or volume, of specific categories of site-applied adhesives and sealants used on the interior side of the building envelope shall comply with the VOC content limits in Table 806.2(1) or alternative VOC emission limits in Table 806.2(2). The VOC content shall be determined in accordance with the appropriate standard being either U.S. EPA Method 24 or SCAQMD Method 304, 316A or 316B. The exempt compound content shall be determined by either SCAQMD Methods 302 and 303 or ASTM D 3960. Table 806.2(1) adhesives and sealants regulatory category and VOC content compliance determination shall conform to the SCAQMD Rule 1168 Adhesive and Sealant Applications as amended on 1/7/05. The provisions of this section shall not apply to adhesives and sealants subject to state or federal consumer product VOC regulations. HVAC duct sealants shall be classified as "Other" category within the SCAQMD Rule 1168 sealants table.

Exception: HVAC air duct sealants are not required to meet the emissions or the VOC content requirements when the air temperature in which they are applied is less than 40°F (4.5°C).

Table 806.2(2) adhesive alternative emissions standards compliance shall be determined utilizing test methodology incorporated by reference in the CDPH/EHLB/Standard Method V.1.1, *Standard Method for Testing VOC Emissions From Indoor Sources*, dated February 2010. The alternative emissions testing shall be performed by a laboratory that has the CDPH/EHLB/Standard Method V.1.1 test methodology in the scope of its ISO 17025 Accreditation.

TABLE 806.2(1)
SITE-APPLIED ADHESIVE AND SEALANT VOC LIMITS

ADHESIVE	VOC LIMIT[a, b]
Indoor carpet adhesives	50
Carpet pad adhesives	50
Outdoor carpet adhesives	150
Wood flooring adhesive	100
Rubber floor adhesives	60
Subfloor adhesives	50
Ceramic tile adhesives	65
VCT and asphalt tile adhesives	50
Dry wall and panel adhesives	50
Cove base adhesives	50
Multipurpose construction adhesives	70
Structural glazing adhesives	100
Single-ply roof membrane adhesives	250
Architectural sealants	250
Architectural sealant primer Nonporous Porous	 250 775
Modified bituminous sealant primer	500
Other sealant primers	750
CPVC solvent cement	490
PVC solvent cement	510
ABS solvent cement	325
Plastic cement welding	250
Adhesive primer for plastic	550
Contact adhesive	80
Special purpose contact adhesive	250
Structural wood member adhesive	140

a. VOC limit less water and less exempt compounds in grams/liter.

b. For low-solid adhesives and sealants, the VOC limit is expressed in grams/liter of material as specified in Rule 1168. For all other adhesives and sealants, the VOC limits are expressed as grams of VOC per liter of adhesive or sealant less water and less exempt compounds as specified in Rule 1168.

TABLE 806.2(2)
VOC EMISSION LIMITS

VOC	LIMIT
Individual VOCs	≤ $^1/_2$ CA chronic REL[a]
Formaldehyde	≤ 16.5 μg/m^3 or ≤ 13.5 ppb[b, c]

a. CDPH/EHLB/Standard Method V.1.1 Chronic Reference Exposure Level (CREL).

b. Effective January 1, 2012, limit became less than or equal to the CDPH/EHLB/Standard Method V.1.1 CREL (≤ 9 μg/m^3 or ≤ 7 ppb)

c. Formaldehyde emission levels need not be reported for materials where formaldehyde is not added by the manufacturer of the material.

806.3 Architectural paints and coatings. A minimum of 85 percent by weight or volume, of site-applied interior architectural coatings shall comply with VOC content limits in Table 806.3(1) or the alternate emissions limits in Table 806.3(2). The exempt compound content shall be determined by ASTM D 3960.

Table 806.3(2) architectural coating alternate emissions standards compliance shall be determined utilizing test methodology incorporated by reference in the CDPH/EHLB/Standard Method V.1.1, *Standard Method for Testing VOC Emissions From Indoor Sources*, dated February 2010. The alternative emissions testing shall be performed by a laboratory that has the CDPH/EHLB/Standard Method V.1.1 test methodology in the scope of its ISO 17025 Accreditation.

806.4 Flooring. A minimum of 85 percent of the total area of flooring installed within the interior of the building shall comply with the requirements of Table 806.4(2). Where flooring with more than one distinct product layer is installed, the emissions from each layer shall comply with these requirements. The test methodology used to determine compliance shall be from CDPH/EHLB/Standard Method V.1.1, *Standard Method for Testing VOC Emissions From Indoor Sources*, dated February 2010. The emissions testing shall be performed by a laboratory that has the CDPH/EHLB/Standard Method V.1.1 test methodology in the scope of its ISO 17025 Accreditation.

Where post-manufacture coatings or surface applications have not been applied, the flooring listed in Table 806.4(1) shall be deemed to comply with the requirements of Table 806.4(2).

806.5 Acoustical ceiling tiles and wall systems. A minimum of 85 percent of acoustical ceiling tiles and wall systems, by square feet, shall comply with the requirements of Table 806.5(2). Where ceiling and wall systems with more than one distinct product layer are installed, the emissions from each layer shall comply with these requirements. The test methodology used to determine compliance shall be from CDPH/EHLB/Standard Method V.1.1, *Standard Method for Testing VOC Emissions From Indoor Sources*, dated February 2010. The emissions testing shall be performed by a laboratory that has the CDPH/EHLB/Standard Method V.1.1 test methodology in the scope of its ISO 17025 Accreditation.

Where post-manufacture coatings or surface applications have not been applied, the ceiling or wall systems listed in Table 806.5(1) shall be deemed to comply with the requirements of Table 806.5(2).

806.6 Insulation. A minimum of 85 percent of insulation shall comply with the requirements of Table 806.6(1) or Table 808.6(2). The test methodology used to determine compliance shall be from CDPH/EHLB/Standard Method V.1.1, *Standard Method for Testing VOC Emissions From Indoor Sources*, dated February 2010. The emissions testing shall be performed by a laboratory that has the CDPH/EHLB/Standard Method V.1.1 test methodology in the scope of its ISO 17025 Accreditation.

TABLE 806.3(2)
ARCHITECTURAL COATINGS VOC EMISSION LIMITS

VOC	LIMIT
Individual	≤ $^1/_2$ CA chronic REL[a]
Formaldehyde	≤ 16.5 μg/m^3 or ≤ 13.5 ppb[b]

a. CA Chronic Reference Exposure Level (CREL).

b. Formaldehyde emission levels need not be reported for materials where formaldehyde is not added by the manufacturer of the material.

TABLE 806.3(1)
VOC CONTENT LIMITS FOR ARCHITECTURAL COATINGS[c, d, e]

CATEGORY	Effective: January 1, 2010 LIMIT[a] g/l	Effective: January 1, 2012 LIMIT[a] g/l
Flat coatings	50	
Nonflat coatings	100	
Nonflat – High-gloss coatings	150	
Specialty coatings:		
Aluminum roof coatings	400	
Basement specialty coatings	400	
Bituminous roof coatings	50	
Bituminous roof primers	350	
Bond breakers	350	
Concrete curing compounds	350	
Concrete/masonry sealers	100	
Driveway sealers	50	
Dry fog coatings	150	
Faux finishing coatings	350	
Fire-resistive coatings	350	
Floor coatings	100	
Form-release compounds	250	
Graphic arts coatings (Sign paints)	500	
High-temperature coatings	420	
Industrial maintenance coatings	250	
Low solids coatings	120[b]	
Magnesite cement coatings	450	
Mastic texture coatings	100	
Metallic pigmented coatings	500	
Multi-color coatings	250	
Pretreatment wash primers	420	
Primers, sealers, and undercoaters	100	
Reactive penetrating sealers	350	
Recycled coatings	250	
Roof coatings	50	
Rust-preventative coatings	400	250
Shellacs, clear	730	
Shellacs, opaque	550	
Specialty primers, sealers, and undercoaters	350	100
Stains	250	
Stone consolidants	450	
Swimming pool coatings	340	
Traffic marking coatings	100	
Tub and tile refinish coatings	420	
Waterproofing membranes	250	
Wood coatings	275	
Wood preservatives	350	
Zinc-rich primers	340	

a. Limits are expressed as VOC Regulatory (except as noted), thinned to the manufacturer's maximum thinning recommendation, excluding any colorant added to tint bases.

(continued)

TABLE 806.3(1)—continued
VOC CONTENT LIMITS FOR ARCHITECTURAL COATINGS[c, d, e]

b. Limit is expressed as VOC actual.

c. The specified limits remain in effect unless revised limits are listed in subsequent columns in the table.

d. Values in this table are derived from those specified by the California Air Resources Board *Suggested Control Measure for Architectural Coatings*, dated February 1, 2008.

e. Table 806.3(1) architectural coating regulatory category and VOC content compliance determination shall conform to the California Air Resources Board *Suggested Control Measure for Architectural Coatings*, dated February 1, 2008.

TABLE 806.4(1)
FLOORING DEEMED TO COMPLY WITH VOC EMISSION LIMITS

Ceramic and concrete tile
Organic-free, mineral-based
Clay pavers
Concrete pavers
Concrete
Metal

TABLE 806.4(2)
FLOORING VOC EMISSION LIMITS

VOC	LIMIT
Individual	≤ $^1/_2$ CA chronic REL[a]
Formaldehyde	≤ 16.5 $\mu g/m^3$ or ≤ 13.5 ppb

a. CA Chronic Reference Exposure Level (CREL).

TABLE 806.5(1)
CEILING AND WALL SYSTEMS DEEMED TO COMPLY WITH VOC EMISSION LIMITS

Ceramic and concrete tile
Organic-free, mineral-based
Gypsum plaster
Clay masonry
Concrete masonry
Concrete
Metal

TABLE 806.5(2)
ACOUSTICAL CEILING TILES AND WALL SYSTEMS VOC EMISSION LIMITS

VOC	LIMIT
Individual	≤ $^1/_2$ CA chronic REL[a]
Formaldehyde	≤ 16.5 $\mu g/m^3$ or ≤ 13.5 ppb

a. CA Chronic Reference Exposure Level (CREL).

TABLE 806.6(1)
INSULATION VOC EMISSION LIMITS

VOC	LIMIT
Individual	≤ $^1/_2$ CA chronic REL[a]
Formaldehyde	≤ 16.5 $\mu g/m^3$ or ≤ 13.5 ppb

a. CA Chronic Reference Exposure Level (CREL).

TABLE 806.6(2)
INSULATION MANUFACTURED WITHOUT FORMALDEHYDE VOC EMISSION LIMITS

VOC	LIMIT
Individual	≤ $^1/_2$ CA chronic REL[a]

a. CA Chronic Reference Exposure Level (CREL).

SECTION 807
ACOUSTICS

807.1 Sound transmission and sound levels. Where required by Table 302.1, buildings and tenant spaces shall comply with the minimum sound transmission class and maximum sound level requirements of Sections 807.2 through 807.5.2.

Exception: The following buildings and spaces need not comply with this section:

1. Building or structures that have the interior environment open to the exterior environment.
2. Parking structures.
3. Concession stands and toilet facilities in Group A-4 and A-5 occupancies.

807.2 Sound transmission. Sound transmission classes established by laboratory measurements shall be determined in accordance with ASTM E 413 based on measurements in accordance with ASTM E 90. Sound transmission classes for concrete masonry and clay masonry assemblies shall be calculated in accordance with TMS 0302 or determined in accordance with ASTM E 413 based on measurements in accordance with ASTM E 90. Field measurements of completed construction, if conducted, shall be in accordance with ASTM E 336 where conditions regarding room size and absorption required in ASTM E 336 are met.

807.2.1 Interior sound transmission. Wall and floor-ceiling assemblies that separate Group A and F occupancies from one another or from Group B, I, M or R occupancies shall have a sound transmission class (STC) of not less than 60 or an apparent sound transmission class (ASTC) of not less than 55 if the completed construction is field tested. Wall and floor-ceiling assemblies that separate Group B, I, M or R occupancies from one another shall have a sound transmission class (STC) of not less than 50 or an apparent sound transmission class (ASTC) of not less than 45 if the completed construction is field tested. Wall and floor-ceiling assemblies that separate Group R condominium occupancies from one another or from other Group B, I, M or R occupancies shall have a sound transmission class (STC) of not less than 55 or an apparent sound transmission class (ASTC) of not less than 50 if the completed construction is field tested.

Exception: This section shall not apply to wall and floor-ceiling assemblies enclosing:

1. Public entrances to tenants of covered and open mall buildings.
2. Concession stands and lavatories in Group A-4 and A-5 occupancies.
3. Spaces and occupancies that are accessory to the main occupancy.

807.2.2 Mechanical and emergency generator equipment and systems. Wall and floor-ceiling assemblies that separate a mechanical equipment room or space from the remainder of the building shall have a sound transmission class (STC) of not less than 50 or an apparent sound transmission class (ASTC) of not less than 45 if the completed construction is field tested. Wall and floor-ceiling assemblies that separate a generator equipment room or space from the remainder of the building shall have a sound transmission class (STC) of not less than 60 or an apparent sound transmission class (ASTC) of not less than 55 if the completed construction is field tested.

807.3 Sound levels. The design and construction of mechanical and electrical generator systems and of walls and floor-ceilings separating such equipment from the outdoors or other building space shall achieve sound levels not greater than specified in Sections 807.3.1 and 807.3.2 during the normal operation of mechanical equipment and generators. Electrical generators used only for emergencies are exempt from the limits on sound levels within the building and need only meet daytime limits for sound-reaching boundaries. Where necessary, walls and floor-ceiling assemblies with sound transmission class (STC) ratings greater than specified in Section 807.2.2 shall be used to meet this requirement.

807.3.1 Sound of mechanical and electrical generator equipment outside of buildings. Where mechanical equipment or electrical generators are located outside of the building envelope or their sound is exposed to the exterior environment, the sound reaching adjacent properties shall comply with all applicable ordinances and zoning performance standards. In the absence of an ordinance or zoning performance standard specifying sound limits at the boundary, or a law specifying different limits if limits are imposed, an adjacent property at the boundary shall not be subjected to a sound level greater than indicated in Table 807.3.1 because of the sound of the equipment. Where a generator is used only for providing emergency power and all periodic operational testing is done during the daytime period of Table 807.3.1, the sound of a generator during the night-time hours shall meet the daytime limits.

807.3.2 Sound of HVAC and mechanical systems within buildings. Sound levels within rooms generated by HVAC and mechanical systems within the building, including electrical generators used regularly but excluding emergency generators, for all modes of operation shall not exceed the limits shown in Table 807.3.2.

807.4 Structure-borne sounds. Floor and ceiling assemblies between dwelling rooms or dwelling units and between dwelling rooms or dwelling units and public or service areas within the structure in occupancies classified as Group A1, A2, A3, B, E, I, M or R shall have an impact insulation classification (IIC) rating of not less than 50 where laboratory-tested and 45 where field-tested when tested in accordance with ASTM E 492. New laboratory tests for impact insulation class (IIC) of an assembly are not required where the IIC has been established by prior tests.

807.5 Special inspections for sound levels. An approved agency, funded by the building owner, shall furnish report(s) of test findings indicating that the sound level results are in compliance with this section, applicable laws and ordinances, and the construction documents. Discrepancies shall be brought to the attention of the design professional and *code official* prior to the completion of that work. A final testing report documenting required testing and corrections of any discrepancies noted in prior tests shall be submitted at a point in time agreed upon by the building owner, or building owner's agent, design professional, and the *code official* for purposes of demonstrating compliance.

807.5.1 Testing for mechanical and electrical generator equipment outside of buildings. Special inspections shall be conducted in accordance with Section 903.1 to demonstrate compliance with the requirements of Section 807.3.1. Testing shall be conducted following the complete installation of the equipment or generators, the installation of sound reduction barriers, and balancing and operation of the equipment or generators. Testing shall be at locations representing the four cardinal directions from the face of the project building. Such testing shall demonstrate that the equipment is capable of compliance with the night-time limits under normal night-time operating conditions, and if higher sound levels are possible during the daytime, compliance with the daytime limits shall also be demonstrated.

807.5.2 Testing for building system background noise. Special inspections shall be conducted in accordance with Section 903.1 to demonstrate compliance with the requirements of Section 807.3.2. Testing shall be executed within not less than 50 percent of the total number of rooms contained in a building or structure of the types listed in Table 807.3.2 for the given occupancy in accordance with Table 903.1. Testing shall occur following the complete installation of the equipment and systems, the installation of any sound reduction barriers, and balancing and operation of the equipment and systems.

807.5.3 Separating assemblies. Wall and floor-ceiling assemblies that separate a mechanical or emergency generator equipment room or space from the remainder of the building shall have a sound transmission class (STC) of not less than 60 determined in accordance with ASTM E 90 and ASTM E 413, or for concrete masonry and clay masonry assemblies as calculated in accordance with TMS 0302 or as determined in accordance with ASTM E 90 and ASTM E 413.

807.5.4 HVAC background sound. HVAC system caused background sound levels for all modes of operation within rooms shall be in accordance with the lower and upper noise criteria (NC) limits as shown in Table 807.3.2. Special inspections shall be required and conducted in accordance with Section 903.1 in order to demonstrate compliance.

807.6 Special inspections for sound transmission. An approved agency, employed by the building owner, shall furnish report(s) of test findings indicating that the results are in compliance with this section and the construction documents. Discrepancies shall be brought to the attention of the design professional and *code official* prior to the completion of that work. A final testing report documenting required testing and corrections of any discrepancies noted in prior tests shall be submitted at a point in time agreed upon by the building owner, or building owner's agent, design professional, and the *code official* for purposes of demonstrating compliance.

Exception: Test reports are not required for *approved* assemblies with an established sound transmission class (STC) rating.

807.6.1 Testing for mechanical and emergency generator equipment outside of buildings. In accordance with Section 807.3.1, all mechanical and emergency generator equipment shall be field tested in accordance with Table 903.1. Testing shall be conducted following the complete installation of the equipment or generators, the installation of sound reduction barriers, and balancing and operation of the equipment or generators. Testing shall be at locations representing the four cardinal directions from the face of the project building. Such testing shall occur on a Tuesday, Wednesday or Thursday at both the day and night times within the periods shown in Table 807.3.1.

807.6.2 Testing for building system background noise. Testing shall be executed in accordance with Section 807.3.1 within not less than 50 percent of the total number of rooms contained in a building or structure, exclusive of closets and storage rooms less than 50 square feet (4.65 m^2) in area, and exclusive of toilet facilities in accordance with Table 903.1. Testing shall occur following the complete installation of the equipment and systems, the installation of any sound reduction barriers, and balancing and operation of the equipment and systems.

SECTION 808 DAYLIGHTING

808.1 General. Fenestration shall be provided in building roofs and walls in accordance with Sections 808.2 and 808.3. Interior spaces shall be planned to benefit from exposure to the natural light offered by the fenestration in accordance with this section.

808.1.1 Fenestration obstructions. Advertisements or displays affixed or applied to a fenestration, or supported

TABLE 807.3.1
MAXIMUM PERMISSIBLE OUTDOOR A-WEIGHTED SOUND LEVELS

INITIATING PROPERTY	ADJACENT PROPERTY	MAXIMUM A-WEIGHTED SOUND LEVEL (dB)	
		Day Time 7:00 AM to 10:00 PM	Night Time 10:00 PM to 7:00 AM
All, except factory, industrial, or storage	All, except factory, industrial, or storage	65	55
Factory, industrial, or storage	All other, except factory, industrial, or storage	65	55
Factory, industrial, or storage	Factory, industrial, or storage	75	75

TABLE 807.3.2
MAXIMUM PERMISSIBLE INDOOR BACKGROUND SOUND IN ROOMS

OCCUPANCY TYPE	ROOM	NOISE CRITERIA (NC) LIMITS
Assembly A-1	Symphony, concert, recital halls	30
	Motion picture theaters	40
Assembly A-3	Places of religious worship, lecture halls not part of educational facilities	35
	Art gallery, exhibit hall, funeral parlor, libraries, and museums	40
	Courtroom	35
	Educational occupancies above 12th grade	(See Educational)
Assembly A-4	Gymnasiums, natatoriums and arenas with seating areas	45
Business B	Office—enclosed greater than 300 square feet	35
	Office—enclosed less than or equal 300 square feet	40
	Office—open plan	45
	Corridors and lobbies	45
	Conference rooms	35
	Educational occupancies above 12th grade	(See Educational)
Educational E	Core learning lecture and classrooms that are less than or equal to 20,000 cubic feet in volume Core learning lecture and classrooms that are greater than 20,000 cubic feet in volume Open plan classrooms Administrative offices and rooms Music teaching studios Music practice rooms	ANSI/ASA S12.60-2010/Part 1 or ANSI/ASA S12.60-2009/Part 2
Institutional I-2	Wards Private and semi-private patient rooms Operating rooms Corridors and public areas	2010 FGI-ASHE Guidelines for Design and Construction of Healthcare Facilities
	Rooms or suites	25 to 35
	Bathroom, kitchen, utility room	40
Residential R-1 and R-2	Meeting rooms	35
	Corridors and lobbies	45
	Service areas	45

For SI: 1 square foot = 0.093 m^2, 1 cubic foot = 28.31 L.

by the building shall not reduce daylighting below the levels prescribed herein.

Exception: The ground floor and the story immediately above the ground floor.

808.2 Applicability. Daylighting of building spaces in accordance with Section 808.3 shall be required for the following occupancies:

1. A Group A-3 occupancy where the specific use of the room or space is for reading areas in libraries, waiting areas in transportation terminals, exhibition halls, gymnasiums, and indoor athletic areas.
2. A Group B occupancy where the specific use of the room or space is for educational facilities for students above the 12th grade, laboratories for testing and research, post offices, print shops, offices, and training and skill development not within a school or academic program.
3. Group E, F and S occupancies.
4. Those portions of Group M occupancies located directly underneath a roof, where the net floor area of the entire occupancy is 10,000 square feet (929 m^2) or greater.

Exception: Daylighting is not required in the following rooms and spaces:

1. Building spaces where darkness is required for the primary use of the space, including, but not limited to, light-sensitive material handling and darkrooms.
2. Building spaces that are required to be cooled below 50°F (10°C).
3. Unconditioned buildings that are equipped with exterior doors that, when opened, provide equivalent daylighting.
4. Alteration, repair, movement, or change of occupancy of existing buildings.

808.3 Daylit area of building spaces. In buildings not greater than two stories above grade, not less than 50 percent of the net floor area shall be located within a daylit area. In buildings three or more stories above grade, not less than 25 percent of the net floor area shall be located within a daylit area. Buildings required to have more than 25,000 square feet (2323 m^2) of daylit area shall comply with Section 808.3.2. All other buildings shall comply with either Section 808.3.1 or Section 808.3.2.

Exception: For buildings not less than three stories above grade with obstructed exterior walls or shaded roofs, the required daylit area shall be modified in accordance with Equation 8-1.

Required daylit area ≥ 25% × TDP **(Equation 8-1)**

The total daylight potential (TDP) is a weighted average of the individual daylight potentials for each floor:

$$TDP = \Sigma(DP_1 \div FA_1/TF) + (DP_2 \div FA_2/TF) + \ldots$$

For floors with roof area immediately above:

$$DP_{1,2\ldots} = 1 - [(OW_1/TW_1) \times (OR_1/TR_1)]$$

For floors without roof area immediately above:

$$DP_{1,2\ldots} = 1 - (OW_1/TW_1)$$

$OW_{1,2\ldots}$ = The length of obstructed exterior wall for each floor that does not face a public way or a yard or court complying with Section 1206 of the *International Building Code* or where the distance to any buildings, structures, or geological formations in front of the wall is less than two times the height of the buildings, structures, or geological formations. For the purposes of this determination, the maximum allowed heights of buildings or structures on adjacent property under existing zoning regulations is permitted to be considered.

$TW_{1,2\ldots}$ = The total length of exterior wall for each floor.

$OR_{1,2\ldots}$ = The roof area immediately above each floor that is shaded during the peak sun angle on the summer solstice by permanent features of the building, or by permanent features of adjacent buildings.

$TR_{1,2\ldots}$ = The total roof area immediately above each floor.

$FA_{1,2\ldots}$ = The *total floor area* of each floor.

TF = The *total building floor area.*

808.3.1 Daylight prescriptive requirements. Daylit areas shall comply with Section 808.3.1.1 or 808.3.1.2. For determining the total daylit area, any overlapping daylit areas shall be counted only once.

The total daylight area shall be the sum of the area of all sidelighting daylight zones and the area of all toplighting zones, except that sidelighting daylight zones shall not be included in the calculation of the area of toplighting daylight areas.

808.3.1.1 Sidelighting. The daylit area shall be illuminated by fenestration that complies with Table 808.3.1.1 and Figure 808.3.1.1(4). Where fenestration is located in a wall, the daylit area shall extend laterally to the nearest 56-inch-high (1422 mm) partition, or up to 1.0 times the height from the floor to the top of fenestration facing within 45 degrees (0.785 rad) of east or west or up to 1.5 times the height from the floor to the top of all other fenestration, whichever is less, and longitudinally from the edge of the fenestration to the nearest 56-inch-high (1422 mm) partition, or up to 2 feet (610 mm), whichever is less, as indicated in Figure 808.3.1.1(1). Where fenestration is located in a rooftop monitor, the daylit area shall extend laterally to the nearest 56-inch-high (1422 mm) partition, or up to 1.0 times the height from the floor to the bottom of the fenestration, whichever is less, and longitudinally from the edge of the fenestration to the nearest 56-inch-high (1422 mm) partition, or up to 0.25 times the height from the floor to the bottom of the fenestration, whichever is less, as indicated in Figures 808.3.1.1(2) and 808.3.1.1(3).

$$EA = (AF \times VT)/DA \quad \textbf{(Equation 8-2)}$$

where:

EA = Effective aperture.

AF = Area of fenestration.

VT = Visible transmittance of the fenestration.

DA = Daylit area.

TABLE 808.3.1.1
MINIMUM EFFECTIVE APERTURE

SKY TYPE	MINIMUM EFFECTIVE APERTURE (percentage)		
	Sidelighting from fenestration in a wall [see Figure 808.3.1.1(1)]	Sidelighting from rooftop monitor [see Figures 808.3.1.1(2) and 808.3.1.1(3)]	Toplighting (see Figure 808.3.1.2)
A[a]	10.0	5.0	1.0
B[b]	12.0	6.0	1.2
C[c]	16.0	8.0	2.2

a. Sky Type A – more than 75 percent mean sunshine, in accordance with the NOAA Annual Mean Sunshine Percentage Table.
b. Sky Type B – 45 percent to 75 percent mean sunshine, in accordance with the NOAA Annual Mean Sunshine Percentage Table.
c. Sky Type C – less than 45 percent mean sunshine, in accordance with the NOAA Annual Mean Sunshine Percentage Table.

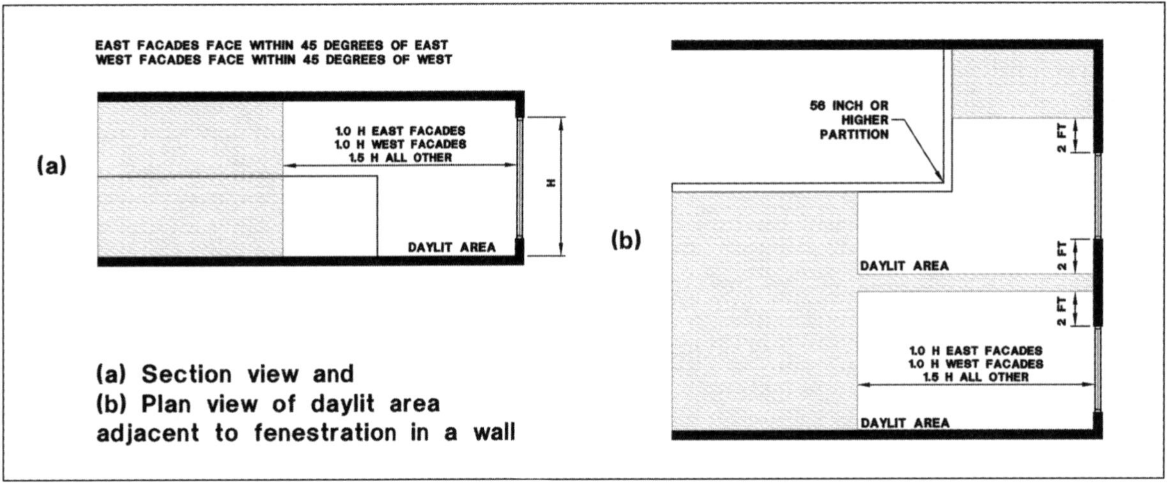

For SI: 1 inch = 25.4 mm, 1 foot = 304.8 mm, 1 degree = 0.017 rad.

FIGURE 808.3.1.1(1)
DAYLIT AREA ADJACENT TO FENESTRATION IN A WALL

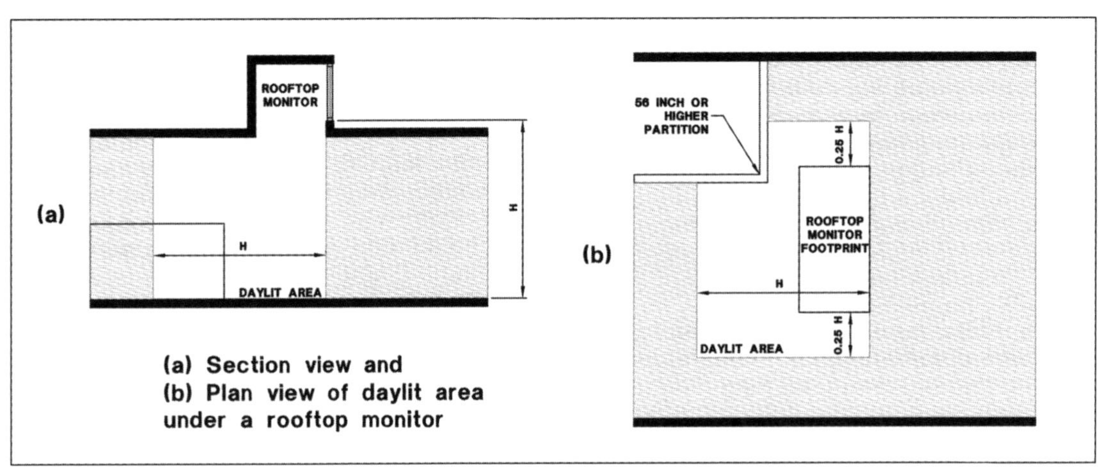

For SI: 1 inch = 25.4 mm.

FIGURE 808.3.1.1(2)
DAYLIT AREA ADJACENT UNDER A ROOFTOP MONITOR

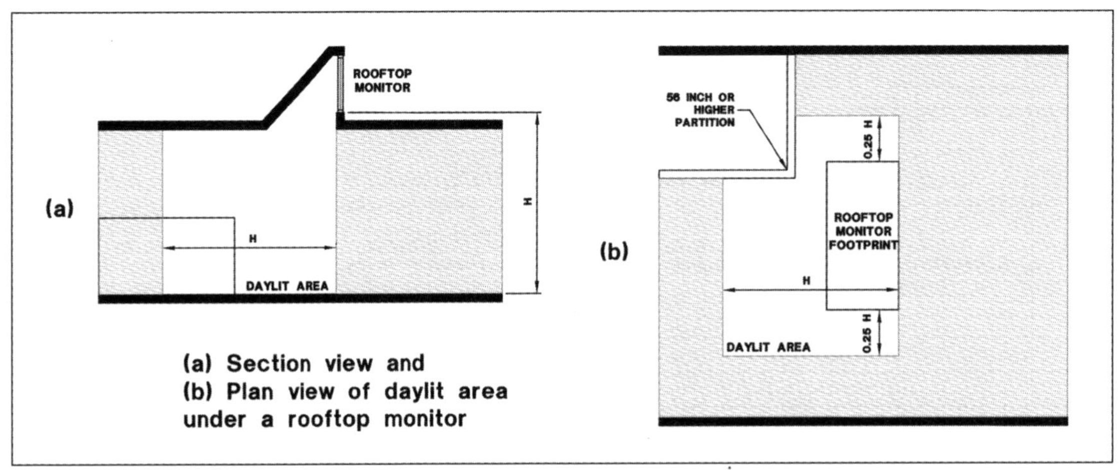

For SI: 1 inch = 25.4 mm.

FIGURE 808.3.1.1(3)
DAYLIT AREA ADJACENT UNDER A ROOFTOP MONITOR

808.3.1.2 Toplighting. The daylit area shall be illuminated by a roof fenestration assembly such as a skylight, sloped glazing or tubular daylighting device that complies with Table 808.3.1.1 and Figure 808.3.1.2. The daylit area extends laterally and longitudinally beyond the glazed opening of the roof fenestration assembly to the nearest 56-inch-high (1422 mm) partition, or up to 0.7 times the height from the floor to the bottom of the rough opening of the daylighting well, whichever is less, as indicated in Figure 808.3.1.2.

808.3.2 Daylight performance requirements. Each daylit area shall comply with the requirements of either Section 808.3.2.1 or 808.3.2.2. Daylight analysis shall be conducted in accordance with Section 808.3.2.3.

808.3.2.1 Morning illumination. Not less than 28 foot-candles (300 lux) and not more than 418 foot-candles (4500 lux) of natural light shall be available at a height of 30 inches (750 mm) above the floor 3 hours before the peak solar angle on the spring equinox.

808.3.2.2 Afternoon illumination. Not less than 28 foot-candles (300 lux) and not more than 418 foot-candles (4500 lux) of natural light shall be available at a height of 30 inches (750 mm) above the floor 3 hours after the peak solar angle on the spring equinox.

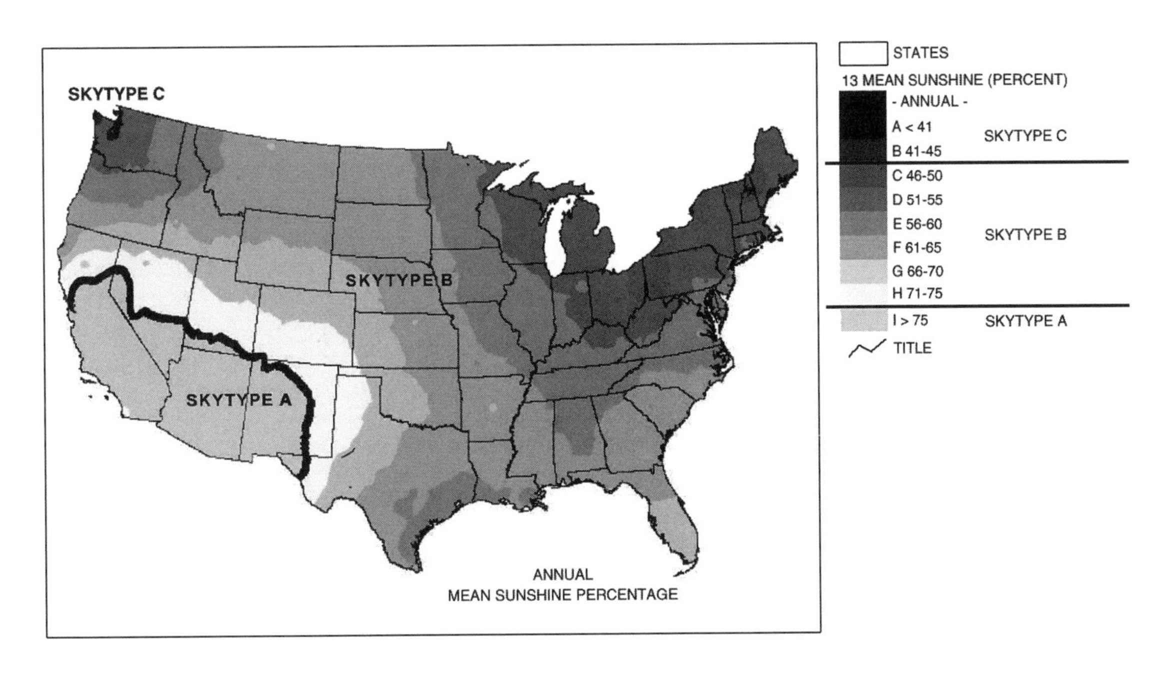

FIGURE 808.3.1.1(4)
SKY TYPES

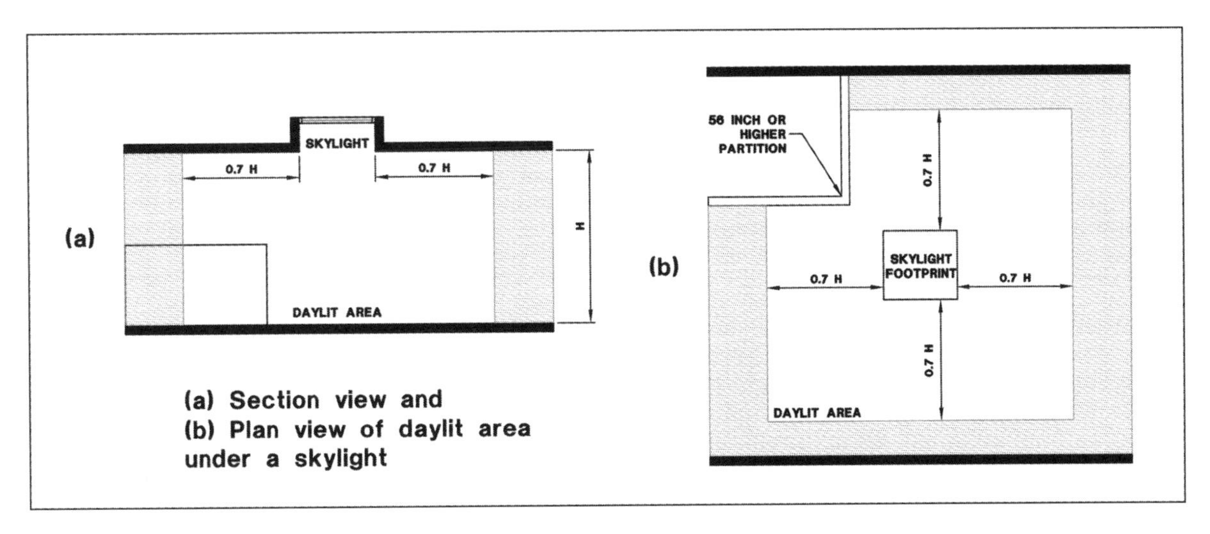

For SI: 1 inch = 25.4 mm.

FIGURE 808.3.1.2
DAYLIT AREA UNDER A SKYLIGHT

808.3.2.3 Daylight analysis. A daylight analysis shall be performed that complies with the following:

1. Sky conditions shall be assumed to be clear.
2. Address the effects of exterior shading devices, buildings, structures, and geological formations on the fenestration of the proposed building and on the ground and other light reflecting surfaces. Include the effects of movable exterior fenestration shading devices. The configuration of fenestration with automatically controlled variable transmittance shall be adjusted to accurately represent the control system operation.
3. Exclude the effects of interior furniture systems, shelving, and stacks.
4. Use the actual reflectance characteristics of all materials.
5. Where blinds, shades and other movable interior fenestration shading devices are included in the analysis and the exact properties of such devices cannot be accurately modeled, such devices shall be assumed to be completely diffusing, with a visible transmittance of 5 percent for fabric shades, and 20 percent for horizontal or vertical blinds.
6. Calculation points shall be spaced not more than 39.4 inches (1 m) by 39.4 inches (1 m). The calculation grid shall start within 20 inches (508 mm) of each wall or partition.
7. Where details about the window framing, mullions, wall thickness and well depth cannot be included in the model, the visible transmittance of all fenestration shall be reduced by 20 percent.

808.4 Sky types. Sky types as described in Section 808.4.1 or 808.4.2 shall be used in determining the applicable effective aperture in Table 808.3.1.1

808.4.1 United States sky types. All states, counties, and territories shall be sky type B, except as named herein. The states and counties in sky type A shall be: all of Arizona; in Nevada the counties of Churchill, Lincoln, Nye, Washoe, and counties south; in New Mexico the counties of Lincoln, Otero, Sandoval, San Juan, Santa Fe, Torrance and counties south; in Texas the counties of Hudspeth, El Paso, and Jeff Davis; in Utah the counties of Iron, Kane, and Washington; and in California all counties except Del Norte, Siskiyou, Modoc, Humboldt, Trinity, and Mendocino. Alaska shall be sky type C.

808.4.2 International sky types. All international locations shall be sky type B, except as follows: locations with an annual average of more than 75 percent sunshine during daytime hours shall be sky type A, and locations with an annual average of less than 45 percent sunshine during daytime hours shall be sky type C.

CHAPTER 9

COMMISSIONING, OPERATION AND MAINTENANCE

SECTION 901 GENERAL

901.1 Scope. The provisions of this chapter are intended to facilitate the pre- and post-occupancy commissioning, operation and maintenance of buildings constructed in accordance with this code in a manner that is consistent with the intent of other provisions of this code, and to further that goal through the education of building owners and maintenance personnel with regard to related best operating and management practices.

SECTION 902 APPROVED AGENCY

902.1 Approved agency. An approved agency shall provide all of the information necessary for the *code official* to determine that the agency meets the applicable requirements. The *code official* shall be permitted to be the approved agency.

902.1.1 Independence. An approved agency shall be objective, competent and independent from the contractor responsible for the work being inspected. The agency shall also disclose possible conflicts of interest so that objectivity can be confirmed.

902.1.2 Equipment. An approved agency shall have adequate equipment to perform the required commissioning. The equipment shall be periodically calibrated.

902.1.3 Personnel. An approved agency shall employ experienced personnel educated in conducting, supervising and evaluating tests and commissioning.

SECTION 903 COMMISSIONING

903.1 General. Where application is made for construction as described in this section, the registered design professional in responsible charge or approved agency shall perform commissioning during construction and after occupancy as required by Table 903.1. Where Table 903.1 specifies that commissioning is to be done on a periodic basis, the registered design professional in responsible charge shall provide a schedule of periodic commissioning with the submittal documents that shall be reviewed and *approved* by the *code official*.

The approved agency shall be qualified and shall demonstrate competence, to the satisfaction of the *code official*, for the commissioning of the particular type of construction or operation. The registered design professional in responsible charge and engineers of record involved in the design of the project are permitted to act as the approved agency provided those personnel meet the qualification requirements of this section to the satisfaction of the *code official*. The approved agency shall provide written documentation to the *code official* demonstrating competence and relevant experience or training. Experience or training shall be considered relevant where the documented experience or training is related in complexity to the same type of commissioning activities for projects of similar complexity and material qualities.

903.1.1 Preoccupancy report requirement. The approved agency shall keep records of the commissioning required by Table 903.1. The approved agency shall furnish commissioning reports to the owner and the registered design professional in responsible charge and, upon request, to the *code official*. Reports shall indicate that work was or was not completed in conformance to *approved* construction documents. Discrepancies shall be brought to the immediate attention of the contractor for correction. Where discrepancies are not corrected, they shall be brought to the attention of the owner, *code official* and to the registered design professional in responsible charge prior to the completion of that phase of the work. Prior to the issuance of a Certificate of Occupancy, a final commissioning report shall be submitted to and accepted by the *code official*.

903.1.2 Post-occupancy report requirement. Post-occupancy commissioning shall occur as specified in the applicable sections of this code. A post-occupancy commissioning report shall be provided to the owner within 30 months after the Certificate of Occupancy is issued for the project and shall be made available to the *code official* upon request.

SECTION 904 BUILDING OPERATIONS AND MAINTENANCE

904.1 General. Building operations and maintenance documents in accordance with Section 904.3 shall be submitted to the owner prior to the issuance of the Certificate of Occupancy. Record documents shall be in accordance with Section 904.2. The building owner shall file a letter with the *code official* certifying the receipt of record documents and building operations and maintenance documents. At least one copy of these materials shall be in the possession of the owner and at least one additional copy shall remain with the building throughout the life of the structure.

904.2 Record documents. The cover sheet of the record documents for the project shall clearly indicate that at least one copy of the materials shall be in the possession of the owner. Record documents shall include all of the following:

1. Copies of the *approved* construction documents, including plans and specifications.
2. As-built plans and specifications indicating the actual locations of piping, ductwork, valves, controls, equipment, access panels, lighting and other similar components where they are concealed or are installed in

(continued on page 89)

TABLE 903.1
COMMISSIONING PLAN

CONSTRUCTION OR SYSTEM REQUIRING VERIFICATION	PREOCCUPANCY	POST-OCCUPANCY	METHOD	OCCURRENCE		SECTION/ REFERENCED STANDARD
				Preoccupancy	Post-occupancy	
Chapter 4: Site Development and Land Use						
Natural resources and base line conditions of building site	X	None	Report	With permit submittal	None	401.2
Landscape irrigation systems	X	None	Field inspection	Installation	None	404.1, 405.1.1
Topsoil and vegetation protection measures; setbacks from protected areas	X	None	Field inspection and report	Installation of measures, prior to other site disturbance	None	405.1.1
Imported soils	X	None	Field inspection and report	With permit submittal; after all-fill operations complete	None	405.1.3
Soil restoration and reuse	X	None	Field inspection and report	Preparation and replacement of soils	None	405.1.4
Stormwater management system operation	None	X	Field inspection	—	24 months	403.1
Erosion and sediment control	X	X	Field inspection	During construction activities	Periodic for 24 months	405.1.1
Hardscape and shading provided by structures and vegetation	X	X	Field inspection and report	During construction and installation	24 months	408.2
Vegetative roofs	X	X	Field inspection and report	Installation of protective membranes, base materials, soils and vegetation	24 months	408.3.2
Site lighting	X	None	Testing and report	Installation	None	409
Chapter 5: Material Resource Conservation and Efficiency						
Moisture control (Section 507.1)						
1. Foundation sub-soil drainage system.	X	None	Field inspection and verification	Periodic inspection for entire sub-soil drainage system	None	507.1 and IBC Ch 18
2. Foundation waterproofing	X	None	Field inspection and verification	Periodic inspection for the entire foundation	None	507.1 and IBC Ch 18
3. Foundation dampproofing	X	None	Field inspection and verification	Periodic inspection for the entire foundation	None	507.1 and IBC Ch 18
4. Under slab water vapor protection	X	None	Field inspection and verification	Periodic inspection for entire slab footprint	None	507.1, IBC Ch 19 and ASTM E 1643
5. Flashing at: exterior windows, doors, skylights, wall flashing and drainage systems	X	None	Field inspection and verification	Periodic inspection for not less than 25 percent of all flashing locations.	None	507.1 and IBC Ch 14
6. Exterior wall coverings	X	None	Field inspection and verification	Periodic inspection for not less than 25 percent of exterior wall cladding systems.	None	507.1 and IBC Ch 14
7. Roof coverings, roof drainage, and flashings	X	None	Field inspection and verification	Periodic inspection for not less than 25 percent of roof covering, roof drainage and flashings.	None	507.1 and IBC Ch 15

(continued)

**TABLE 903.1—(continued)
COMMISSIONING PLAN**

CONSTRUCTION OR SYSTEM REQUIRING VERIFICATION	PREOCCUPANCY	POST-OCCUPANCY	METHOD	OCCURRENCE		SECTION/ REFERENCED STANDARD
				Preoccupancy	Post-occupancy	
Chapter 6: Energy						
Energy consumption, monitoring, targeting and reporting						
a. Monitoring system	X	None	Inspection and verification	During construction and prior to occupancy	None	603, 610.5
b. Calibration	X	X	Testing and review and evaluation or test reports	During commissioning	Annually	603, 610.5
Mechanical systems completion – all buildings						
a. Air system balancing – provide the means for system balancing	X	None	Inspection and verification	During construction and prior to occupancy	None	611.1.2.1 and through reference to IECC
b. Hydronic system balancing – provide means for system balancing	X	None	Inspection and verification	During construction and prior to occupancy	None	611.1.2.2 and through reference to IECC
c. Mechanical system manuals – construction documents to require O&M manual	X	None	Verification of construction documents	Plan review	None	611.1.5.2
Mechanical systems – buildings over 5,000 square feet total building floor area						
a. Commissioning required and noted in plans and specifications	X	None	Verification of construction documents	Plan review	None	611.1
b. Documentation of required commissioning outcomes	X	None	Verification with the building owner	Subsequent to completion of all commissioning activities	None	611.1
c. Preparation and availability of a commissioning plan	X	None	Verification with the RDP or commissioning agent	Between plan review and commissioning initiation	None	611.1.1
d. Balance HVAC systems (both air and hydronic)	X	X	HVAC system installer/contractor or commissioning agent	After installation of HVAC systems and prior to occupancy	TBD	611.1.2
e. Functional performance testing of HVAC equipment	X	X	HVAC system installer/contractor or commissioning agent	After installation of HVAC systems and prior to occupancy	TBD	611.1.3
f. Functional performance testing of HVAC controls and control systems	X	X	HVAC system installer/contractor or commissioning agent	After installation of HVAC systems and prior to occupancy	TBD	611.1.3.2
g. Preparation of preliminary commissioning report	None	X	HVAC system installer/contractor or commissioning agent	None	Subsequent to commissioning	611.1.4

(continued)

TABLE 903.1—(continued)
COMMISSIONING PLAN

CONSTRUCTION OR SYSTEM REQUIRING VERIFICATION	PREOCCUPANCY	POST-OCCUPANCY	METHOD	OCCURRENCE		SECTION/ REFERENCED STANDARD
				Preoccupancy	Post-occupancy	
h. Acceptance of HVAC systems and equipment/system verification report	None	X	Building owner	None	Letter verifying receipt of the commissioning report	611.1.4.1
i. Preparation and distribution of final HVAC system completion—Documentation that construction documents require drawings, manuals, balancing reports and commissioning report be provided to the owner and that they have been provided	None	X	RDP, contractor or commissioning authority	None	90 days after final certificate of occupancy	611.1.5
Chapter 6: Lighting						
Auto demand reduction control system functionality	X	X	Functional testing	Final inspection	18-24 months	604.4
Plug load controls	X	None	Functional testing	Final inspection	None	608.6
Connection of appliances to switched receptacles	—	X	Field inspection	None	18-24 months	608.6
Specified transformer nameplate efficiency rating	X	None	Field inspection	Final inspection	None	608.8.1.1
Verification of lamp	X	X	Field inspection	Final inspection	18-24 months	608.10
Verification of ballast	X	None	Field inspection	Final inspection	None	608.10
Lighting controls						
a. Installation	X	None	Field inspection	Post-installation	None	608.11
b. Calibration	X	X	System installer/contractor or commissioning agent	Post-installation	18-24 months	611.3.3
Chapter 7: Water Resource Conservation, Quality and Efficiency						
Appliances	X	None	—	—	—	702.6
Hot water distribution	X	None	—	—	—	702.8
Cooling tower performance	—	X	—	—	—	703.7.7
Metering	X	None	—	—	—	705.1.1
Rainwater system water quality	None	X	Field testing and verification	None	707.15.1	707.15.1
Gray water system water quality	None	X	Field testing and verification	None	708.13.8	708.13.8
Soil percolation test	X	None	Field inspection and report	Prior to installation of gray water irrigation system	None	708.14.2

(continued)

TABLE 903.1—(continued)
COMMISSIONING PLAN

CONSTRUCTION OR SYSTEM REQUIRING VERIFICATION	PREOCCUPANCY	POST-OCCUPANCY	METHOD	OCCURRENCE		SECTION/ REFERENCED STANDARD
				Preoccupancy	Post-occupancy	
Chapter 8: Indoor Environmental Quality and Comfort						
Building construction, features, operations and maintenance facilitation						
Air-handling system access	X	X	Field inspection and verification	During construction and prior to occupancy	18 - 24 months	802.2
Air-handling system filters	X	X	Field inspection and verification	During construction and prior to occupancy	18 - 24 months	802.3
HVAC systems						
Temperature and humidity in occupied spaces	—	X	Field inspection and verification	—	18 - 24 months	803.2
Specific indoor air quality & pollutant control measures						
Listing, installation and venting of fireplaces and combustion appliances	X	—	Field inspection and verification	During construction and prior to occupancy	—	804.1
Sound transmission						
Mechanical and emergency generator equipment located outside buildings or located where exposed to exterior environment.	X	None	Field testing and verification	See Section 807.5.1	None	807.5.1
HVAC background sound	X	None	Field testing and verification	See Section 807.5.2	None	807.5.2

For SI: 1 square foot = 0.0929 m^2.

locations other than those indicated on the *approved* construction documents.

3. For sites that have previously been a *brownfield*, or required environmental corrective action, remediation or restoration at the federal, state or local level, copies of engineering and institutional control information shall be provided.
4. A copy of the Certificate of Occupancy.

904.3 Building operations and maintenance documents. The building operations and maintenance documents shall consist of manufacturer's specifications and recommendations, programming procedures and data points, narratives, and other means of illustrating to the owner how the building, site and systems are intended to be maintained and operated. The following information shall be included in the materials, as applicable to the specific project:

1. Directions to the owner or occupant on the manual cover sheet indicating that at least one copy of the materials shall be in the possession of the owner or occupant.
2. Operations and maintenance manuals for equipment, products and systems installed under or related to the provisions of Chapter 4 including, but not limited to, the following, as applicable:
   - 2.1. Vegetative shading, vegetative roofs and natural resource protections and setbacks.
   - 2.2. Water-conserving landscape and irrigation systems.
   - 2.3. Stormwater management systems.
   - 2.4. Permanent erosion control measures.
   - 2.5. Landscape or tree management plans.
3. Operations and maintenance documents for materials, products, assemblies and systems installed under or related to the provisions of this code for material resource conservation in accordance with Chapter 5 including, but not limited to, the following, as applicable:
   - 3.1. Care and maintenance instructions and recommended replacement schedule for flooring, including, but not limited to, carpeting, walk-off mats and tile.
   - 3.2. Care and maintenance instructions for natural materials including, but not limited to, wood, bio-based materials and stone.
   - 3.3. Available manufacturer's instructions on maintenance for:
     - 3.3.1. Exterior wall finishes.
     - 3.3.2. Roof coverings.
     - 3.3.3. Exterior doors, windows and skylights.
   - 3.4. Information and recommended schedule for required routine maintenance measures, including, but not limited to, painting and refinishing.

4. Operations and maintenance documents for equipment, products and systems installed under or related to the provisions of this code for energy conservation in accordance with Chapter 6 including, but not limited to, the following:
   - 4.1. Heating, ventilating and air-conditioning systems including:
     - 4.1.1. Recommended equipment maintenance schedule.
     - 4.1.2. Air filters and fluid filters, including recommended replacement schedule and materials.
     - 4.1.3. Time clocks, including settings determined during commissioning.
     - 4.1.4. Programmable controls and thermostats, including settings determined during commissioning.
   - 4.2. Domestic hot water systems including performance criteria and controls.
   - 4.3. *Building thermal envelope* systems including:
     - 4.3.1. Glazing systems inspection schedule.
     - 4.3.2. Performance criteria for replacements and repairs.
     - 4.3.3. Information and recommended schedule on required routine maintenance measures, including but not limited to, sealants, mortar joints and screens.
   - 4.4. Electrical and lighting systems including:
     - 4.4.1. Technical specifications and operating instructions for installed lighting equipment.
     - 4.4.2. Luminaire maintenance and cleaning plan.
     - 4.4.3. Lamp schedule, recommended relamping plan, and lamp disposal information.
     - 4.4.4. Programmable and automatic controls documentation, including settings determined during commissioning.
     - 4.4.5. Occupant sensor and daylight sensors documentation, including settings determined during commissioning.
   - 4.5. Automatic demand reduction systems.
5. Operations and maintenance documents for equipment, products and systems installed under or related to the provisions of this code for water conservation in accordance with Chapter 7, including, but not limited to the following:
   - 5.1. Domestic fixtures.
   - 5.2. Water-regulating devices including faucets and valves.
   - 5.3. Irrigation and rainwater and gray water catchment.
6. Operations and maintenance documents for equipment products and systems under or related to the provisions of this code for indoor environmental quality in accordance with Chapter 8, including, but not limited to, the following:
   - 6.1. Humidification/dehumidification.
   - 6.2. Green cleaning products, procedures and techniques.
   - 6.3. Recommended window cleaning schedule.
   - 6.4. Ventilation controls.
   - 6.5. Floor finishes.
   - 6.6. Fireplaces and combustion appliances.

CHAPTER 10

EXISTING BUILDINGS

SECTION 1001 GENERAL

1001.1 Scope. The provisions of this chapter shall control the alteration, repair, addition, maintenance and operation and change of occupancy of existing buildings and structures. Relocated existing buildings shall comply with Chapter 10. Existing building sites shall comply with Chapter 11.

1001.2 Building operation and maintenance. Previously commissioned buildings and parts thereof, shall be operated and maintained in conformance to the code edition applicable at the time of construction. The owner shall be responsible for the operation and maintenance of existing buildings. The requirements of this chapter shall not provide the basis for removal or abrogation of fire protection and safety systems and devices in existing structures.

1001.3 Compliance. Alterations, repairs, additions and changes of occupancy to existing structures shall comply with the provisions of this chapter.

Exception: Where a tenant in a multi-tenant building does not have control within that tenant space of a complete system or item, compliance for that complete system or item shall not be required.

1001.4 Existing materials, assemblies, configurations and systems. Materials, assemblies, configurations and systems already in use that conform to requirements or approvals in effect at the time of their erection or installation shall be permitted to remain in use unless determined by the *code official* to be dangerous to life, health or safety. Where such conditions are determined to be dangerous to the environment, life, health or safety, they shall be mitigated or made safe.

SECTION 1002 ADDITIONS

1002.1 General. Additions to any site-built building or structure shall comply with the requirements of this code for new construction. Any addition to a modular building that is relocated within or into a jurisdiction that is in compliance with requirements or approvals in effect at the time of its construction shall comply with Section 1002 of this code.

SECTION 1003 ALTERATIONS TO EXISTING BUILDINGS

1003.1 General. Alterations to existing buildings and building systems shall be in accordance with the provisions of this code for those assemblies, systems and components being altered. Unaltered portions, components and systems of the building, including relocated modular buildings, shall be in accordance with the provisions of the code in force at the time of their construction. Alterations shall not be made to an existing building or structure that will cause the existing building or structure to be in violation of any provisions of this code.

1003.2 Requirements for alterations. Alterations of portions or components of buildings shall comply with Sections 1003.2.1 through 1003.2.7.

Exceptions:

1. The total cost of improvements required by Sections 1003.2.1 through 1003.2.7 shall not be required to exceed 10 percent of the costs of the alterations exclusive of land and building site improvements.
2. This section shall not require compliance that exceeds that required for systems regulated by Chapters 6 through 8 of this code.
3. Materials, assemblies and components regulated by Sections 1003.2.1 through 1003.2.7 that are dependent upon properties of other concealed materials, assemblies or system components to function properly and where the properties of the concealed materials, assemblies or components are unknown or insufficient and will not be revealed during construction.
4. Alterations are not required to comply with the requirements of Sections 1003.2.1 through 1003.2.7 where the *code official* determines the alterations to be *infeasible* based upon the existing configuration of spaces, unless those spaces or portions thereof will be reconfigured as part of the alteration project.
5. Where a tenant in a multi-tenant building does not have control within that tenant space of a complete system or item, compliance for that complete system or item shall not be required.
6. Where the total cost of the alteration to the existing building is less than the percent of the value of the building as indicated in Table 1003.2, compliance with Section 1003.2 shall not be required. The percent value of the building shall be determined by the original construction cost plus completed improvement costs of the building.

TABLE 1003.2 MINIMUM VALUES FOR ADDITIONAL REQUIREMENTS TO ALTERATIONS

BUILDING SIZE (square feet)	PERCENT OF BUILDING VALUE
Less than 5,000	20
5,000 – 50,000	10
50,001 – 500,000	1
over 500,000	0

For SI: 1 square foot = 0.0929 m^2.

1003.2.1 Metering devices. Dedicated individual utility or private metering devices that measure and verify energy and water use within the building or space shall be provided for at least one of the following:

1. Electrical energy consumption for individual tenant spaces.
2. Water consumption for individual tenant spaces.
3. Natural gas or fuel oil consumption for individual tenant spaces.
4. Lighting loads.
5. Motor and drive loads.
6. Chiller part-load efficiency.
7. Cooling loads.
8. Economizer and heat recovery loads.
9. Boiler efficiencies.
10. Building process systems and equipment loads.
11. Water consumption for landscape irrigation.

1003.2.2 Heating, ventilating and air-conditioning. Heating, ventilating and air-conditioning systems and equipment shall be in accordance with the following:

1. Time clock and automatic time switch controls that can turn systems off and on according to building occupancy requirements shall be provided and connected to the following HVAC equipment: chillers and other space-cooling equipment, chilled water pumps, boilers and other space-heating devices, hot water pumps, heat exchanger circulation pumps, supply fans, return fans, and exhaust fans. Where occupant override is provided, it shall be designed with a timer to automatically revert to time clock and automatic time switch controls in not longer than 12 hours.

   Exception: A time clock or automatic time switch controls shall not be required for spaces where any of the following conditions exist:

   1. A time clock is not required by Section C403.2.4.3 of the *International Energy Conservation Code*.
   2. There is 24-hour occupancy materials with special atmospheric requirements dependent on 24-hour space conditioning.
   3. A majority of the areas of the building served by the system are under setback thermostat control.
   4. Manufacturer's specifications stipulate that the system must not be shut off.

2. Functional outside air economizers shall be provided on all cooling systems of more than $4^1/_2$ tons total cooling capability, 54,000 Btu/h, or more than 1800 cfm (9.144 $m^3/s \cdot m^2$) air flow, provided manufacturer's guidelines are available for adding the economizer to the existing system.

   Exception: An outside air economizer shall not be required for buildings or special uses where 100 percent outside air for ventilation is required or where any of the following conditions exist:

   1. Section C403.3.1 of the *International Energy Conservation Code* would not require an economizer.
   2. The existing system has a water-based economizer.
   3. The existing system does not have an outside air intake.
   4. Special economizer operations such as, but not limited to, carefully controlled humidity would require more energy use than is conserved.
   5. There is insufficient space to install necessary equipment.
   6. Installation of an economizer would require major modifications to the building's life safety system.
   7. The existing system is a multi-zone system where the same intake air is used at the same time for either heating or cooling in different parts of the building.

3. HVAC piping and ducts, including those located above suspended ceilings, shall comply with Sections 606.3 and 606.4.

   Exception: Additional insulation shall not be required for piping where any of the following conditions exist:

   1. Additional insulation shall not be required for piping where any of the following conditions exist:

      1.1. It is located within HVAC equipment;

      1.2. It is located within conditioned space that conveys fluids between 60°F (15.6°C) and 105°F (40.6°C);

      1.3. Piping that is already insulated and the insulation is in good condition; or

   2. Where HVAC ducts and piping are installed in a building cavity or interstitial framing space of insufficient width to accommodate the duct or pipe and the insulation required by Section 606.3 and Table 606.4, the insulation thickness shall be permitted to have the maximum thickness that the wall can accommodate, but

shall not be less than $^1/_2$-inch (12.7 mm) thick.

4. Where central heat is intended to be replaced with individual electric space heaters, the application for the electrical permit shall include documentation demonstrating that the new electric heaters will not consume more energy than the existing nonelectric heaters.
5. Boiler systems shall have been cleaned and tuned within one year prior to the alteration. Boilers shall be equipped with an outdoor air lock-out thermostat or a temperature reset control.
6. Chillers shall be equipped with an outdoor air lock-out thermostat and chilled water reset control.
7. A maximum 5-year phase out plan shall be provided for buildings with existing systems that use CFC-based refrigerants.
8. Where mechanical and electrical systems and equipment are joined with microprocessors that communicate with each other or to a computer, a properly integrated building automation system shall be installed to optimize energy, operations, and indoor comfort. The building automation system shall:
    - 8.1. Allow the owner to set up schedules of operation for the equipment and provide equipment optimal start with adaptive learning;
    - 8.2. Provide trim and respond capabilities based on zone demand;
    - 8.3. Offer the ability to monitor energy usage, including the ability to meter electric, gas, water, steam, hot water, chilled water, and fuel oil services;
    - 8.4. Offer economizing based on enthalpy calculation and/or CO_2 set point control;
    - 8.5. Offer load shedding when power companies are at peak demand and need; and
    - 8.6. Offer the ability to send alarms to alert building owner, manager, or operator when problems occur due to system failures.

1003.2.3 Service water systems. Service water systems and equipment shall be in accordance with the following:

1. Water heater and hot water storage tanks shall have a combined minimum total of external and internal insulation value of R-16.
2. Accessible hot and cold water supply and distribution pipes shall comply with Section 607.6. The insulation shall not be required to extend beyond the *building thermal envelope*.
3. Circulating pump systems for hot water supply purposes other than comfort heating shall be controlled as specified in Section 607.7.
4. Showerhead, toilet, urinal and faucet flow rates shall be in accordance with this code.

1003.2.4 Lighting. Lighting systems and equipment shall be in accordance with sections C405.2.2.3 and C405.2.4 of the *International Energy Conservation Code*.

1003.2.5 Swimming pools and spas. Swimming pools and spas and their equipment shall be in accordance with the following:

1. Heated swimming pools and spas shall be equipped with a cover listed and labeled in accordance with ASTM F 1346, or a liquid pool cover feed system, for unoccupied hours.

    Exception: A cover shall not be required for indoor pools or spas in which water temperature is less than 80°F (26.7°C) during time of nonuse.
2. Backwash systems shall be based on pressure drop and shall not be based on a timer.
3. Pool and spa recirculation pumps shall be under timeclock control.

    Exception: Filtration pumps where the public health standard requires 24-hour pump operation.
4. Heaters shall have been cleaned and tuned for efficiency within one year prior to the alteration. Where this has not been done, the heaters shall be cleaned and tuned as part of the alteration work.

1003.2.6 Insulation of unconditioned attics. In buildings with three or fewer stories above grade plane, ceiling insulation with a minimum R-value as required by this code shall be installed in accessible attic spaces that are directly above conditioned spaces. For the purposes of this section, accessible attic space is the space between ceiling joists and roof rafters where the vertical clear height from the top of a ceiling joist or the bottom chord of a truss, to the underside of the roof sheathing at the roof ridge, is greater than 24 inches (610 mm). Where the required R-value insulation cannot fit in the attic space, the maximum amount of insulation compatible with available space and existing uses shall be installed.

1003.2.7 Roof replacement insulation. For roof replacement on an existing building with insulation entirely above the deck and where the roof slope is less than two units vertical in 12 units horizontal (16-percent slope), the insulation shall conform to the energy conservation requirements for insulation entirely above deck in the *International Energy Conservation Code*.

Exception: Where the required R-value cannot be provided due to thickness limitations presented by existing rooftop conditions, including heating, ventilating and air-conditioning equipment, low door or glazing heights, parapet heights, proper roof flashing heights, the maximum thickness of insulation compatible with the available space and existing uses shall be installed.

SECTION 1004 CHANGE OF OCCUPANCY

1004.1 Change of occupancy. Where a change in occupancy of a building or tenant space places it in a different division of the same group of occupancy or in a different group of occupancies, as determined in accordance with the provisions of the *International Building Code*, compliance with Sections 1001.3 and 1001.4 shall be required.

Exception: Historic buildings in accordance with Section 1005 shall not be required to comply with Section 1004.

SECTION 1005 HISTORIC BUILDINGS

1005.1 Historic buildings. The provisions of this code relating to the construction, repair, alteration, addition, restoration and movement of structures, and change of occupancy, where each individual provision is evaluated separately on its own merit, shall not be mandatory for historic buildings for any of the following conditions:

1. Where implementation of such provisions would require a change in the visible configuration of building components in a manner that is not in keeping with the building's historic nature, as determined by the *code official*; or
2. Where compliance with such provisions would produce a conflict with a building function that is fundamental to the historic nature of the building.

SECTION 1006 DEMOLITION

1006.1 Deconstruction and demolition material and waste management plan. Where buildings, structures or portions thereof are deconstructed or demolished, a minimum of 50 percent of materials shall be diverted from landfills. A construction material and waste management plan shall be developed that is in accordance with Section 503.1, that includes procedures for deconstruction, and that documents the total materials in buildings, structures and portions thereof to be deconstructed or demolished and the materials to be diverted.

SECTION 1007 JURISDICTIONAL REQUIREMENTS

1007.1 General. Sections 1007.2 and 1007.3 shall be mandatory and enforced only where specifically indicated by the jurisdiction in Table 302.1.

1007.2 Evaluation and certification of existing buildings and building sites. Where a permit application is accepted by a jurisdiction for the evaluation of an existing building and building site in accordance with the requirements of this code as applicable to a new project, and this code does not otherwise require compliance, evaluation shall be in accordance with the requirements of this section.

1007.2.1 Certificate of compliance. Where compliance with the requirements of this code as applicable to a new building is verified by the *code official* for an existing building and building site, a certificate shall be issued indicating compliance to this code, as modified by the limitations contained in Sections 1007.2.2 through 1007.2.3.2.

1007.2.2 Specific exclusions. Where evidence of compliance is not available, existing buildings evaluated under Section 1007.2 shall not be subject to the requirements of Section 806. Provisions of this code related to the project's construction phase, including Sections 401.2, 406.1, 406.2, 502, 503.1 and 803.1, those portions of Section 405 related to the construction phase, and other sections as *approved* by the *code official*, shall not be required for buildings evaluated under Section 1007.2. Where buildings do not comply with the aforementioned sections, the certification shall specifically list the sections for which compliance has not been required or verified.

1007.2.3 Existing concealed construction. Existing concealed construction in buildings regulated by Section 1007.2 shall be in accordance with Sections 1007.2.3.1 and 1007.2.3.2.

1007.2.3.1 Previously approved documents. Previously *approved* construction documents for the initial construction of an existing building and, where possible, description of changing uses and major upgrades over the building's lifetime for which a certificate of occupancy was previously issued shall be deemed an acceptable indication of materials, assemblies and equipment in concealed spaces, except where field inspection reveals sufficient evidence suggesting noncompliance, subject to the evaluation of the *code official*.

1007.2.3.2 Previously approved documents not available. Where previously *approved* construction documents for the initial construction of an existing project are not available, materials, assemblies and equipment in spaces in existing buildings and existing portions thereof that are concealed, including, but not limited to, materials in spaces within walls and floor/ceiling assemblies, shall be exposed and spot checked in limited areas as determined by the *code official*.

1007.3 Post certificate of occupancy zEPI, energy demand, and CO_2e emissions reporting. Where the jurisdiction indicates in Table 302.1 that ongoing post certificate of occupancy zEPI, energy demand and CO_2e emissions reporting is required, and where the jurisdiction has indicated in Table 302.1 that enhanced energy performance in accordance with Section 302.1 or CO_2e emissions in accordance with Section 602.2 are required, zEPI, energy demand, and CO_2e emissions reporting shall be provided in accordance with this section.

1007.3.1 Purpose. The purpose of this section is to provide for the uniform reporting and display of the total annual net energy use, peak demand for each energy form and emissions associated with building operations and building sites.

1007.3.2 Intent. The intent of these requirements is to provide for the ongoing reporting and display of the total annual net energy use, peak energy demand and emissions associated with operation of the building and its systems to document ongoing compliance with the provisions of Sections 601 and 602.

1007.3.3 Reporting. Reports in accordance with Sections 1007.3.3.1 through 1007.3.3.3 shall be generated.

1007.3.3.1 Annual net energy use. The zEPI associated with the operation of the building and the buildings on the site, as determined in accordance with Section 602.1, shall be reported by the building owner or the owner's registered agent to the [INSERT NAME OF APPROPRIATE STATE OR LOCAL GOVERNMENT AGENCY RESPONSIBLE FOR COLLECTING REPORTED INFORMATION].

Where there are multiple buildings on a building site, each building shall have its zEPI reported separately. Where there are energy uses associated with the building site other than the buildings on the site, the zEPI for the building site shall be reported separately.

Energy use for the previous year shall cover the complete calendar year and be reported on, or before, March 1st of the following year.

1007.3.3.2 Peak monthly energy demand reporting. The peak demand of all energy forms serving each building and the building site shall be reported by the building owner or the owner's registered agent to the [INSERT NAME OF APPROPRIATE STATE OR LOCAL GOVERNMENT AGENCY RESPONSIBLE FOR COLLECTING REPORTED INFORMATION].

Where there are multiple buildings on a building site, each building shall have its energy demand reported separately. Where there are energy uses associated with the building site other than the buildings on the site, the energy demand for the building site shall be reported separately.

Monthly energy demand data for the previous year shall cover the complete calendar year and be reported on, or before, March 1st of the following year.

1007.3.3.3 Annual CO_2e emissions reporting. The annual emissions associated with the operation of the building and its systems, as determined in accordance with Section 602.2, shall be reported by the building owner or the owner's registered agent to the [INSERT NAME OF APPROPRIATE STATE OR LOCAL GOVERNMENT AGENCY RESPONSIBLE FOR COLLECTING REPORTED INFORMATION].

Where there are multiple buildings on a building site, each building shall have its annual emissions reported separately. Where there are energy uses associated with the building site other than the buildings on the site, the annual CO_2e emissions for the building site shall be reported separately.

Emissions reported for the previous year shall cover the complete calendar year and be reported on, or before, March 1st of the following year.

CHAPTER 11

EXISTING BUILDING SITE DEVELOPMENT

SECTION 1101 GENERAL

1101.1 Scope. The provisions of this chapter shall control the alteration, repair, maintenance and operation of existing building sites and the alteration to building site improvements. Chapter 11 applies where building site improvements are being made, or where additions are made to, or changes of occupancy occur within, the existing buildings on the site.

1101.2 Operation and maintenance. Building sites shall be operated and maintained in conformance to the code edition under which the site improvements were installed. The owner or the owner's designated agent shall be responsible for the operation and maintenance of building sites. To determine compliance with this section, the *code official* shall have the authority to require a building site to be reinspected. The requirements of this chapter shall not provide the basis for removal or abrogation of protections or systems from existing building sites.

1101.3 Compliance. Alterations and repairs to building sites shall comply with the provisions of this code unless provided otherwise in this chapter. Where differences occur between the provisions of this code and the provisions of other locally adopted land use, zoning or site development regulations, the provisions of the most restrictive code or regulation shall apply.

1101.4 Building site materials, systems and landscaping. Building materials used for building site development shall comply with the requirements of this section.

1104.4.1 Existing materials, assemblies, configurations and systems. Materials and systems already in use on a building site in compliance with the requirements or approvals in effect at the time of their installation shall be permitted to remain in use unless determined by the *code official* to be dangerous to the environment, life, health or safety. Where such conditions are determined to be dangerous to the environment, life, health or safety, they shall be mitigated or made safe.

Existing buildings and site improvements located within or located closer to protected areas than permitted by Section 402.1 but that are in compliance with the requirements or approvals in effect at the time of their installation shall be permitted to remain in use unless determined by the *code official* to be dangerous to the environment, life, health and safety of the community and the occupants of the building site. Where such conditions are determined to be dangerous to the environment, life, health or safety, they shall be mitigated or made safe.

1101.4.2 New and replacement materials, assemblies, configurations and systems. Except as otherwise required or permitted by this code, materials, assemblies, configurations and systems permitted by the applicable code for new construction shall be used. Like materials shall be permitted for repairs and alterations provided no hazard to the environment, life, health or property is created. Hazardous materials shall not be used where the code for new construction would not permit their use at building sites of similar occupancy, purpose and location.

SECTION 1102 ADDITIONS

1102.1 General. Additions to any building site improvements shall comply with the requirements of this code for new construction. Unaltered portions of a building site shall be in accordance with the provisions of the code in force at the time of their construction.

Where additions to a building, or additions to building site improvements result in the alteration of existing portions or improvements of the building site, those alterations shall comply with this section and Section 1103.

Additions to an existing building site shall be made to ensure the following:

1. Existing building site improvements together with the additional or expanded improvements are not less conforming to the provisions of this code than the existing building site was prior to the addition; and
2. Where additions to any building reduces, or requires alteration to, building site improvements, the alterations to the building site together with unaltered site improvements shall not be less conforming to the provisions of this code prior to the addition to the building or structure.

SECTION 1103 ALTERATIONS TO EXISTING BUILDING SITES

1103.1 General. Alterations to existing portions or site improvements on building sites shall be in accordance with the provisions of this code for those portions or building site improvements being altered. Unaltered portions and site improvements of the building site shall be in accordance with the provisions of the code in force at the time of their construction. Alterations shall be such that the existing building site is no less conforming to the provisions of this code than the existing building site was prior to the alteration.

Unaltered portions and site improvements of a building site shall be in accordance with the provisions of the code in force at the time of their construction or preservation.

Exception: Where, in the opinion of the *code official*, there is no significant compromise of the intent of this code, the *code official* shall have the authority to approve materials and assemblies that perform in a manner that is at least the equivalent of those being replaced.

1103.2 Changes to hardscapes and surface vehicle parking. Where existing hardscapes are altered, the alterations shall comply with the provisions of this code.

Exceptions:

1. Existing hardscapes and vegetation are permitted to be replaced with materials shown in previously *approved* construction documents.
2. Where existing vehicle surface parking lots are altered without changing parking space configuration or increasing the number of parking spaces, the altered parking lot shall not be required to comply with Section 407.4.

SECTION 1104 CHANGE OF OCCUPANCY

1104.1 Conformance. Where a change in the use or occupancy of a building or tenant space places it in a different division of the same group or occupancy or in a different group of occupancies, as determined in accordance with the provisions of the *International Building Code*, compliance with Section 1104.2 shall be required. Altered portions of, and additions to, existing buildings and existing building sites that are not a result of change of occupancy requirements, shall comply with Chapter 10 and this chapter.

1104.2 Building site improvements. Where a change in occupancy results in an increase in the occupant load of the building, bicycle parking shall comply with the following:

1. Short-term bicycle parking spaces shall be provided in accordance with Section 407.3 equivalent to a new building of the new occupancy.
2. Where the existing building and building site have parking for motorized vehicles, long-term bicycle parking shall be provided in accordance with Section 407.3 equivalent to a new building of the new occupancy. Where the existing building does not contain covered parking spaces for vehicles, only 25 percent of the long-term bicycle parking needs to be covered.

SECTION 1105 HISTORIC BUILDING SITES

1105.1 Historic building sites. The provisions of this code relating to the construction, repair, alteration, addition and restoration of building sites and site improvements, where each individual provision is evaluated separately on its own merit, shall not be mandatory for historic building sites for any of the following conditions:

1. Where implementation of that provision would change the visible configuration of building site improvements in a manner that is not in keeping with the building site's historic nature, as determined by the *code official*, in consultation with the authority having jurisdiction over historic buildings or sites;
2. Where compliance with that provision would produce a conflict with a building site function that is fundamental to the historic nature of the building site, as determined by the *code official*, in consultation with the authority having jurisdiction over historic buildings or sites; or
3. Where such building sites are judged by the *code official* in consultation with the authority having jurisdiction over historic buildings or sites to not constitute a distinct environmental hazard.

CHAPTER 12
REFERENCED STANDARDS

This chapter lists the standards that are referenced in various sections of this document. The standards are listed herein by the promulgating agency of the standard, the standard identification, the effective date and title, and the section or sections of this document that reference the standard. The application of the referenced standards shall be as specified in Section 102.4.

AHRI

Air Conditioning, Heating and Refrigeration Institute
2111 Wilson Boulevard, Suite 500
Arlington, VA 22201

Standard reference number	Title	Referenced in code section number
810—2007	Standard for Performance Rating of Automatic Commercial Ice-Makers	Table 609.2.3
870—2009	Direct Geoexchange Heat Pumps	Table 606.2.2.1

AMCA

Air Movement and Control Association International
25 West 43rd Street
Fourth Floor
New York, NY 10036

Standard reference number	Title	Referenced in code section number
205—10	Energy Efficiency Classification for Fans	606.2.2.3
220—05	Laboratory Methods of Testing Air Curtain Units for Aerodynamic Performance Rating	605.1.2.3

AITC

American Institute of Timber Construction
7012 South Revere Parkway, Suite 140
Englewood, CO 80112

Standard reference number	Title	Referenced in code section number
ANSI/AITC 190.1—2007	Structural Glued Laminated Timber	202

ANSI

American National Standards Institute
25 West 43rd Street, Fourth Floor
New York, NY 10036

Standard reference number	Title	Referenced in code section number
Z21.50/CSA 2.22.2007	Vented Gas Fireplaces	804.1.3
Z21.88a/CSA 2.33a—09	ANSI/CSA Standard for Vented Gas Fireplace Heaters	804.1.3

APHA

American Public Health Association
800 I Street NW
Washington, DC 20001

Standard reference number	Title	Referenced in code section number
2005	Standard Methods for Examination of Water and Waste Water-21st Edition	707.15.1.1

ARB

California Air Resource Board
1001 "I" Street, P. O. Box 2815
Sacramento, CA 9512

Standard reference number	Title	Referenced in code section number
February 1, 2008	California Air Resources Board, Architectural Coatings Suggested Control Measures February 1, 2008.	Table 806.3(1)

ASA

Acoustical Society of America
Suite 1N01
2 Huntington Quadrangle
Melville, NY 11747-4502

Standard reference number	Title	Referenced in code section number
ANSI/ASA S12.60-2010/ Pt.1	Acoustical Performance Criteria, Design Requirements, and Guidelines for Schools, Part 1: Permanent Schools	Table 807.4.3
ANSI/ASA S12.60-2009/ Pt.2	Acoustical Performance Criteria, Design Requirements, and Guidelines for Schools, Part 2: Relocatable Classroom Factors	Table 807.4.3

ASABE

American Society of Agricultural and Biological Engineers
2950 Niles Road
St. Joseph, MI 49085

Standard reference number	Title	Referenced in code section number
S313.3-99 (R2009)	Soil Cone Penetrometer	405.1.4.2
EP542-99 (R2009)	Procedures for Using and Reporting Data Obtained with the Soil Cone Penetrometer	405.1.4.2

ASHE

The American Society for Healthcare Engineering of the American Hospital Association
155 N. Wacker Drive, Suite 400
Chicago, IL 60606

Standard reference number	Title	Referenced in code section number
2010 FGI-ASHE	Guidelines for Design and Construction of Healthcare Facilities.	Table 807.3.2

ASME

American Society of Mechanical Engineers
Three Park Avenue
New York, NY 10016-5990

Standard reference number	Title	Referenced in code section number
A112.18.1/ CSA B125.1—2010	Plumbing Supply Fittings	Table 702.1, 702.2

ASHRAE

American Society of Heating, Refrigerating and Air-Conditioning Engineers, Inc.
1791 Tullie Circle
Atlanta, GA 30329-2305

Standard reference number	Title	Referenced in code section number
52.2—2007	Method of Testing General Ventilation Air-Cleaning Devices for Removal Efficiency by Particle Size	803.1.3, 803.5
55—2004	Thermal Environmental Conditions on Human Occupancy	606.5.1, 803.2

ASHRAE—continued

62.1—2010	Ventilation for Acceptable Indoor Air Quality	604.3
72—05	Method of Testing Commercial Refrigerators and Freezers	Table 609.2.3
90.1—2010	Energy Standard for Buildings Except Low-rise Residential Buildings (ANSI/ASHRAE/IESNA 90.1-2007)	602.1.2, 602.1.2.1
189.1—2011	Standard for the Design of High-performance Green Buildings, Except Low-rise Residential Buildings	101.3, 301.1.1

ASSE

American Society of Sanitary Engineering
901 Canterbury Road, Suite A
Westlake, OH 44145

Standard reference number	Title	Referenced in code section number
1016—2010	Performance Requirements for Automatic Compensating, Valves for Individual Showers and Tub/Shower Combinations	Table 702.1

ASTM

ASTM International
100 Barr Harbor
West Conshohocken, PA 19428-2959

Standard reference number	Title	Referenced in code section number
C 1371—04a	Standard Test Method for Determination of Emittance of Materials Near Room Temperature Using Portable Emissometers	408.3.1.1
C 1549—09	Standard Test Method for Determination of Solar Reflectance Near Ambient Temperature Using a Portable Solar Reflectometer	408.2.1, 408.3.1.1
D 1253—08	Standard Test Method for Residual Chlorine in Water	707.11.6, 707.15.1.1.2, 708.12.5
D 2974—07a	Standard Test Methods for Moisture, Ash, and Organic Matter of Peat and other Organic Soils	405.1.4.2
D 3385—09	Standard Test Method for Infiltration Rate of Soils in Field Using Double-Ring Infiltrometer	405.1.4.2
D 3960—05	Standard Practice of Determining Volatile Organic Compound (VOC) Content of Paints & Related Coatings	806.2, 806.3
D 5055—10	Standard Specification for Establishing and Monitoring Structural Capacities of Prefabricated Wood I-Joists	202
D 5093—02 (2008)	Standard Test Method for Field Measurement of Infiltration Rate Using Double-Ring Infiltrometer With Sealed-Inner Ring	405.1.4.2
D 5197—09	Test Method for Determination of Formaldehyde and Other Carbonyl Compounds in Air (Active Sampler Methodology)	804.3
D 5456—10	Standard Specification for Evaluation of Structural Composite Lumber Products	202
D 5464—07	Standard Test Method for pH Measurement of Water of Low Conductivity	707.15.1.1.2
D 5466—01 (2007)	Test Method for Determination of Volatile Organic Chemicals in Atmospheres (Canister Sampling Methodology)	804.2
D 5907—10	Standard Test Methods for Filterable Matter (Total Dissolved Solids) and Nonfilterable Matter (Total Suspended Solids) in Water	707.15.1.1.2
D 6007—02 (2008)	Standard Test Method for Determining Formaldehyde Concentrations in Air from Wood Products Using a Small-Scale Chamber	Table 806.1
D 6196—03 (2009)	Standard Practice for Selection of Sorbents, Sampling, and Thermal Desorption Analysis Procedures for Volatile Organic Compounds in Air	804.3
D 6345—10	Standard Guide for Selection of Methods for Active, Integrative Sampling of Volatile Organic Compounds in Air	804.3
D 6698—07	Standard Test Method for On-Line Measurement of Turbidity Below 5 NTU in Water	707.15.1.1.2
D 6866—11	Standard Test Methods for Determining the Biobased Content of Solid, Liquid, and Gaseous Samples Using Radiocarbon Analysis	505.2.4
D 7612—10	Standard Practice in Categorizing Wood and Wood-Based Products according to their Fiber Sources	202
E 90—04	Test Method for Laboratory Measurement of Airborne Sound Transmission Loss of Building Partitions and Elements	807.2, 807.3 807.4.1, 809.2, 809.3
E 336—2010	Standard Test Method for Measurement of Airborne Sound Attenuation Between Rooms in Buildings	807.2
E 408—71 (2008)	Standard Test Methods for Total Normal Emittance of Surfaces Using Inspection-Meter Techniques	408.3.1.1
E 413—10	Classification for Rating Sound Insulation	807.4.1

ASTM—continued

E 492—09	Standard Test Method for Laboratory Measurement of Impact Sound Transmission Through Floor-Ceiling Assemblies Using the Tapping Machine	807.4
E 779—10	Standard Test Method for Determining Air Leakage Rate by Ton Pressurization	605.1.2.2
E 1332—90 (2003)	Standard Classification for the Determination of Outdoor-Indoor Transmission Class	807.2
E 1333—10	Standard Test Method for Determining Formaldehyde Concentrations in Air and Emission Rates from Wood Products Using a Large Chamber	Table 806.1
E 1509—04	Standard Specification for Room Heaters, Pellet Fuel-Burning Type	804.1.6
E 1643—10	Standard Practice for Selection, Design, Installation, and Inspection of Water Vapor Retarders Used in Contact with Earth or Granules Fill Under Concrete Slabs	804.2.2, Table 903.1
E 1918—06	Standard Test Method for Measuring Solar Reflectance of Horizontal and Low-Sloped Surfaces in the Field	408.2.1, 408.3.1.1
E 1980—11	Standard Practice for Calculating Solar Reflectance Index of Horizontal and Low-Sloped Opaque Surfaces	408.3.1.2
E 2399—11	Standard Test Method for Maximum Media Density for Dead Load Analysis of Vegetative (Green) Roof Systems	408.3.2
E 2635—08	Standard Practice for Water Conservation in Buildings Through In-Situ Water Reclamation	709.9
E 2727—10	Standard Practice for Assessment of Rainwater Quality	707.12.10
F 1275—03 (2008)	Standard Test Method for Performance of Griddles	Table 609.2.3
F 1346—91 (2010)	Standard Performance Specification for Safety Covers and Labeling Requirements for All Covers for Swimming Pools, Spas and Hot Tubs	1003.2.5
F 1361—07	Standard Test Method for Performance of Open Deep Fat Fryers	Table 609.2.3
F 1496—99 (2005)e1	Standard Test Method for Performance of Convection Ovens	Table 609.2.3
F 1484—05	Standard Test Methods for Performance of Steam Cookers	Table 609.2.3
F 1605—95 (2007)	Standard Test Method for Performance of Double-Sided Griddles	Table 609.2.3
F 1639—05	Standard Test Method for Performance of Combination Ovens	702.17
F 1696—07	Standard Test Method for Energy Performance of Single-Rack, Door-Type Commercial Dishwashing Machines	Table 609.2.3
F 1920—11	Standard Test Method for Performance of Rack Conveyor, Commercial Dishwashing Machines	Table 609.2.3
F 2140—11	Standard Test Method for Performance of Hot Food Holding Cabinets	Table 609.2.3
F 2144—09	Standard Test Method for Performance of Large Open Vat Fryers	Table 609.2.3
F 2861—10	Standard Test Method for Enhanced Performance of Combination Oven in Various Modes	Table 609.2.3

CCR

California Code of Regulations
Department of Industrial Relations
Office of the Director
455 Golden Gate Avenue
San Francisco, CA 94102

Standard reference number	Title	Referenced in code section number
Section 93120—Title 17	California Code Regulations, Airborne Toxic Control Measure to Reduce Formaldehyde Emissions from Composite Wood Products	806.1

CDPH

California Department of Public Health
1615 Capitol Avenue
Sacramento, CA 95814

Standard reference number	Title	Referenced in code section number
CDPH Section 01350	EHLB Standard Method for Testing VOC Emissions From Indoor Sources	806.2

CRRC

Cool Roof Rating Council
1610 Harrison Street
Oakland, CA 94612

Standard reference number	Title	Referenced in code section number
CRRC—1 2010	Cool Roof Rating Council, CRRC-1 Standard	404.3.1.1

CSA

Canadian Standards Association
5060 Spectrum Way
Mississauga, Ontario, Canada L4N 5N6

Standard reference number	Title	Referenced in code section number
CAN/CSA B366.1—2009	Solid-Fuel-Fired Central Heating Appliances	804.1.6
CSA Z21.50/CSA 2.22—07	Vented Gas Fireplaces	804.1.3
CSA Z21.88a/CSA 2.33a—09	ANSI/CSA Standard for Vented Gas Fireplace Heaters	804.1.3

DCHS

California Department of Health Services
Office of Regulations
P.O. Box 997413, MS 0015
Sacramento, CA 95899-7413

Standard reference number	Title	Referenced in code section number
CA/DHS/EHLB/ R-174—2010	Standard Method for the Testing and Evaluation of Volatile Organic Chemical Emissions from Indoor Sources using Environmental Chambers Version 1.1 February 2010	806.2, Table 806.2(2), 806.3, Table 806.2(2), 806.5, 806.6, 809.2.4

DOC

U.S. Department of Commerce
National Institute of Standards and Technology
1401 Constitution Avenue NW
Washington, DC 20230

Standard reference number	Title	Referenced in code section number
PS1—09	Structural Plywood	202
PS2—10	Performance Standard for Wood-Based Structural-Use Panels	202

DOE

U.S. Department of Energy
c/o Superintendent of Documents
U.S. Government Printing Office
Washington, DC 20402-9325

Standard reference number	Title	Referenced in code section number
10 CFR Part 431	Sub-Part K, Appendix C	Table 609.8.1.1(1), Table 609.8.1.1(2), Table 609.8.1.1(3)

EPA

Environmental Protection Agency
Ariel Rios Building
1200 Pennsylvania Avenue, NW
Washington, DC 20460

Standard reference number	Title	Referenced in code section number
40 CFR, Part 60 Subpart AAA	EPA Standards of Performance for New Residential Wood Heaters	804.1.5
40 CFR 300	Small Business Liability Relief and Brownfield Revitalization Act-Public Law 107-118	202
EPAct 2005	Energy Policy Act 2005	608.8.1.1
EPA eGRID 2007	Version 1.1; 2005 data; EPA eGrid Data	602.1.2.3, Table 602.1.2.1, 602.2.1, Table 602.2.1
ENERGY STAR	Energy Star	202, 702.6.1, 702.6.2, 702.6.4
US EPA Method 24	Determination of Volatile Matter Content, Water Content, Density, Volume Solids and Weight Solids of Surface Coatings	806.2
Water Sense February 2007	High Efficiency Toilet Specification	Table 702.1
Water Sense October 2007	Lavatory Faucet Specification	Table 702.1
Water Sense October 2009	Flushing Urinal Specification	Table 702.1
Water Sense March 2010	Showerhead Specification	Table 702.1

FSC

Forest Stewardship Council
212 Third Avenue, North, Suite 504
Minneapolis, MN 55401

Standard reference number	Title	Referenced in code section number
STD-40-004 V2-1EN—2011	Standard for Chain of Custody Certification	505.2.4

ICC

International Code Council, Inc.
500 New Jersey Avenue, NW
6th Floor
Washington, DC 20001

Standard reference number	Title	Referenced in code section number
IBC—12	International Building Code®	101.3, 101.3.1, 102.4, 102.4.1, 102.6, 103.1, 104.1, 201.3, 202, 303.1, 402.1.1.3, 402.2.3, 407.3.2, 407.4, 410.2.1.2, 605.1.1.1, 608.3, 608.4.1, 608.5, 609.2, 610.2.2, 703.7.1, 710.14.1, 803.1.2.1, 804.1.7, Table 903.1, 1004.1, 1104.1
ICCPC—12	International Code Council Performance Code®	102.4
IEBC—12	International Existing Building Code®	102.4, 102.6
IECC—12	International Energy Conservation Code®	102.4, 102.6, 201.3, 202, 604.4, 605.1, 605.1.1, 605.1.1.1, 605.1.2, 605.1.2.1, 605.1.2.3, 606.1, 606.2.1, 606.3, 606.5, 606.8, 606.8.1, 607.1, 607.2.1, 607.3, 607.4, 608.1, 608.8.1.1, 608.1.1.2, 608.1.1.3, 608.5, 608.7, 608.9, 608.11, 609.2, 611.1, 611.1.4.3, 611.3, 611.3.2, 611.3.3.1, 611.3.3.3, 611.4, 611.4.2, Table 903.1, 1003.2.2, 1003.2.4, 1003.2.7, 1104.1
IFC—12	International Fire Code®	102.4, 102.6, 201.3, 408.3.2, 608.3, 608.4.1, 610.2.2
IFGC—12	International Fuel Gas Code®	102.4, 201.3, 603.3.1, 609.2, 804.1.2
IMC—12	International Mechanical Code®	102.4, 201.3, 603.3.2, 604.3, 606.7, 606.8.1, 609.2, 611.1.2.1, 703.7, 803.1.2.1, 803.3, 804.1.2
IPC—12	International Plumbing Code®	102.4, 201.3, 609.2, 704.1.3, 704.2, 705.2, 707.4, 707.11.3.2, 707.11.4.1, 707.11.4.2, 707.11.4.3, 707.11.7.2, 707.11.7.4, 707.11.7.6, 707.11.7.11, 707.11.10, 707.11.12.1, 707.11.12.1, 707.11.12.2, 707.11.12.3, 707.12.1, 707.12.2, 707.12.3, 707.12.4, 708.3, 708.5, 708.10, 708.12.1.1, 708.12.2, 708.12.3.1, 708.12.3.2, 708.12.3.3, 708.12.6.4, 708.12.6.5, 708.12.6.7, 708.12.6.9, 708.12.8, 708.12.9, 708.12.9.3, 708.12.10.1, 708.12.10.2, 708.12.10.3, 708.13.1, 708.13.2, 708.13.3, 708.13.4, 709.3, 709.4, 709.6, 709.7, 709.8, 709.9.1.1, 709.9.1.2, 709.9.1.3, 709.10.1, 709.10.2
IPMC—12	International Property Maintenance Code®	102.4, 102.6
IRC—12	International Residential Code®	102.4, 201.3, 804.2
ICC-700—2008	National Green Building Standard	101.3.1, 302.1, Table 302.1

IEC

The International Electrotechnical Commission
Central Office
3, rue de Varembe'
P. O. Box 131
Ch-1211 Geneva 20, Switzerland

Standard reference number	Title	Referenced in code section number
EN60034-30—2009	Standard on Efficiency Classes for Low Voltage AC Motors	609.2.1.2.1, 609.2.2.2

IESNA

Illuminating Engineering Society of North America
120 Wall Street, 17th Floor
New York, NY 10005-4001

Standard reference number	Title	Referenced in code section number
TM-15—07	Luminaire Classification System for Outdoor Luminaires	Table 405.3(2), Table 409.2, Table 409.3(1), Table 409.3(2)

ISO

International Organization for Standardization
ISO Central Secretariat
1 ch, de la Voie-Creuse, Case Postale 56
CH-1211 Geneva 20, Switzerland

Standard reference number	Title	Referenced in code section number
7708—1995	Air quality – Particle Size Fraction Definitions for Health-related Sampling	804.3
13256-1—1998	Water-source Heat Pumps – Testing and Rating for Performance – Part 1: Water-to-air and Brine-to-air Heat Pumps	Table 606.2.2.1
13256-2—1998	Water-source Heat Pumps – Testing and Rating for Performance – Part 2: Water-to-water and Brine-to-water Heat Pumps	Table 606.2.2.1
14044—2006	Environmental Management – Lifecycle Assessment—Requirements and Guidelines	303.1
ISO/IEC 17025—2005 2004—11	General Requirements for the Competence of Testing and Calibration Laboratories	806.2, Table 806.2(2), 806.3, 806.4, 806.5, 806.6, 809.2.4

NEMA

National Electrical Manufacturers Association
1300 North 17th Street, Suite 1752
Rosslyn, VA 22209

Standard reference number	Title	Referenced in code section number
LSD 23—2010	Recommended Practice –Lamp Seasoning for Fluorescent Dimming Systems	611.3.3.4

NFPA

National Fire Protection Association
1 Batterymarch Park
Quincy, MA 02269

Standard reference number	Title	Referenced in code section number
NFPA 70—2011	National Electrical Code	603.3.4, 610.2.2, 610.3
NFPA 72—2010	National Alarm and Signaling Code	710.6.2

NREL

National Renewable Energy Laboratory
1617 Cole Boulevard
Golden, CO 80401-3305

Standard reference number	Title	Referenced in code section number
SERI TR-642-761	A Simplified Clear Sky Model for Direct and Diffuse Insolation on Horizontal Surfaces	610.2.1

NSF

NSF International
789 Dixboro Road
Ann Arbor, MI 48105

Standard reference number	Title	Referenced in code section number
NSF/ANSI 3—10	Commercial Warewashing Equipment	Table 609.2.3
NSF/ANSI 44—09	Residential Cation Exchange Water	704.1.2, 704.1.4
NSF/ANSI 50—09	Equipment for Swimming Pools, Spas, Hot Tubs, and other Recreational Water Facilities	708.12.7.1
NSF/ANSI 58—09	Reverse Osmosis Drinking Water Treatment Systems	704.2
NSF/ANSI 61—09	Drinking Water Systems Components – Health Effects	707.15
NSF/P151—95	Health Effects from Rain Water Catchment Systems Components	707.11.1.1
NSF 350—11	Onsite Residential and Commercial Water Reuse Treatment Systems	704.3

SCAQMD

South Coast Air Quality Management District
21865 Capley Drive
Diamond Bar, CA 91765

Standard reference number	Title	Referenced in code section number
SCAQMD Method 302—91 (Revised 1993)	Distillation of Solvents from Paints, Coatings and Inks, South Coast Air Quality Management District	803.2
SCAQMD Method 303—91 (Revised 1993)	Determination of Exempt Compounds, South Coast Air Quality Management District	803.2
SCAQMD Method 304—91 (Revised February 1996)	Determination of Volatile Organic Compounds (VOC) in Various Materials, South Coast Air Quality Management District	803.2
SCAQMD Method 316A—92	Determination of Volatile Organic Compounds (VOC) in Materials Used for Pipes and Fittings	803.2
SCAQMD Method 316B—92	Determination of Volatile Organic Compounds (VOC) In Adhesives containing Cyanoacrylates	803.2
SCAQMD Rule 1168	Adhesives and Sealant Applications	806.2

SFI

Sustainable Forest Initiative, Inc.
900 17th Street, NW, Suite 700
Washington, DC 20006

Standard reference number	Title	Referenced in code section number
SFI—2010-2014	Sustainable Forest Initiative 2010-2014	505.2.4

SMACNA

Sheet Metal & Air Conditioning Contractors National Assoc., Inc.
4021 Lafayette Center Road
Chantilly, VA 22021

Standard reference number	Title	Referenced in code section number
2010	SMACNA HVAC Air Duct Leakage Test Manual (1st Edition)	606.3.1

TCIA

Tree Care Industry Association
136 Harvey Road, Suite 101
Londonderry, NH 03053

Standard reference number	Title	Referenced in code section number
ANSI A300 Part 5—2005	Tree Shrub and Other Woody Plt Mgmt-Management of Trees and Shrubs during Site Planning, Site Development, and Construction	405.2.1.1

TMS

The Masonry Society
3970 Broadway, Unit 201-D
Boulder, CO 80304-1135

Standard reference number	Title	Referenced in code section number
0302—2011	Standard Method for Determining the Sound Transmission Class Rating for Masonry Walls	809.3, 809.5.1

UL

Underwriters Laboratories Inc.
333 Pfingsten Road
Northbrook, IL 60062

Standard reference number	Title	Referenced in code section number
UL 1482—2011	Room Heaters, Solid Fuel Type	804.1.6
UL 1993—2009	Standard for Safety of Self-Ballasted Lamps and Lamp Adapters	506.3
UL 2523—2009	Solid Fuel-Fired Hydronic Heating Appliances, Water Heaters and Boilers	804.1.6

USCC

US Composting Council
1 Comac Loop 14 B1
Rokonkoma, NY 11779

Standard reference number	Title	Referenced in code section number
TMECC 05.7a	Test Method for the Examination of Composting and Compost	405.1.2

USDA

United States Department of Agriculture
Office of Energy Policy and New Uses
Room 361, Reporters Bldg.
300 Seventh Street, SW
Washington, DC 20024

Standard reference number	Title	Referenced in code section number
7 CFR Part 2902 Rev. 1/1/06	Guidelines for Designating Bio-based Products for Federal Procurement	505.2.4
MP 1475—90	USDA Plant Hardiness Zone Map, Miscellaneous Publication 1475	408.3.2

APPENDIX A

PROJECT ELECTIVES

The provisions contained in this appendix are not mandatory unless specifically referenced in the adopting ordinance.

SECTION A101 GENERAL

A101.1 Scope. The provisions of this appendix are designed to offer conservation practices that achieve greater benefit than the minimum requirements of the *International Green Construction Code*™ (IgCC™).

A101.2 Intent. This appendix shall provide a basis by which a jurisdiction can implement measures to increase natural resource conservation, material resource conservation, energy conservation, water conservation and environmental comfort and mitigate impacts of building site development.

SECTION A102 APPLICABILITY AND CONFORMANCE

A102.1 General. Project electives shall be applicable to building, structures and building sites constructed under the provisions of this code.

A102.2 Required number of and selection of project electives. The jurisdiction shall indicate the number of project electives required in the blank provided in the row that references Section A102.2 in Tables A104, A105, A106, A107 and A108. Each project constructed in the jurisdiction shall be required to comply with this number of project electives. A total of not less than this number of project electives shall be selected by the owner from each table. Selected project electives shall be applied as mandatory requirements for the project. Selected project electives shall be communicated to the *code official* by means of checking the appropriate boxes in the tables and providing a copy of the tables, or by inclusion of a list of selected project electives, with the construction documents.

SECTION A103 DEFINITIONS

A103.1 Definitions. The following words and terms shall, for the purposes of this appendix, have the meanings shown herein. Refer to Chapter 2 of this code for general definitions.

DESIGN LIFE. The intended service life or the period of time that a building or its component parts are expected to meet or exceed the performance requirements.

GEOTHERMAL ENERGY. Renewable energy generated from the interior of the Earth and used to produce energy for heating buildings or serving building commercial or industrial processes.

PROJECT ELECTIVE. The minimum total number of project electives that must be selected and complied with as indicated in Section A102.2 and Tables A104, A105, A106, A107 and A108.

SERVICE LIFE. The period of time after installation during which a building or its component parts meets or exceeds the performance requirements.

VOCs, TOTAL (TVOCs). Sum of the concentrations of all identified and unidentified *volatile organic compounds* between and including n-hexane through n-hexadecane (i.e., C_6 - C_{16}) as measured by gas chromatography/mass spectrometry total ion-current chromatogram method and are quantified by converting the total area of the chromatogram in that analytical window to toluene equivalents.

SECTION A104 SITE PROJECT ELECTIVES

A104.1 Flood hazard area project elective. Where Section 402.2.1 is not listed in Table 302.1 as a mandatory requirement, and in specific *flood hazard areas* if Section 402.2.2 is not a mandatory requirement, projects seeking *flood hazard area* project electives in accordance with Section A102.2 shall comply with one of the project electives identified in Sections A104.1.1 through A104.1.3.

A104.1.1 Flood hazard area preservation. Where less than 25 percent of a building site is located within a *flood hazard area*, buildings and building site improvements shall be located on portions of the building site that are located outside of the *flood hazard area*. The building site shall not be filled or regraded to raise the elevation of the site to remove areas from the *flood hazard area*.

A104.1.2 Flood hazard area minimization. Where 25 percent or more of a building site is located within a *flood hazard area*, the lowest floors of buildings that are located within the *flood hazard area* shall be not less than 1 foot (305 mm) above the design flood elevation as established by the *International Building Code*, or not less than the height, as established by the jurisdiction, above the design flood elevation, whichever is higher. The placement of fill on a building site shall not be used to achieve the required height above the design flood elevation.

A104.1.3 Flood hazard area, existing building. Where additions, alterations, or repairs are made to an existing building located in a *flood hazard area*, and the cost of the work equals or exceeds 40 percent of the market value of the structure before the improvement or repair is started, the entire building shall be brought into compliance with the flood-resistant construction requirements in the *International Building Code* for new buildings and structures.

TABLE A104
SITE PROJECT ELECTIVES

SECTION	DESCRIPTION	MINIMUM NUMBER OF ELECTIVES REQUIRED AND ELECTIVES SELECTED	
A102.2	The jurisdiction shall indicate a number between and including 0 and up to and including 6 to establish the minimum total number of project electives that must be satisfied.	—	
A104.1.1	Flood hazard area preservation	☐ Yes	☐ No
A101.1.2	Flood hazard area minimization	☐ Yes	☐ No
A101.1.3	Flood hazard area, existing building	☐ Yes	☐ No
A104.2	Wildlife corridor	☐ Yes	☐ No
A104.3	Infill site	☐ Yes	☐ No
A104.4	Brownfield site	☐ Yes	☐ No
A104.5	Site restoration	☐ Yes	☐ No
A104.6	Mixed use development	☐ Yes	☐ No
A104.7	Changing and shower facilities	☐ Yes	☐ No
A104.8	Long-term bicycle parking and storage	☐ Yes	☐ No
A104.9	Heat island	☐ Yes	☐ No
A104.9.1	Site hardscape project elective 1	☐ Yes	☐ No
A104.9.2	Site hardscape project elective 2	☐ Yes	☐ No
A104.9.3	Site hardscape project elective 3	☐ Yes	☐ No
A104.9.4	Roof covering project elective	☐ Yes	☐ No

A104.2 Wildlife corridor project elective. Site development that restores a wildlife corridor, connecting wildlife corridors on adjacent lots, shall be recognized as a project elective.

A104.3 Infill site project elective. The development of a building site that is an infill site with a new building and associated site improvements shall be recognized as a project elective.

A104.4 Brownfield site project elective. The development of a building site that is a *brownfield* site with a new building with associated site improvements shall be recognized as a project elective. The development shall be in accordance with the following:

1. Phase I and II Environmental Assessment and, as necessary, the documentation of the site remediation plan and completion of the plan, as *approved* by the jurisdictional agency in charge of environmental regulations.
2. Where contamination levels are above risk-based standards for intended reuse and remediation is required, building and site development shall provide effective remediation *approved* by the local, state or federal government agency which classified the site as a *brownfield*, by one of the following:
   - 2.1. The effective remediation is completed in the manner described in the remediation plan *approved* by the agency which classified the site as a *brownfield.*
   - 2.2. A remediation commensurate with the initial *approved* plan which the agency approves upon completion by issuing a letter stating that no further remediation action is required.
3. The *brownfield* site project elective fully accomplishes the applicable state and local *brownfields* program cleanup goals, with all supporting documentation as required by the state, tribal or other responsible authority.

A104.5 Site restoration project elective. Previously developed sites that restore 25 percent or more of the nonbuilding footprint building site area with native or adaptive vegetation shall be recognized as a project elective.

A104.6 Mixed-use development project elective. Development of a mixed-use building shall be recognized as a project elective. The building shall be in accordance with all of the following:

1. It shall have not less than two stories.
2. Eight or more dwelling units of Group R-1 or R-2 occupancy shall be located above the first story.
3. The first story shall contain one or more of the following occupancies: A-1, A-2, A-3, B, M, Group E daycare, or Group R-2 live/work units.

A104.7 Changing and shower facilities project elective. Where a new building is less than 10,000 square feet (929 m^2) in *total building floor area*, providing changing and shower facilities in accordance with Section 407.2 shall be recognized as a project elective.

A104.8 Long-term bicycle parking and storage project elective. The development of a new building and associated site improvements where additional long-term bicycle parking is provided in accordance with all of the following shall be recognized as a single project elective:

1. Provide long-term bicycle parking that is twice the number of parking spaces required by Table 407.3;

2. Provide spaces in accordance with Section 407.3.2; and
3. Locate not less than 90 percent of long-term bicycle parking within a building or provide the parking with a permanent cover including, but not limited to, roof overhangs, awnings, or bicycle storage lockers.

A104.9 Heat island. Project electives related to heat island impact shall comply with Sections A104.9.1 through A104.9.4. Compliance with multiple electives shall be recognized.

A104.9.1 Site hardscape project elective 1. In climate zones 1 through 6, as established in the *International Energy Conservation Code*, the development of a new building and associated site improvements where a minimum of 75 percent of the site hardscape is in accordance with one or any combination of options in Sections 408.2.1 through 408.2.4, shall be recognized as a project elective.

A104.9.2 Site hardscape project elective 2. In climate zones 1 through 6, as established in the *International Energy Conservation Code,* the development of a new building and associated site improvements where a minimum of 100 percent of the site hardscape is in accordance with one or any combination of options in Sections 408.2.1 through 408.2.4, shall be recognized as a project elective.

A104.9.3 Site hardscape project elective 3. In climate zones 7 and 8, as established in the *International Energy Conservation Code,* the development of a new building and associated site improvements where a minimum of 50 percent of the site hardscape is in accordance with one or any combination of options in Sections 408.2.1 through 408.2.4, shall be recognized as a project elective.

A104.9.4 Roof covering project elective. In climate zones 4 through 8, as established in the *International Energy Conservation Code,* the development of a new building with roof coverings in accordance with Section 408.3, shall be recognized as a project elective.

SECTION A105 MATERIAL RESOURCE CONSERVATION AND EFFICIENCY

A105.1 Waste management project elective. Projects seeking a waste management project elective shall comply with Section 503.1, except that the nonhazardous construction waste materials required to be diverted from landfills shall be increased by 20 percent. Where another percentage is indicated by the jurisdiction in Table 302.1, projects seeking this credit shall increase diversion by 20 percent above the percentage indicated in Table 302.1.

A105.2 Construction waste landfill maximum project elective. Projects seeking a construction waste landfill maximum project elective in accordance with Table A105 and Section A102.2 shall comply with Section 503.1 except that not more than 4 pounds (1.814 kg) of construction waste, excluding hardscape, per square foot (0.0929 m^2) of building area shall be disposed of in a landfill. Building construction waste and hardscape waste shall be measured separately.

A105.3 Material selection project electives. Each of the following shall be considered a separate material selection project elective. The project electives are cumulative and compliance with each item shall be recognized individually.

1. Compliance with this project elective shall require compliance with Section 505.2, except that buildings and structures shall contain used, recycled content, recyclable, bio-based and indigenous materials that comply with Sections 505.1 through 505.2.5 such that the aggregate total materials compliant with those sections constitute at least 70 percent of the total building products and materials used, based on mass, volume or cost, used singularly or in combination.
2. Compliance with Item 1 except that such materials shall be used for at least 85 percent of the total mass, volume or cost of materials in the project.

A105.4 Building service life plan project electives. Projects seeking a building service life plan project elective shall comply with this section. The building service life plan (BSLP) in accordance with Section A105.4.1 shall be included in the construction documents.

TABLE A105
MATERIAL RESOURCE CONSERVATION AND EFFICIENCY

SECTION	DESCRIPTION	MINIMUM NUMBER OF ELECTIVES REQUIRED AND ELECTIVES SELECTED	
A102.2	The jurisdiction shall indicate a number between and including 0 and up to and including 4 to establish the minimum total number of project electives that must be satisfied.	—	
A105.1	Waste management	☐ Yes	☐ No
A105.2	Construction waste landfill maximum	☐ Yes	☐ No
A105.3(1)	Reused, recycled content, recyclable, bio-based and indigenous materials (70%)	☐ Yes	☐ No
A105.3(2)	Reused, recycled content, recyclable, bio-based and indigenous materials (85%)	☐ Yes	☐ No
A105.4	Service life plan	☐ Yes	☐ No
A105.5	Design for deconstruction and building reuse	☐ Yes	☐ No
A105.6	Existing building reuse	☐ Yes	☐ No
A105.7	Historic building reuse	☐ Yes	☐ No

A105.4.1 Plan and components. The building service life plan (BSLP) shall indicate the intended length in years of the design service life for the building as determined by the building owner or *registered design professional*, and shall include a maintenance, repair, and replacement schedule for each of the following components. The maintenance, repair and replacement schedule shall be based on manufacturer's reference service life data or other *approved* sources for the building components. The manufacturer's reference service life data or data from other *approved* sources shall be included in the documentation.

1. Structural elements and concealed materials and assemblies.
2. Materials and assemblies where replacement is cost prohibitive or impractical.
3. Major materials and assemblies that are replaceable.
4. Roof coverings.
5. Mechanical, electrical and plumbing equipment and systems.
6. Site hardscape.

A105.5 Design for deconstruction and building reuse project elective. Projects seeking a design for deconstruction and building reuse project elective shall be designed for deconstruction of not less than 90 percent of the total components, assemblies, or modules to allow essentially the entire building to be reused. Design for deconstruction shall be documented on the building's plans and construction documents.

A105.6 Existing building reuse project elective. The development of a building site on which an existing building is already located and in which not less than 75 percent of the existing core and shell of the structure will be reused shall be recognized as a project elective.

A105.7 Historic building reuse project elective. The development of a building site on which an existing building is already located and in which not less than 75 percent of the existing core and shell of a locally or nationally designated historic structure will be reused shall be recognized as a project elective.

SECTION A106 ENERGY CONSERVATION, EFFICIENCY AND EARTH ATMOSPHERIC QUALITY

A106.1 zEPI reduction project electives. Project electives for buildings pursuing performance-based compliance in accordance with Section 601.3.1 shall be in accordance with the portions of Table A106 that reference Section A106.1, Equation 6-1 and the calculation procedures specified in Section 602.1.2.1.

A106.2 Mechanical systems project elective. Buildings seeking a mechanical systems project elective in accordance with Sections A102.2 and A106 shall comply with Sections A106.2.1 through A106.2.5.

A106.2.1 Prescriptive path. The building shall be designed prescriptively in accordance with Section 601.3.2.

TABLE A106
ENERGY CONSERVATION AND EFFICIENCY

SECTION	DESCRIPTION	MINIMUM NUMBER OF ELECTIVES REQUIRED AND ELECTIVES SELECTED
A102.2	The jurisdiction shall indicate a number between and including 0 and up to and including 10 to establish the minimum total number of project electives that must be satisfied.	—
A106.1	zEPI reduction project electives	☐ Yes ☐ No
A106.1	Project zEPI is at least 5 points lower than required by Table 302.1	☐ 1 elective
A106.1	Project zEPI is at least 10 points lower than required by Table 302.1	☐ 2 electives
A106.1	Project zEPI is at least 15 points lower than required by Table 302.1	☐ 3 electives
A106.1	Project zEPI is at least 20 points lower than required by Table 302.1	☐ 4 electives
A106.1	Project zEPI is at least 25 points lower than required by Table 302.1	☐ 5 electives
A106.1	Project zEPI is at least 30 points lower than required by Table 302.1	☐ 6 electives
A106.1	Project zEPI is at least 35 points lower than required by Table 302.1	☐ 7 electives
A106.1	Project zEPI is at least 40 points lower than required by Table 302.1	☐ 8 electives
A106.1	Project zEPI is at least 45 points lower than required by Table 302.1	☐ 9 electives
A106.1	Project zEPI is at least 51 points lower than required by Table 302.1	☐ 10 electives
A106.2	Mechanical systems project elective	☐ Yes ☐ No
A106.3	Service water heating	☐ Yes ☐ No
A106.4	Lighting systems	☐ Yes ☐ No
A106.5	Passive design	☐ Yes ☐ No
A106.6	Renewable energy systems—5 percent	☐ Yes ☐ No
A106.6	Renewable energy systems—10 percent	☐ Yes ☐ No
A106.6	Renewable energy systems—20 percent	☐ Yes ☐ No

A106.2.2 Mechanical equipment. Mechanical equipment shall comply with Sections A106.2.2.1 through A106.2.2.4 to achieve the mechanical systems project elective.

A106.2.2.1 Heating equipment. For heating equipment, the part-load efficiency of the equipment shall be not less than 10 percent greater than the part-load efficiencies shown in the applicable tables of the *International Energy Conservation Code*, or ASHRAE 90.1, or the equipment shall be ENERGY STAR qualified, as applicable.

A106.2.2.2 Cooling equipment. For cooling equipment, the part-load efficiency of the equipment shall be not less than 10 percent greater than the part-load efficiencies shown in the applicable tables of the *International Energy Conservation Code*, or ASHRAE 90.1, or the equipment shall be ENERGY STAR qualified.

A106.2.2.3 Ground source heat pumps. Ground source heat pumps shall meet the provisions of Table A106.2.2.3 based on the applicable referenced test procedure.

A106.2.2.4 Multi-stage ground source heat pumps. The efficiency of multi-stage ground source heat pumps shall meet the provisions of Table A106.2.2.3 based on the applicable referenced test procedure.

TABLE A106.2.2.3
ENERGY-EFFICIENCY CRITERIA FOR GROUND SOURCE HEAT PUMPS

PRODUCT TYPE	*MINIMUM* EER	*MINIMUM* COP
Water-to-Air Closed loop **TEST PROCEDURE - ISO 13256-1**	14.1	3.3
Water-to-Air Open loop **TEST PROCEDURE - ISO 13256-1**	16.2	3.6
Water-to-Water Closed loop **TEST PROCEDURE - ISO 13256-2**	15.1	3.0
Water-to-Water Open loop **TEST PROCEDURE - ISO 13256-2**	19.1	3.4
Direct Expansion (DX) or Direct GeoExchange (DGX) **TEST PROCEDURE - AHRI 870**	15.0	3.5

EER = energy-efficiency ratio; COP = coefficient of performance

A106.2.3 Duct insulation. Ducts shall be insulated to R-8 or greater where located in unconditioned spaces and R-11 minimum where located outside of the building structure. Where located within a building envelope assembly, the duct or plenum shall be separated from the building exterior or unconditioned or exempt spaces by R-8 insulation or greater.

A106.2.4 Duct system testing. Duct systems shall be leak-tested in accordance with the SMACNA *HVAC Air Duct Leakage Test Manual* and shall have a rate of air leakage (CL) less than or equal to 4 as determined in accordance with Equation 4-5 of the *International Energy Conservation Code*.

A106.2.4.1 Documentation. Documentation shall be furnished by the designer demonstrating that representative sections totaling not less than 50 percent of the duct area have been tested and that all tested sections meet the requirements of Section A106.2.4.

A106.2.5 Service water heating equipment. The efficiency of the service water heating equipment shall be not less than 10 percent greater than the efficiencies shown in the *International Energy Conservation Code* and ASHRAE 90.1 or the service water heating equipment shall be ENERGY STAR qualified.

A106.3 Service water heating project elective. Buildings seeking a service water heating project elective in accordance with Sections A102.2 and A106.3 shall comply with Sections A106.3.1 through A106.3.3.

A106.3.1 Prescriptive path. The building shall be designed prescriptively in accordance with Section 601.3.2.

A106.3.2 Occupancy. The building shall be designed to serve one of the following occupancies:

1. Group A-2, restaurants and banquet halls;
2. Group F, laundries;
3. Group R-1, boarding houses (transient), hotels (transient), motels (transient);
4. Group R-2 buildings;
5. Group A-3, health clubs and spas; and
6. Group I-2, hospitals, mental hospitals and nursing homes.

A106.3.3 Service water heating efficiency. The efficiency of the service water heating equipment shall be at least 10 percent greater than the efficiencies shown in the *International Energy Conservation Code* and ASHRAE 90.1 or the service water heating equipment shall be ENERGY STAR qualified.

A106.4 Lighting system efficiency project elective. Buildings seeking a lighting system efficiency project elective in accordance with Sections A102.2 and A106.4 shall comply with Sections A106.4.1 through A106.4.3.

A106.4.1 Prescriptive path. The building shall be designed prescriptively in accordance with Section 602.3.1.

A106.4.2 Interior lighting system efficiency. The interior connected lighting power shall be 10 percent less than the allowance determined in accordance with Section C405.5 of the *International Energy Conservation Code*.

A106.4.3 Exterior lighting system efficiency. The exterior connected lighting power shall be 10 percent less than the allowance determined in accordance with Section C405.6 of the *International Energy Conservation Code*.

A106.5 Passive design project elective. Buildings seeking a passive design project elective in accordance with Sections A102.2 and A106. 5 shall comply with Sections A106.5.1 and A106.5.2.

A106.5.1 Performance path. The building shall be designed using the performance path in accordance with Section 601.3.1.

A106.5.2 Passive design provisions. The simulation of energy use performed pursuant to Section 602 shall document that not less than 40 percent of the annual energy use reduction realized by the proposed design has been achieved through passive heating, cooling, and ventilation design, as compared to the standard reference design. Passive heating and cooling shall use strategies including, but not limited to, building orientation, fenestration provisions, material selection, insulation choices, overhangs, shading means, microclimate vegetation and water use, passive cooling towers, natural heat storage, natural ventilation, and thermal mass.

A106.6 Renewable energy system project electives. Buildings seeking a renewable energy system project elective or electives shall be equipped with one or more renewable energy systems in accordance with Section 610.1 that have the capacity to provide the percent of annual energy used within the building as selected in Table A106. Capacity shall be demonstrated in accordance with Section 610.1.1 or 610.1.2.

SECTION A107 WATER RESOURCE CONSERVATION AND EFFICIENCY

A107.1 Indoor water use. This section contains project electives related to indoor water use.

A107.2 Onsite waste water treatment project elective. Where projects are intended to qualify for an onsite waste water treatment project elective in accordance with Section A107.2, all waste water from the building shall be treated to meet the quality requirements appropriate for its intended use and as required by law.

A107.3 Alternate onsite nonpotable water for outdoor hose connections project elective. Where projects are intended to qualify for an alternate onsite nonpotable for outdoor hose connections project elective in accordance with Section A107.3, sillcocks, hose bibs, wall hydrants, yard hydrants, and other outdoor outlets shall be supplied by nonpotable water. Such outlets shall be located in a locked vault or shall be operable only by means of a removable key.

A107.3.1 Signage. Each outlet shall be provided with signage in accordance with Section 706.2.

A107.4 Alternate onsite nonpotable water for plumbing fixture flushing water project elective. Where projects are intended to qualify for an *alternate onsite nonpotable water* for plumbing fixture flushing project elective in accordance with Section A107.4, nonpotable water shall be used for flushing water closets and urinals.

A107.4.1 Water quality. Nonpotable water for water closet and urinal flushing shall meet minimum water quality requirements as established for indoor flushing applications by local codes and regulations. Where chlorine is used for disinfection, the nonpotable water shall contain not more than 4 mg/L of chloramines or free chlorine. Where ozone is used for disinfection, the nonpotable water shall not contain gas bubbles having elevated levels of ozone at the point of use.

A107.4.2 Filtration required. Nonpotable water utilized for water closet and urinal flushing applications shall be filtered by a 100 micron or finer filter.

A107.4.3 Signage. The entries to rooms having water closets or urinals that are supplied with nonpotable water shall be provided with signage in accordance with Section 706.2.

A107.5 Automatic fire sprinkler system project elective. Where projects are intended to qualify for an automatic fire sprinklers system project elective in accordance with Section A107.5, automatic fire sprinkler systems shall be supplied with nonpotable water from an onsite rainwater collection system. Such rainwater collection system shall comply with Section 707. The requirements of Sections A107.5.1 and A107.5.3 shall apply to the fire sprinkler system and the onsite rainwater collection system.

A107.5.1 Emergency power. An emergency power system complying with Chapter 27 of the *International Building Code* shall be provided for powering the pump and controls for the onsite rainwater collection system.

TABLE A107
WATER RESOURCE CONSERVATION AND EFFICIENCY

SECTION	DESCRIPTION	MINIMUM NUMBER OF ELECTIVES REQUIRED AND ELECTIVES SELECTED	
A102.2	The jurisdiction shall indicate a number between and including 0 and up to and including 6 to establish the minimum total number of project electives that must be satisfied.	—	
A107.2	Onsite waste water treatment	☐ Yes	☐ No
A107.3	Alternate onsite nonpotable water for outdoor hose connections	☐ Yes	☐ No
A107.4	Alternate onsite nonpotable water for plumbing fixture flushing	☐ Yes	☐ No
A107.5	Automatic fire sprinkler system	☐ Yes	☐ No
A107.6	Alternate onsite nonpotable water to fire pumps	☐ Yes	☐ No
A107.7	Alternate onsite nonpotable water for industrial process makeup water	☐ Yes	☐ No
A107.8	Alternate onsite nonpotable water for cooling tower makeup water	☐ Yes	☐ No
A107.9	Gray water collection	☐ Yes	☐ No

A107.5.2 Source volume indication. The fire command center for the building shall be equipped with a device that indicates the volume of nonpotable water contained in the collection reservoir. The indicator shall be *approved* and shall be in compliance with NFPA 72.

A107.5.3 Quality of water used for fire suppression. The required quality and treatment of the nonpotable water stored and used for fire suppression shall be determined by authority(s) having jurisdiction.

A107.6 Alternate onsite nonpotable water to fire pumps project elective. Where projects are intended to qualify for an *alternate onsite nonpotable water* to fire pumps project elective in accordance with Sections A107.6, one or more fire pumps shall be located within 200 feet (60 960 mm) of a nonpotable water collection system of sufficient quality, pressure, and capacity for fire pump applications and the fire pumps shall be connected to such source of nonpotable water. The connections shall be in accordance with Section 403.3.2 of the *International Building Code*.

A107.6.1 Quality of water used for fire suppression. The required quality and treatment of the nonpotable water stored and used for fire suppression shall be determined by the authority having jurisdiction.

A107.6.2 Signage. Fire pumps connected to a nonpotable water supply shall have signage in accordance with Section 706.2 provided at the building's fire command center and at each fire pump.

A107.7 Alternate onsite nonpotable water for industrial process makeup water project elective. Where projects are intended to qualify for an *alternate onsite nonpotable water* for industrial process makeup water project elective in accordance with Section A107.7, industrial processes requiring makeup water shall utilize nonpotable water except where the process requires potable water for proper functioning.

A107.7.1 Signage. Rooms containing process equipment supplied with nonpotable water shall be provided with signage in accordance with Section 706.2.

A107.8 Alternate onsite nonpotable water for cooling tower makeup water project elective. Where projects are intended to qualify for an *alternate onsite nonpotable water* for cooling tower makeup water project elective in accordance with Section A107.7, nonpotable water shall be utilized for cooling tower makeup water in accordance with the requirements of Section 706.3.

A107.9 Gray water collection project elective. Where projects are intended to qualify for a gray water collection project elective in accordance with Section A107.8, waste water from lavatories, showers, bathtubs, clothes washers, and laundry trays shall be collected for reuse onsite in accordance with Section 708.

SECTION A108 INDOOR ENVIRONMENTAL QUALITY AND COMFORT

A108.1 VOC emissions project electives. Sections A108.2 through A108.5 shall be considered to be separate project electives. The electives shall be cumulative and compliance with each project elective shall be recognized individually.

A108.2 Flooring material project elective. Where projects are intended to qualify for a "flooring material" project elective, all flooring installed within the interior of the building shall comply with Section 806.4 or shall be one or more of the following flooring materials that are deemed to comply with VOC emission limits:

1. Ceramic and concrete tile
2. Clay pavers
3. Concrete
4. Concrete pavers
5. Metal
6. Organic-free, mineral-based

A108.3 Ceiling materials project elective. Where projects are intended to qualify for a "ceiling materials" project elective, all ceiling systems shall comply with Section 806.5 or shall be one or more of the following ceiling systems that are deemed to comply with VOC emission limits:

1. Ceramic tile
2. Clay masonry
3. Concrete
4. Concrete masonry
5. Metal
6. Organic-free, mineral-based

TABLE A108
INDOOR ENVIRONMENTAL QUALITY AND COMFORT

SECTION	DESCRIPTION	MINIMUM NUMBER OF ELECTIVES REQUIRED AND ELECTIVES SELECTED	
A102.2	The jurisdiction shall indicate a number between and including 0 and up to and including 3 to establish the minimum total number of project electives that must be satisfied.	—	
A108.2	VOC emissions—flooring	☐ Yes	☐ No
A108.3	VOC emissions—ceiling systems	☐ Yes	☐ No
A108.4	VOC emissions—wall systems	☐ Yes	☐ No
A108.5	Total VOC limit	☐ Yes	☐ No
A108.6	Views to building exterior	☐ Yes	☐ No

A108.4 Wall materials project elective. Where projects are intended to qualify for a "wall materials" project elective, all wall systems shall comply with Section 806.5 or shall be one or more of the following wall systems that are deemed to comply with VOC emission limits:

1. Ceramic tile
2. Clay masonry
3. Concrete
4. Concrete masonry
5. Metal
6. Organic-free, mineral-based

A108.5 Total VOC limit project elective. Where projects are intended to qualify for a "total VOC limit" project elective in accordance with a minimum of 50 percent of all adhesives and sealants, architectural paints and coatings, flooring, acoustical ceiling tiles and wall systems and Insulation shall have a Total *Volatile Organic Compounds* (TVOCs) emission limit of ≤ 500 ug/m³. The test methodology used to determine compliance shall be from CDPH/EHLB/Standard Method V.1.1. The emissions testing shall be performed by a laboratory that has the CDPH/EHLB/Standard Method V.1.1 test methodology in the scope of its ISO 17025 Accreditation.

A108.6 Views to building exterior project elective. Where projects are intended to qualify for a "views to building exterior" project elective in accordance with Section A108.6, not less than 50 percent of the net floor area shall have a direct line of sight to the exterior through clear vision glazing. A total of not less than 45 square feet (4.18 m^2) of clear vision glazing in the exterior wall or roof shall be visible. The direct line of sight shall originate at a height of 42 inches (1067 mm) above the finished floor of the space, shall terminate at the clear vision glazing in the exterior wall or roof, and shall be less than 40 feet (12 192 mm) in length.

Exception: Where the direct line of sight is less than 25 feet (7620 mm) in length, a total of not less than 18 square feet (1.67 m^2) of clear vision glazing in the exterior wall or roof shall be visible.

SECTION A109 REFERENCED STANDARDS

Agency	Standard	Title	Referenced in
AHRI		Certified Direct GEO Exchange Certification 870-2011	Table A106.2.2.3
ASHRAE	90.1-2010	Energy Standard for Buildings Except Low-Rise Residential Buildings	A106.2.2.1 A106.2.2.2 A106.2.5 A106.3.3
CDPH		Standard Method for the Testing and Evaluation of Volatile Organic Chemical Emissions From Indoor Sources Using Environmental Chambers, Version 1.1 - 2010	A108.5
EPA		Energy Star	A106.2.2.1 A106.2.2.2 A106.2.5 A106.3.3
ICC	IBC—12	International Building Code	A104.1.2, A104.1.3, A107.5.1
	IECC—12	International Energy Conservation Code	A104.9.1 A104.9.2 A104.9.3 A104.9.4 A106.2.2.1 A106.2.2.2 A106.2.4 A106.2.5 A106.3.3 A106.4.2 A106.4.3
	IPC—12	International Plumbing Code	A104.1.2, A104.1.3
ISO	13256-1: 1998	Water-source Heat Pumps—Testing and Rating for Performance – Part 1: Water-to-air and Brine-to-air Heat Pumps	Table A106.2.2.3
	13256-1: 1998	Water-source Heat Pumps—Testing and Rating for Performance – Part 2: Water-to-air and Brine-to-air Heat Pumps	Table A106.2.2.3
	17025-2005	General Requirements for the Competence of Testing and Calibration Laboratories	A107.5
NFPA	72-10	National Fire Alarm Code	A107.5.2
SMACNA		HVAC Air Duct Leakage Test Manual -1985	A106.2.4

APPENDIX B
RADON MITIGATION

SECTION B101
GENERAL

B101.1 Radon mitigation. Buildings in areas of High and Moderate Radon Potential (Zone 1 and 2), as determined by Figure B101.1 and Table B101.1 shall comply with Sections B201.1 through B201.10.

SECTION B102
DEFINITIONS

DRAIN TILE LOOP. A continuous length of drain tile or perforated pipe extending around all or part of the internal or external perimeter of a basement or crawl space footing.

RADON GAS. A naturally occurring, chemically inert, radioactive gas found in soil that is not detectable by human senses.

SOIL-GAS-RETARDER. A continuous membrane of 6-mil (0.15 mm) polyethylene or other equivalent material used to retard the flow of soil gases into a building.

SUBMEMBRANE DEPRESSURIZATION SYSTEM. A system designed to achieve lower-submembrane air pressure relative to crawl space air pressure by use of a vent drawing air from beneath the soil-gas-retarder membrane.

SUBSLAB DEPRESSURIZATION SYSTEM (Active). A system designed to achieve lower subslab air pressure relative to indoor air pressure by use of a fan-powered vent drawing air from beneath the slab.

SUBSLAB DEPRESSURIZATION SYSTEM (Passive). A system designed to achieve lower subslab air pressure relative to indoor air pressure by use of a vent pipe routed through the conditioned space of a building and connecting the subslab area with outdoor air, thereby relying on the convective flow of air upward in the vent to draw air from beneath the slab.

SECTION B201
MITIGATION PROCEDURES

B201.1 Subfloor preparation. A layer of gas-permeable material shall be placed under all concrete slabs and other floor systems that directly contact the ground and that are within the walls of the occupied spaces of the building, as a prerequisite for passive and active subslab depressurization systems. The gas-permeable layer shall consist of one of the following:

1. A uniform layer of clean aggregate, not less than 4 inches (102 mm) in thickness. The aggregate shall consist of material that will pass through a 2-inch (51 mm) sieve and be retained by a $^{1}/_{2}$-inch (12.7mm) sieve. Size 5, 56 or 6 aggregate shall be used and shall meet the specifications of ASTM C 33. Where compaction is required or practiced, a geotextile fabric or reinforced vapor retarder shall be used beneath the aggregate to prevent fines and soil from being introduced into the aggregate.
2. A uniform layer of sand (native or fill), not less than 4 inches (102 mm) in thickness, overlain by a layer or strips of geotextile drainage matting designed to allow the lateral flow of soil gases.
3. Geotextile drainage matting, or other materials, systems or floor designs with demonstrated capability to permit depressurization across the entire subfloor area.

B201.2 Subslab radon suction pit. A radon suction pit without aggregate shall be installed in the center of each 100,000 square feet (9390 m^2) of floor area that is in contact with the earth and that has no subslab barriers. The suction pit void area shall be not less than 4 square feet (0.371 m^2) and the pit shall be not less than 8 inches (203 mm) in depth. The resulting suction pit void to aggregate interface shall be 7 square feet (0.65 m^2), or 30 times the cross sectional area of a 6-inch (157.4 mm) radon vent pipe. Alternatively, a concrete drainage distribution box or similar structure meeting the 30:1 ratio shall be employed. The suction pit shall be covered with $^{3}/_{4}$-inch-thick (19.05 mm) pressure-treated plywood or an equivalent material prior to pouring the slab. The section of slab covering the suction pit shall be reinforced.

B201.3 Radon vent piping. Radon vent piping shall be not less than 6 inches (157.4 mm) in diameter and constructed of PVC or equivalent gas-tight pipe.

B201.3.1 Subslab suction pit horizontal vent pipe. A section of vent pipe not less than of 5 feet (1.52 m) in length shall be placed in the aggregate and shall enter the suction pit horizontally. One end of the vent pipe shall be placed so as to terminate midway in the suction pit. The vent pipe shall be supported at the boundary of the aggregate-void space so as to maintain its position. The horizontal pipe run shall provide positive condensation drainage to the suction pit with a pitch of not less than $^{1}/_{8}$ inch per foot (13 mm per meter).

B201.3.2 Subslab suction pit vertical vent pipe. A 90-degree (1.57 rad) elbow shall be installed on the end of the vent pipe in the aggregate. A section of vent pipe shall be connected to the elbow and shall pass vertically through and above the slab to a height of not less than 2 feet (610 mm), and shall be covered with a temporary cap. A pipe sleeve or coupling extending through the full depth of the slab shall be used to protect the vent pipe where it passes through the slab, and the slab penetration shall be sealed in accordance with Section B201.5.

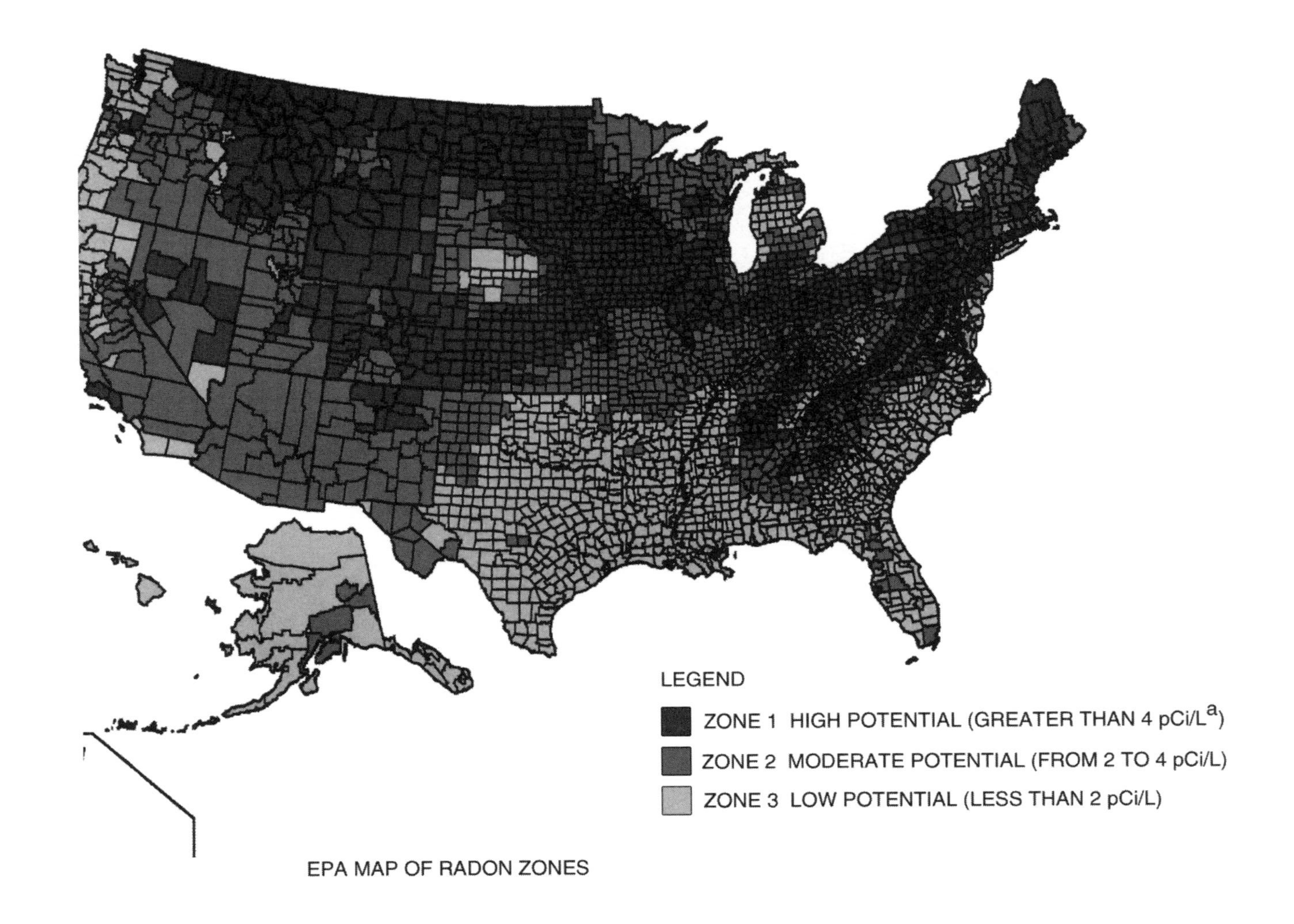

FIGURE B101.1
EPA MAP OF RADON ZONES

a. pCi/L standard for picocuries per liter of radon gas. EPA recommends that all homes that measure 4 pCi/L and greater be mitigated

The United States Environmental Protection Agency and the United States Geological Survey have evaluated the radon potential in the United States and have developed a map of radon zones designed to assist *code officials* in deciding whether radon-resistant features are applicable in new construction.

The map assigns each of the 3,141 counties in the United States to one of three zones based on radon potential. Each zone designation reflects the average short-term radon measurement result that can be expected to be measured in a building without the implementation of radon control methods. The radon zone designation of highest priority is Zone 1 and Zone 2. Table 804.2 lists the Zone 1 counties illustrated on the map. The counties are listed by state and alphabetically within each zone. More detailed information can be obtained from state-specific booklets (EPA-402-R-93-021 through 070) available through the U.S. EPA National Service Center for Environmental Publications (NSCEP), U.S. EPA Regional Offices or State Radon Offices.

TABLE B101.1
EPA RADON ZONE 1 and 2 COUNTIES BY STATE[a]

ALABAMA
Zone 1
Calhoun
Clay
Cleburne
Colbert
Coosa
Franklin
Jackson
Lauderdale
Lawrence
Limestone
Madison
Morgan
Talladega

Zone 2
Autauga
Barbour
Bibb
Blount
Bullock
Cherokee
Chilton
Cullman
Dallas
DeKalb
Elmore
Etowah
Fayette
Greene
Hale
Jefferson
Lamar
Lee
Lowndes
Macon
Marion
Marshall
Montgomery
Perry
Pickens
Randolph
Russell
Shelby
St Clair
Sumter
Tuscaloosa
Walker
Winston

ALASKA
Zone 2
Anchorage Municipality
Dillingham Census Area
Fairbanks North Star
Borough
Kenai Peninsula
Borough
Matanuska-Susitna
Borough
Southeast Fairbanks
Census Area

ARIZONA
Zone 2
Apache
Cochise
Coconino
Gila
Graham
Greenlee
La Paz
Maricopa
Mohave
Navajo
Pima
Pinal
Santa Cruz
Yavapai
Yuma

ARKANSAS
Zone 2
Baxter
Benton
Boone
Carroll
Fulton
Garland
Independence
Izard
Marion
Montgomery
Randolph
Searcy
Sharp
Stone

CALIFORNIA
Zone 1
Santa Barbara
Ventura

Zone 2
Alameda
Alpine
Amador
Calaveras
Contra Costa
El Dorado
Fresno
Inyo
Kern
Los Angeles
Madera
Mariposa
Mono
Monterey
Nevada
Placer
Plumas
Riverside
San Benito
San Bernardino
San Francisco
San Luis Obispo
San Mateo
Santa Clara
Santa Cruz
Sierra
Tulare
Tuolumne
Yuba

COLORADO
Zone 1
Adams
Arapahoe
Baca
Bent
Boulder
Broomfield
Chaffee
Cheyenne
Clear Creek
Crowley
Custer
Delta
Denver
Dolores
Douglas
El Paso
Elbert
Fremont
Garfield
Gilpin
Grand
Gunnison
Huerfano
Jackson
Jefferson
Kiowa
Kit Carson
La Plata
Larimer
Las Animas
Lincoln
Logan
Mesa
Moffat
Montezuma
Montrose
Morgan
Otero
Ouray
Park
Phillips
Pitkin
Prowers
Pueblo
Rio Blanco
San Miguel
Sedgwick
Summit
Teller
Washington
Weld
Yuma

Zone 2
Alamosa
Archuleta
Conejos
Costilla
Eagle
Hinsdale
Lake
Mineral
Rio Grande
Routt
Saguache
San Juan

CONNECTICUT
Zone 1
Fairfield
Middlesex
New Haven
New London

Zone 2
Litchfield
Tolland
Windham

DELAWARE
Zone 2
New Castle

FLORIDA
Zone 2
Alachua
Citrus
Columbia
Hillsborough
Leon
Marion
Miami-Dade
Polk
Union

GEORGIA
Zone 1
Cobb
DeKalb
Fulton
Gwinnett

Zone 2
Banks
Barrow
Bartow
Butts
Carroll
Catoosa
Cherokee
Clarke
Clayton
Coweta
Dawson
Douglas
Elbert
Fannin
Fayette
Floyd
Forsyth
Franklin
Gilmer
Greene

(continued)

TABLE B101.1—continued
EPA RADON ZONE 1 and 2 COUNTIES BY STATE[a]

Habersham
Hall
Haralson
Harris
Hart
Heard
Henry
Jackson
Jasper
Lamar
Lumpkin
Madison
Meriwether
Monroe
Morgan
Newton
Oconee
Oglethorpe
Paulding
Pickens
Pike
Rabun
Richmond
Rockdale
Spalding
Stephens
Talbot
Towns
Troup
Union
Upson
Walker
Walton
White
Whitfield

IDAHO
Zone 1
Benewah
Blaine
Boise
Bonner
Boundary
Butte
Camas
Clark
Clearwater
Custer
Elmore
Fremont
Gooding
Idaho
Kootenai
Latah
Lemhi
Shoshone
Valley

Zone 2
Ada
Bannock
Bear Lake
Bingham
Bonneville
Canyon
Caribou
Cassia
Franklin
Jefferson
Jerome
Lincoln
Madison
Minidoka
Oneida
Owyhee
Payette
Power
Teton
Twin Falls

ILLINOIS
Zone 1
Adams
Boone
Brown
Bureau
Calhoun
Carroll
Cass
Champaign
Coles
De Kalb
De Witt
Douglas
Edgar
Ford
Fulton
Greene
Grundy
Hancock
Henderson
Henry
Iroquois
Jersey
Jo Daviess
Kane
Kendall
Knox
LaSalle
Lee
Livingston
Logan
Macon
Marshall
Mason
McDonough
McLean
Menard
Mercer
Morgan
Moultrie
Ogle
Peoria
Piatt
Pike
Putnam
Rock Island
Sangamon
Schuyler
Scott
Stark
Stephenson
Tazewell
Vermilion
Warren
Whiteside
Winnebago
Woodford

Zone 2
Bond
Christian
Clark
Clay
Clinton
Cook
Crawford
Cumberland
DuPage
Edwards
Effingham
Fayette
Franklin
Gallatin
Hamilton
Hardin
Jackson
Jasper
Jefferson
Johnson
Kankakee
Lake
Lawrence
Macoupin
Madison
Marion
McHenry
Monroe
Montgomery
Perry
Pope
Randolph
Richland
Saline
Shelby
St Clair
Union
Wabash
Washington
Wayne
White
Will
Williamson

INDIANA
Zone 1
Adams
Allen
Bartholomew
Benton
Blackford
Boone
Carroll
Cass
Clark
Clinton
De Kalb
Decatur
Delaware
Elkhart
Fayette
Fountain
Fulton
Grant
Hamilton
Hancock
Harrison
Hendricks
Henry
Howard
Huntington
Jay
Jennings
Johnson
Kosciusko
LaGrange
Lawrence
Madison
Marion
Marshall
Miami
Monroe
Montgomery
Noble
Orange
Putnam
Randolph
Rush
Scott
Shelby
St Joseph
Steuben
Tippecanoe
Tipton
Union
Vermillion
Wabash
Warren
Washington
Wayne
Wells
White
Whitley

Zone 2
Brown
Clay
Crawford
Daviess
Dearborn
Dubois
Floyd
Franklin
Gibson
Greene
Jackson
Jasper
Jefferson
Knox

(continued)

TABLE B101.1—continued
EPA RADON ZONE 1 and 2 COUNTIES BY STATE[a]

Lake
LaPorte
Martin
Morgan
Newton
Ohio
Owen
Parke
Perry
Pike
Porter
Posey
Pulaski
Ripley
Spencer
Starke
Sullivan
Switzerland
Vanderburgh
Vigo
Warrick

IOWA
Zone 1
Adair
Adams
Allamakee
Appanoose
Audubon
Benton
Black Hawk
Boone
Bremer
Buchanan
Buena Vista
Butler
Calhoun
Carroll
Cass
Cedar
Cerro Gordo
Cherokee
Chickasaw
Clarke
Clay
Clayton
Clinton
Crawford
Dallas
Davis
Decatur
Delaware
Des Moines
Dickinson
Dubuque
Emmet
Fayette
Floyd
Franklin
Fremont
Greene
Grundy
Guthrie
Hamilton
Hancock
Hardin
Harrison
Henry
Howard
Humboldt
Ida
Iowa
Jackson
Jasper
Jefferson
Johnson
Jones
Keokuk
Kossuth
Lee
Linn
Louisa
Lucas
Lyon
Madison
Mahaska
Marion
Marshall
Mills
Mitchell
Monona
Monroe
Montgomery
Muscatine
O'Brien
Osceola
Page
Palo Alto
Plymouth
Pocahontas
Polk
Pottawattamie
Poweshiek
Ringgold
Sac
Scott
Shelby
Sioux
Story
Tama
Taylor
Union
Van Buren
Wapello
Warren
Washington
Wayne
Webster
Winnebago
Winneshiek
Woodbury
Worth
Wright

KANSAS
Zone 1
Atchison
Barton
Brown
Cheyenne
Clay
Cloud
Decatur
Dickinson
Douglas
Ellis
Ellsworth
Finney
Ford
Geary
Gove
Graham
Grant
Gray
Greeley
Hamilton
Haskell
Hodgeman
Jackson
Jewell
Johnson
Kearny
Kingman
Kiowa
Lane
Leavenworth
Lincoln
Logan
Marion
Marshall
McPherson
Meade
Mitchell
Nemaha
Ness
Norton
Osborne
Ottawa
Pawnee
Phillips
Pottawatomie
Pratt
Rawlins
Republic
Rice
Riley
Rooks
Rush
Russell
Saline
Scott
Sheridan
Sherman
Smith
Stanton
Thomas
Trego
Wallace
Washington
Wichita
Wyandotte

Zone 2
Allen
Anderson
Barber
Bourbon
Butler
Chase
Chautauqua
Cherokee
Clark
Coffey
Comanche
Cowley
Crawford
Doniphan
Edwards
Elk
Franklin
Greenwood
Harper
Harvey
Jefferson
Labette
Linn
Lyon
Miami
Montgomery
Morris
Morton
Neosho
Osage
Reno
Sedgwick
Seward
Shawnee
Stafford
Stevens
Sumner
Wabaunsee
Wilson
Woodson

KENTUCKY
Zone 1
Adair
Allen
Barren
Bourbon
Boyle
Bullitt
Casey
Clark
Cumberland
Fayette
Franklin
Green
Harrison
Hart
Jefferson
Jessamine
Lincoln
Marion
Mercer
Metcalfe
Monroe
Nelson
Pendleton
Pulaski

(continued)

TABLE B101.1—continued
EPA RADON ZONE 1 and 2 COUNTIES BY STATE[a]

Robertson
Russell
Scott
Taylor
Warren
Woodford

Zone 2
Anderson
Bath
Bell
Boone
Boyd
Bracken
Breathitt
Breckinridge
Butler
Caldwell
Campbell
Carroll
Carter
Christian
Clay
Clinton
Crittenden
Daviess
Edmonson
Elliott
Estill
Fleming
Floyd
Gallatin
Garrard
Grant
Grayson
Greenup
Hancock
Hardin
Harlan
Henderson
Henry
Hopkins
Jackson
Johnson
Kenton
Knott
Knox
Larue
Laurel
Lawrence
Lee
Leslie
Letcher
Lewis
Livingston
Logan
Lyon
Madison
Magoffin
Martin
Mason
McCreary
McLean
Meade
Menifee
Montgomery
Morgan
Muhlenberg
Nicholas
Ohio
Oldham
Owen
Owsley
Perry
Pike
Powell
Rockcastle
Rowan
Shelby
Simpson
Spencer
Todd
Trigg
Trimble
Union
Washington
Wayne
Webster
Whitley
Wolfe

MAINE
Zone 1
Androscoggin
Aroostook
Cumberland
Franklin
Hancock
Kennebec
Lincoln
Oxford
Penobscot
Piscataquis
Somerset
York

Zone 2
Knox
Sagadahoc
Waldo
Washington

MARYLAND
Zone 1
Baltimore
Calvert
Carroll
Frederick
Harford
Howard
Montgomery
Washington

Zone 2
Allegany
Anne Arundel
Baltimore City
Cecil
Charles
Garrett
Prince George's
Somerset

MASSACHUSETTS
Zone 1
Essex
Middlesex
Worcester

Zone 2
Barnstable
Berkshire
Bristol
Dukes
Franklin
Hampden
Hampshire
Nantucket
Norfolk
Plymouth

MICHIGAN
Zone 1
Branch
Calhoun
Cass
Hillsdale
Jackson
Kalamazoo
Lenawee
St Joseph
Washtenaw

Zone 2
Alcona
Alger
Alpena
Antrim
Baraga
Barry
Charlevoix
Clinton
Dickinson
Eaton
Emmet
Genesee
Gogebic
Houghton
Ingham
Ionia
Iron
Kent
Keweenaw
Lapeer
Leelanau
Livingston
Marquette
Menominee
Monroe
Montcalm
Montmorency
Oakland
Otsego
Presque Isle
Sanilac
Shiawassee

MINNESOTA
Zone 1
Becker
Big Stone
Blue Earth
Brown
Carver
Chippewa
Clay
Cottonwood
Dakota
Dodge
Douglas
Faribault Count
Fillmore
Freeborn
Goodhue
Grant
Hennepin
Houston
Hubbard
Jackson
Kanabec
Kandiyohi
Kittson
Lac qui Parle
Le Sueur
Lincoln
Lyon
Mahnomen
Marshall
Martin
McLeod
Meeker
Mower
Murray
Nicollet
Nobles
Norman
Olmsted
Otter Tail
Pennington
Pipestone
Polk
Pope
Ramsey
Red Lake
Redwood
Renville
Rice
Rock
Roseau
Scott
Sherburne
Sibley
Stearns
Steele
Stevens
Swift
Todd
Traverse
Wabasha
Wadena
Waseca

(continued)

TABLE B101.1—continued
EPA RADON ZONE 1 and 2 COUNTIES BY STATE[a]

Washington
Watonwan
Wilkin
Winona
Wright
Yellow Medicine

Zone 2
Aitkin
Anoka
Beltrami
Benton
Carlton
Cass
Chisago
Clearwater
Cook
Crow Wing
Isanti
Itasca
Koochiching
Lake
Lake of the Woods
Mille Lacs
Morrison
Pine
St Louis

MISSISSIPPI
Zone 2
Alcorn
Chickasaw
Clay
Lee
Lowndes
Noxubee
Pontotoc
Rankin
Union
Washington

MISSOURI
Zone 1
Andrew
Atchison
Buchanan
Cass
Clay
Clinton
Holt
Iron
Jackson
Nodaway
Platte

Zone 2
Adair
Audrain
Barry
Barton
Bates
Benton
Bollinger
Boone
Caldwell
Callaway
Camden
Cape Girardeau
Carroll
Carter
Cedar
Chariton
Christian
Clark
Cole
Cooper
Crawford
Dade
Dallas
Daviess
DeKalb
Dent
Douglas
Franklin
Gasconade
Gentry
Greene
Grundy
Harrison
Henry
Hickory
Howard
Howell
Jasper
Jefferson
Johnson
Knox
Laclede
Lafayette
Lawrence
Lewis
Lincoln
Linn
Livingston
Macon
Madison
Maries
Marion
McDonald
Mercer
Miller
Moniteau
Monroe
Montgomery
Morgan
Newton
Oregon
Osage
Ozark
Perry
Pettis
Phelps
Pike
Polk
Pulaski
Putnam
Ralls
Randolph
Ray
Reynolds
Ripley
Saline
Schuyler
Scotland
Shannon
Shelby
St Charles
St Clair
St Francois
St Louis city
St Louis
Ste Genevieve
Stone
Sullivan
Taney
Texas
Vernon
Warren
Washington
Wayne
Webster
Worth
Wright

MONTANA
Zone 1
Beaverhead
Big Horn
Blaine
Broadwater
Carbon
Carter
Cascade
Chouteau
Custer
Daniels
Dawson
Deer Lodge
Fallon
Fergus
Flathead
Gallatin
Garfield
Glacier
Granite
Hill
Jefferson
Judith Basin
Lake
Lewis and Clark
Liberty
Lincoln
Madison
McCone
Meagher
Mineral
Missoula
Park
Phillips
Pondera
Powder River
Powell
Prairie
Ravalli
Richland
Roosevelt
Rosebud
Sanders
Sheridan
Silver Bow
Stillwater
Teton
Toole
Valley
Wibaux

Zone 2
Golden Valley
Musselshell
Petroleum
Sweet Grass
Treasure
Wheatland
Yellowstone

NEBRASKA
Zone 1
Adams
Boone
Boyd
Burt
Butler
Cass
Cedar
Clay
Colfax
Cuming
Dakota
Dixon
Dodge
Douglas
Fillmore
Franklin
Frontier
Furnas
Gage
Gosper
Greeley
Hamilton
Harlan
Hayes
Hitchcock
Jefferson
Johnson
Kearney
Knox
Lancaster
Madison
Nance
Nemaha
Nuckolls
Otoe
Pawnee
Phelps
Pierce
Platte
Polk
Red Willow
Richardson
Saline

(continued)

TABLE B101.1—continued
EPA RADON ZONE 1 and 2 COUNTIES BY STATE[a]

Sarpy
Saunders
Seward
Stanton
Thayer
Thurston
Washington
Wayne
Webster
York

Zone 2
Antelope
Banner
Box Butte
Buffalo
Chase
Cheyenne
Custer
Dawes
Dawson
Deuel
Dundy
Hall
Howard
Keith
Keya Paha
Kimball
Merrick
Morrill
Perkins
Scotts Bluff
Sheridan
Sherman
Sioux
Valley

NEVADA
Zone 1
Carson City
Douglas
Eureka
Lander
Lincoln
Lyon
Mineral
Pershing
White Pine

Zone 2
Churchill
Elko
Esmeralda
Humboldt
Nye
Storey
Washoe

NEW HAMPSHIRE
Zone 1
Carroll

Zone 2
Belknap
Cheshire
Coos
Grafton
Hillsborough
Merrimack
Rockingham
Strafford
Sullivan

NEW JERSEY
Zone 1
Hunterdon
Mercer
Monmouth
Morris
Somerset
Sussex
Warren

Zone 2
Bergen
Burlington
Camden
Cumberland
Essex
Gloucester
Hudson
Middlesex
Passaic
Salem
Union

NEW MEXICO
Zone 1
Bernalillo
Colfax
Mora
Rio Arriba
San Miguel
Santa Fe
Taos

Zone 2
Catron
Chaves
Cibola
Curry
De Baca
Dona Ana
Eddy
Grant
Guadalupe
Harding
Hidalgo
Lea
Lincoln
Los Alamos
Luna
McKinley
Otero
Quay
Roosevelt
San Juan
Sandoval
Sierra
Socorro
Torrance
Union
Valencia

NEW YORK
Zone 1
Albany
Allegany
Broome
Cattaraugus
Cayuga
Chautauqua
Chemung
Chenango
Columbia
Cortland
Delaware
Dutchess
Erie
Genesee
Greene
Livingston
Madison
Onondaga
Ontario
Orange
Otsego
Putnam
Rensselaer
Schoharie
Schuyler
Seneca
Steuben
Sullivan
Tioga
Tompkins
Ulster
Washington
Wyoming
Yates

Zone 2
Clinton
Jefferson
Lewis
Monroe
Montgomery
Niagara
Oneida
Orleans
Oswego
Saratoga
Schenectady
St Lawrence
Wayne

NORTH CAROLINA
Zone 1
Alleghany
Buncombe
Cherokee
Henderson
Mitchell
Rockingham
Transylvania
Watauga

Zone 2
Alexander
Ashe
Avery
Burke
Caldwell
Caswell
Catawba
Clay
Cleveland
Forsyth
Franklin
Gaston
Graham
Haywood
Iredell
Jackson
Lincoln
Macon
Madison
McDowell
Polk
Rutherford
Stokes
Surry
Swain
Vance
Wake
Warren
Wilkes
Yadkin
Yancey

NORTH DAKOTA
Zone 1
Adams
Barnes
Benson
Billings
Bottineau
Bowman
Burke
Burleigh
Cass
Cavalier
Dickey
Divide
Dunn
Eddy
Emmons
Foster
Golden Valley
Grand Forks
Grant
Griggs
Hettinger
Kidder
LaMoure
Logan
McHenry
McIntosh
McKenzie
McLean
Mercer

(continued)

TABLE B101.1—continued
EPA RADON ZONE 1 and 2 COUNTIES BY STATE[a]

Morton
Mountrail
Nelson
Oliver
Pembina
Pierce
Ramsey
Ransom
Renville
Richland
Rolette
Sargent
Sheridan
Sioux
Slope
Stark
Steele
Stutsman
Towner
Traill
Walsh
Ward
Wells
Williams

OHIO
Zone 1
Adams
Allen
Ashland
Auglaize
Belmont
Butler
Carroll
Champaign
Clark
Clinton
Columbiana
Coshocton
Crawford
Darke
Delaware
Fairfield
Fayette
Franklin
Greene
Guernsey
Hamilton
Hancock
Hardin
Harrison
Holmes
Huron
Jefferson
Knox
Licking
Logan
Madison
Marion
Mercer
Miami
Montgomery
Morrow
Muskingum
Perry
Pickaway
Pike
Preble
Richland
Ross
Seneca
Shelby
Stark
Summit
Tuscarawas
Union
Van Wert
Warren
Wayne
Wyandot

Zone 2
Ashtabula
Athens
Brown
Clermont
Cuyahoga
Defiance
Erie
Fulton
Gallia
Geauga
Henry
Highland
Hocking
Jackson
Lake
Lawrence
Lorain
Lucas
Mahoning
Medina
Meigs
Monroe
Morgan
Noble
Ottawa
Paulding
Portage
Putnam
Sandusky
Scioto
Trumbull
Vinton
Washington
Williams
Wood

OKLAHOMA
Zone 2
Adair
Beaver
Cherokee
Cimarron
Delaware
Ellis
Mayes
Sequoyah
Texas

OREGON
Zone 2
Baker
Clatsop
Columbia
Crook
Gilliam
Grant
Harney
Hood River
Jefferson
Klamath
Lake
Malheur
Morrow
Multnomah
Sherman
Umatilla
Union
Wasco
Washington
Wheeler
Yamhill

PENNSYLVANIA
Zone 1
Adams
Allegheny
Armstrong
Beaver
Bedford
Berks
Blair
Bradford
Bucks
Butler
Cameron
Carbon
Centre
Chester
Clarion
Clearfield
Clinton
Columbia
Cumberland
Dauphin
Delaware
Franklin
Fulton
Huntingdon
Indiana
Juniata
Lackawanna
Lancaster
Lebanon
Lehigh
Luzerne
Lycoming
Mifflin
Monroe
Montgomery
Montour
Northampton
Northumberland
Perry
Schuylkill
Snyder
Sullivan
Susquehanna
Tioga
Union
Venango
Westmoreland
Wyoming
York

Zone 2
Cambria
Crawford
Elk
Erie
Fayette
Forest
Greene
Jefferson
Lawrence
McKean
Mercer
Pike
Potter
Somerset
Warren
Washington
Wayne

RHODE ISLAND
Zone 1
Kent
Washington

Zone 2
Newport
Providence

SOUTH CAROLINA
Zone 1
Greenville

Zone 2
Abbeville
Anderson
Cherokee
Laurens
Oconee
Pickens
Spartanburg
York

SOUTH DAKOTA
Zone 1
Aurora
Beadle
Bon Homme
Brookings
Brown
Brule
Buffalo
Campbell
Charles Mix
Clark

(continued)

TABLE B101.1—continued
EPA RADON ZONE 1 and 2 COUNTIES BY STATE[a]

Clay
Codington
Corson
Davison
Day
Deuel
Douglas
Edmunds
Faulk
Grant
Hamlin
Hand
Hanson
Hughes
Hutchinson
Hyde
Jerauld
Kingsbury
Lake
Lincoln
Lyman
Marshall
McCook
McPherson
Miner
Minnehaha
Moody
Perkins
Potter
Roberts
Sanborn
Spink
Stanley
Sully
Turner
Union
Walworth
Yankton

Zone 2
Bennett
Butte
Custer
Dewey
Fall River
Gregory
Haakon
Harding
Jackson
Jones
Lawrence
Meade
Mellette
Pennington
Shannon
Todd
Tripp
Ziebach

TENNESSEE
Zone 1
Anderson
Bedford
Blount
Bradley
Claiborne
Davidson
Giles
Grainger
Greene
Hamblen
Hancock
Hawkins
Hickman
Humphreys
Jackson
Jefferson
Knox
Lawrence
Lewis
Lincoln
Loudon
Macon
Madison
Marshall
McMinn
Meigs
Monroe
Moore
Perry
Roane
Rutherford
Smith
Sullivan
Trousdale
Union
Washington
Wayne
Williamson
Wilson

Zone 2
Benton
Cannon
Carter
Cheatham
Chester
Clay
Cocke
Coffee
Decatur
DeKalb
Dickson
Fentress
Hamilton
Hardin
Henderson
Houston
Johnson
Marion
McNairy
Montgomery
Overton
Pickett
Polk
Putnam
Robertson
Sevier
Stewart
Sumner
Unicoi
Van Buren
Warren
White

TEXAS
Zone 2
Armstrong
Bailey
Brewster
Carson
Castro
Crosby
Culberson
Dallam
Deaf Smith
Donley
Floyd
Garza
Gray
Hale
Hansford
Hartley
Hemphill
Hockley
Hudspeth
Hutchinson
Jeff Davis
Lamb
Lipscomb
Llano
Lubbock
Lynn
Mason
Moore
Ochiltree
Oldham
Parmer
Potter
Presidio
Randall
Reeves
Roberts
Sherman
Swisher
Terrell

UTAH
Zone 1
Carbon
Duchesne
Grand
Piute
Sanpete
Sevier
Uintah

Zone 2
Beaver
Box Elder
Cache
Daggett
Davis
Emery
Garfield
Iron
Juab
Kane
Millard
Morgan
Rich
Salt Lake
San Juan
Summit
Tooele
Utah
Wasatch
Washington
Wayne
Weber

VERMONT
Zone 2
Addison
Bennington
Caledonia
Essex
Franklin
Lamoille
Orange
Orleans
Rutland
Washington
Windham
Windsor

VIRGINIA
Zone 1
Alleghany
Amelia
Appomattox
Augusta
Bath
Bland
Botetourt
Brunswick
Buckingham
Campbell
Chesterfield
Clarke
Craig
Cumberland
Dinwiddie
Fairfax
Fluvanna
Frederick
Giles
Goochland
Henry
Highland
Lee
Louisa
Montgomery
Nottoway
Orange
Page
Patrick
Pittsylvania
Powhatan
Pulaski

(continued)

TABLE B101.1—continued
EPA RADON ZONE 1 and 2 COUNTIES BY STATE[a]

Roanoke
Rockbridge
Rockingham
Russell
Scott
Shenandoah
Smyth
Spotsylvania
Stafford
Tazewell
Warren
Washington
Wythe

Zone 2
Albemarle
Amherst
Arlington
Bedford
Buchanan
Carroll
Charlotte
Culpeper
Dickenson
Fauquier
Floyd
Franklin
Grayson
Greene
Halifax
Loudoun
Lunenburg
Madison
Mecklenburg
Nelson
Prince Edward
Prince William
Rappahannock
Wise

WASHINGTON
Zone 1
Clark
Ferry
Okanogan
Pend Oreille
Skamania
Spokane
Stevens

Zone 2
Adams
Asotin
Benton
Columbia
Douglas
Franklin
Garfield
Grant
Kittitas
Klickitat
Lincoln
Walla Walla
Whitman
Yakima

WEST VIRGINIA
Zone 1
Berkeley
Brooke
Grant
Greenbrier
Hampshire
Hancock
Hardy
Jefferson
Marshall
Mercer
Mineral
Monongalia
Monroe
Morgan
Ohio
Pendleton
Pocahontas
Preston
Summers
Wetzel

Zone 2
Barbour
Braxton
Cabell
Calhoun
Clay
Doddridge
Fayette
Gilmer
Harrison
Jackson
Lewis
Lincoln
Marion
Mason
Nicholas
Pleasants
Putnam
Raleigh
Randolph
Ritchie
Roane
Taylor
Tucker
Tyler
Upshur
Wayne
Webster
Wirt
Wood

WISCONSIN
Zone 1
Buffalo
Crawford
Dane
Dodge
Door
Fond du Lac
Grant
Green
Green Lake
Iowa
Jefferson
Lafayette
Langlade
Marathon
Menominee
Pepin
Pierce
Portage
Richland
Rock
Shawano
St Croix
Vernon
Walworth
Washington
Waukesha
Waupaca
Wood

Zone 2
Adams
Ashland
Barron
Bayfield
Brown
Burnett
Calumet
Chippewa
Clark
Columbia
Douglas
Dunn
Eau Claire
Florence
Forest
Iron
Jackson
Juneau
Kenosha
Kewaunee
La Crosse
Lincoln
Manitowoc
Marinette
Marquette
Milwaukee
Monroe
Oconto
Oneida
Outagamie
Ozaukee
Polk
Price
Racine
Rusk
Sauk
Sawyer
Sheboygan
Taylor
Trempealeau
Vilas
Washburn
Waushara
Winnebago

WYOMING
Zone 1
Albany
Big Horn
Campbell
Carbon
Converse
Crook
Fremont
Goshen
Hot Springs
Johnson
Laramie
Lincoln
Natrona
Niobrara
Park
Sheridan
Sublette
Sweetwater
Teton
Uinta
Washakie

Zone 2
Platte
Weston

B201.4 Soil-gas-retarder. A minimum 6-mil (0.15 mm) [or 3-mil (0.075 mm) cross-laminated] polyethylene or equivalent flexible sheeting material that conforms to ASTM E 1643 shall be placed on top of the gas-permeable layer prior to casting the slab or placing the floor assembly to serve as a soil-gas-retarder by bridging any cracks that develop in the slab or floor assembly and to prevent concrete from entering the void spaces in the aggregate base material. The sheeting shall cover the entire floor area with separate sections of sheeting lapped at least 12 inches (305 mm). The sheeting shall fit closely around any pipe, wire or other penetrations of the material. All punctures or tears in the material shall be sealed or covered with additional sheeting having an overlap of not less than 12 inches (305 mm) on all sides.

B201.5 Entry routes. Potential radon entry routes shall be sealed or closed in accordance with Sections B201.5.1 through B201.5.10.

B201.5.1 Floor openings. Piping and other penetrations through concrete slabs or other floor assemblies shall be filled or sealed with a polyurethane caulk or equivalent sealant that complies with ASTM C 920 Class 25 or greater and is applied in accordance with the manufacturer's recommendations. Prior to sealing, backer rods shall be used to fill gaps greater than $^1/_2$ inch (12.7 mm).

B201.5.2 Concrete joints. Slab joints, control saw joints, isolation joints, construction joints, pour joints, floor and wall intersection joints, and any other joints in concrete slabs or between slabs and foundation walls shall be sealed with a caulk or sealant. Gaps and joints shall be cleared of loose material and filled with a polyurethane caulk or other elastomeric sealant that complies with ASTM C 920 Class 25 or greater and is applied in accordance with the manufacturer's recommendations. Prior to sealing, backer rods shall be used to fill gaps that are greater than $^1/_2$ inch (12.7 mm) in depth.

B201.5.3 Drains. Where floor, condensate and other drains discharge to the soil and not a sewer, such drains shall be provided with a water-seal trap or shall be water trapped or routed through nonperforated pipe to a point above grade.

B201.5.4 Sumps. Sump pits open to soil or serving as the termination point for subslab or exterior drain tile loops shall be covered with a gasketed or otherwise sealed lid. Sumps used as a floor drain shall have a lid equipped with a trapped inlet.

B201.5.5 Foundation walls. Hollow block masonry foundation walls shall be constructed with either a continuous course of solid masonry, one course of masonry grouted solid, or a solid concrete beam at or above finished ground surface to prevent passage of air from the interior of the wall into the living space. Where a brick veneer or other masonry ledge is installed, the course immediately below that ledge shall be sealed. Joints, cracks and other openings around all penetrations of both exterior and interior surfaces of masonry block or wood foundation walls below the ground surface shall be filled with a polyurethane caulk or other equivalent sealant that complies with ASTM C 920 Class 25 or greater and is applied in accordance with the manufacturer's recommendations. Penetrations of concrete walls shall be filled.

B201.5.6 Dampproofing. The exterior surfaces of portions of concrete and masonry block walls below the ground surface shall be dampproofed.

B201.5.7 Air-handling units. Air-handling units in crawl spaces shall be sealed to prevent air from being drawn into the unit.

Exception: Units with gasketed seams or units that are otherwise sealed by the manufacturer to prevent leakage.

B201.5.8 Ducts. Ductwork for supply or return air shall not be located in crawl spaces or beneath a slab in areas with high or moderate radon potential. Where ductwork passes through or beneath a slab, it shall be of seamless material or sealed water tight. Joints in such ductwork shall be sealed water tight.

B201.5.9 Crawl space floors. Openings around all penetrations through floors above crawl spaces shall be caulked or otherwise filled to prevent air leakage.

B201.5.10 Crawl space access. Access doors and other openings or penetrations between basements and adjoining crawl spaces shall be closed, gasketed or otherwise filled to prevent air leakage.

B201.6 Passive submembrane depressurization system. In buildings with crawl space foundations, the following components of a passive submembrane depressurization system shall be installed during construction.

Exception: Buildings in which an *approved* mechanical crawl space ventilation system or other equivalent system is installed.

B201.6.1 Ventilation. Crawl spaces shall be provided with vents to the exterior of the building.

B201.6.2 Soil-gas-retarder. The soil in crawl spaces shall be covered with a continuous layer of minimum 6-mil (0.15 mm) polyethylene soil-gas-retarder that conforms to ASTM E 1643. The ground cover shall be lapped a minimum of 12 inches (305 mm) at joints and shall extend to all foundation walls enclosing the crawl space area.

B201.6.3 Vent pipe. A plumbing tee or other *approved* connection shall be inserted horizontally beneath the sheeting and connected to a 3- or 4-inch-diameter (76 mm or 102 mm) fitting with a vertical vent pipe installed through the sheeting. The vent pipe shall be extended up through the building floors, terminate at least 12 inches (305 mm) above the roof in a location at least 10 feet (3048 mm) away from any window or other opening into the conditioned spaces of the building that is less than 2 feet (610 mm) below the exhaust point, and 10 feet (3048 mm) from any window or other opening in adjoining or adjacent buildings.

B201.7 Passive subslab depressurization system. In basement or slab-on-grade buildings, the following components of a passive subslab depressurization system shall be installed during construction.

B201.7.1 Vent pipe. A minimum 3-inch-diameter (76 mm) ABS, PVC or equivalent gas-tight pipe shall be embedded vertically into the subslab aggregate or other permeable material before the slab is cast. A "T" fitting or equivalent method shall be used to ensure that the pipe opening remains within the subslab permeable material. Alternatively, the 3-inch (76 mm) pipe shall be inserted directly into an interior perimeter drain tile loop or through a sealed sump cover where the sump is exposed to the subslab aggregate or connected to it through a drainage system.

The pipe shall be extended up through the building floors, terminate at least 12 inches (305 mm) above the surface of the roof in a location at least 10 feet (3048 mm) away from any window or other opening into the conditioned spaces of the building that is less than 2 feet (610 mm) below the exhaust point, and 10 feet (3048 mm) from any window or other opening in adjoining or adjacent buildings.

B201.7.2 Multiple vent pipes. In buildings where interior footings or other barriers separate the subslab aggregate or other gas-permeable material, each area shall be fitted with an individual vent pipe. Vent pipes shall connect to a single vent that terminates above the roof or each individual vent pipe shall terminate separately above the roof.

B201.8 Vent pipe drainage. All components of the radon vent pipe system shall be installed to provide positive drainage to a suction pit beneath the slab, or to the ground beneath the slab or soil-gas-retarder. The slope of vent piping shall be not less than $^1/_8$ unit vertical in 12 units horizontal.

B201.9 Vent pipe accessibility. Radon vent pipes shall be accessible for future fan installation through an attic or other area outside the habitable space.

Exception: The radon vent pipe need not be accessible in an attic space where an *approved* roof-top electrical supply is provided for future use.

B201.10 Vent pipe identification. All exposed and visible interior radon vent pipes shall be identified with at least one marking on each floor and in accessible attics. The marking shall read: "Radon Reduction System."

B201.11 Combination foundations. Combination basement/crawl space or slab-on-grade/crawl space foundations shall have separate radon vent pipes installed in each type of foundation area. Each radon vent pipe shall terminate above the roof or shall be connected to a single vent that terminates above the roof.

B201.12 Power source. To provide for future installation of an active submembrane or subslab depressurization system, an electrical circuit terminated in an *approved* box shall be installed during construction in the attic or other anticipated location of vent pipe fans. An electrical supply shall also be accessible in anticipated locations of system failure alarms.

SECTION B202 REFERENCED STANDARDS

ASTM		
C 920-11	Standard Specification for Elastomeric Joint Sealants	B201.5.1 B201.5.2 B201.5.5
E 1643	Standard Practice for Selection, Design, Installation, and Installation, and Inspection of Water Vapor Retarder Used in Contact with Earth or Granules Fill Under Concrete Slabs	B201.6.2

APPENDIX C
OPTIONAL ORDINANCE

The provisions contained in this appendix are not mandatory unless specifically referenced in the adopting ordinance.

The *International Codes* are designed and promulgated to be adopted by reference by ordinance. Jurisdictions wishing to adopt the *International Green Construction Code*™ (IgCC™) as enforceable regulations of sustainable construction practice governing structures and premises should ensure that certain factual and fiscal information is included in the adopting ordinance at the time adoption is being considered by the appropriate governmental body.

The following sample adoption ordinance addresses several key elements of a code adoption ordinance, including the information required for insertion into the code text and an evidentiary-based adoption structure contain bonding requirements tied to the issuance of building permits, certificates of occupancy and the compliance verification process, a concept already familiar to jurisdictions' master development plans for larger-scale, Planned-Unit Developments (PUD's).

Most importantly, this Optional Ordinance intends to open the dialogue among stakeholders, and give jurisdictions a place to start an fiscal and evidentiary-based adoption structure utilizing performance bonding requirements tied to the compliance verification process. The bonding requirement is designed to ensure that the project complies with the IgCC. The bond is held by the jurisdiction. Bond amounts are set at a percentage of total cost of the building, based on local economic and geo-centric requirements overseen by jurisdictional authorities, and tied to square footage.

SAMPLE ORDINANCE FOR ADOPTION OF THE *INTERNATIONAL GREEN CONSTRUCTION CODE* ORDINANCE NO.________

An ordinance of the **[JURISDICTION]** adopting the *International Green Construction Code*™, regulating and governing the impact of buildings and structures on the environment in the **[JURISDICTION]**; providing for the issuance of permits and collection of fees therefore; repealing Ordinance No. ______ of the **[JURISDICTION]** and all other ordinances and parts of the ordinances in conflict therewith.

The **[GOVERNING BODY]** of the **[JURISDICTION]** does ordain as follows:

Section 1. That a certain document, three (3) copies of which are on file in the office of the **[TITLE OF JURISDICTION'S KEEPER OF RECORDS]** of **[NAME OF JURISDICTION]**, being marked and designated as the *International Green Construction Code*, 2012 edition, including Appendix Chapters **[FILL IN THE APPENDIX CHAPTERS BEING ADOPTED]**, as published by the International Code Council, be and is hereby adopted as the Green Construction Code of the **[JURISDICTION]**, in the State of **[STATE NAME]** for regulating and governing the impact of buildings and structures on the environment as herein provided; providing for the issuance of permits and collection of fees thereof; and each and all of the regulations, provisions, penalties, conditions and terms of said Green Construction Code on file in the office of the **[JURISDICTION]** are hereby referred to, adopted, and made a part hereof, as if fully set out in this ordinance, with the additions, insertions, deletions and changes, if any, prescribed in Section 2 of this ordinance.

Section 2. The following sections are hereby revised:

Section 101.1. Insert: **[NAME OF JURISDICTION]**

Table 302.1. Insert: **[JURISDICTIONAL REQUIREMENTS]**.

Section 1007.3.3.1. Insert: **[AGENCY RESPONSIBLE]** where Section 1007.3 is selected in Table 302.1.

Section 1007.3.3.2. Insert: **[AGENCY RESPONSIBLE]** where Section 1007.3 is selected in Table 302.1.

Section 1007.3.3.3. Insert: **[AGENCY RESPONSIBLE]** where Section 1007.3 is selected in Table 302.1.

Section 3. That Ordinance No. ______ of **[JURISDICTION]** entitled **[FILL IN HERE THE COMPLETE TITLE OF THE ORDINANCE OR ORDINANCES IN EFFECT AT THE PRESENT TIME SO THAT THEY WILL BE REPEALED BY DEFINITE MENTION]** and all other ordinances or parts of ordinances in conflict herewith are hereby repealed.

Section 4. That if any section, subsection, sentence, clause or phrase of this ordinance is, for any reason, held to be unconstitutional, such decision shall not affect the validity of the remaining portions of this ordinance. The **[GOVERNING BODY]** hereby declares that it would have passed this ordinance, and each section, subsection, clause or phrase thereof, irrespective of the fact that any one or more sections, subsections, sentences, clauses and phrases be declared unconstitutional.

Section 5. That nothing in this ordinance or in the Green Construction Code hereby adopted shall be construed to affect any suit or proceeding impending in any court, or any rights acquired, or liability incurred, or any cause or causes of action acquired or existing, under any act or ordinance hereby repealed as cited in Section 3 of this ordinance; nor shall any just or legal right or remedy of any character be lost, impaired or affected by this ordinance.

Section 6. That the **[GOVERNING BODY]** hereby directs and causes for all privately owned nonresidential projects of at least **[INSERT]** square feet, a performance bond, irrevocable letter of credit from a financial institution authorized to do business in the jurisdiction, or evidence of cash deposited in an escrow account in a financial institution in the jurisdiction, to be provided to the jurisdiction, with the bond, LOC or escrow, "due and payable prior to receipt of certificate of occupancy."

(a) A commercial applicant who applies for an incentive described in Section 7 shall provide a performance bond which shall be due and payable upon approval of the first building construction permit application.

(b) On or before **[EFFECTIVE DATE]**, all applicants for construction governed by Section 1 shall provide a performance bond, which shall be due and payable prior to issuance of a certificate of occupancy.

(c) For the purpose of compliance with subsections (a) and (b) of this section, in lieu of the bond required by this section, the **[GOVERNING BODY]** may accept an irrevocable letter of credit from a financial institution authorized to do business in the **[JURISDICTION]** or evidence of cash deposited in an escrow account in a financial institution in the **[JURISDICTION]** in the name of the licensee and the **[JURISDICTION]**. The letter of credit or escrow account shall be in the amounts required by subsection (d) of this section.

(d) The amount of the required performance bond under subsection (a) of this section shall be 1 percent of the incentive provided.

(e) The amount of the required performance bond under subsection (b) of this section shall be:

(1) For a project not exceeding 150,000 square feet of gross floor area, 2 percent of the total cost of the building;

(2) For a project from 150,001 to 250,000 square feet of gross floor area, 3 percent of the total cost of the building; and.

(3) For a project exceeding 250,000 square feet building of gross floor area, 4 percent of the total cost of the building.

(f) The maximum amount of a performance bond shall be $3 million.

(g) All or part of the performance bond shall be forfeited to the **[JURISDICTION]** and deposited in a Green Building Fund if the building fails to meet the verification requirements described in sub-parts (1) and (2) below.

(1) Publicly-owned, private leasing of public property, publicly financed buildings, and tenant improvements.

(2) Privately-owned buildings. Any new construction or substantial improvement of a nonresidential privately-owned project with **[INSERT]** square feet of gross floor area or more shall:

a. On or before **[EFFECTIVE DATE]**, submit to the Department of Buildings, as part of any building construction permit application, a green building checklist documenting the green building elements to be pursued in the building construction permit.

b. Within 365 days (1 year) of the receipt of a certificate of occupancy, be verified "in compliance" with the requirements of this ordinance through:

i. An agency of the **[JURISDICTION]** government; or

ii. Third-party entities which meet criteria to be established by the **[GOVERNING BODY]** by rulemaking within 180 days of the effective date of this ordinance;

as having fulfilled or exceeded the requirements of the *International Green Construction Code*.

(h) The **[JURISDICTION]** shall draw down on the bond funds if the required green building verification is not provided within 730 days (2 years) after issuing the first certificate of occupancy.

(i) The **[GOVERNING BODY]** shall promulgate rules to establish additional requirements for the drawing down or return of performance bonds.

Section 7. Incentives.

(a) Within 180 days of the effective date of this ordinance, the **[GOVERNING BODY]** shall establish an incentive program to promote early adoption of green building practices by applicants for building construction permits for buildings regulated by this code. The incentive program shall be funded by funds deposited in the Green Building Fund, subject to the availability of funds. As part of the incentive program, the **[GOVERNING BODY]** shall establish a **[PROPERTY TAX INCENTIVE OR**

INCENTIVES PROGRAM] for Qualifying Green Building Properties, and **[MAY PROVIDE GRANTS]** to help defray costs associated with the early adoption of the green building practices of the *International Green Construction Code*.

Section. 8. Green Building Fund.

(a) There is established a fund designated as the Green Building Fund, which shall be ENROLLED ORIGINAL Codification **[JURISDICTION]** of Columbia Official Code, 2001 Edition 8 West Group Publisher, 1-800-328-9378, separate from the General Fund of the **[JURISDICTION]** of Columbia. All additional monies obtained pursuant to sections 6 and 9, and all interest earned on those funds, shall be deposited into the Fund without regard to fiscal year limitation pursuant to an act of Congress, and used solely to pay the costs of operating and maintaining the Fund and for the purposes stated in subsection (c) of this section. All funds, interest, and other amounts deposited into the Fund shall not be transferred or revert to the General Fund of the **[JURISDICTION]** of Columbia at the end of any fiscal year or at any other time, but shall continually be available for the uses and purposes set forth in this section, subject to authorization by Congress in an appropriations act.

(b) The **[GOVERNING BODY]** shall administer the monies deposited in the Fund.

(c) The Fund shall be used as follows:

(1) Staffing and operating costs to provide technical assistance, plan review, and inspections and monitoring of green buildings;

(2) Education, training and outreach to the public and private sectors on green building practices; and

(3) Incentive funding for private buildings as provided for in Section 7.

Section. 9. Green building fee.

(a) A green building fee is established to fund the implementation this ordinance and the Green Building Fund.

(b) Upon the effective date of this ordinance, the green building fee shall be established by increasing the building construction permit fees in effect at the time in accordance with the following schedule of additional fees:

(1) New construction – an additional $0.0020 per square foot.

(2) Alterations and repairs exceeding $1,000 but not exceeding $1 million – an additional 0.13 percent of construction value; and

(3) Alterations and repairs exceeding $1 million - an additional 0.065 percent of construction value.

Section. 10. Establishment of a **[JURISDICTION]** Green Building Advisory Council.

(a) The Department of the Environment shall provide the central coordination and technical assistance to **[JURISDICTION]** agencies and instrumentalities in the implementation of the provisions of this ordinance.

(b) Within 90 days after the effective date of this ordinance, the **[GOVERNING BODY]** shall establish a Green Building Advisory Council to monitor the **[JURISDICTION]**'s compliance with the requirements of this ordinance and to make policy recommendations designed to continually improve and update the ordinance.

(c) The **[JURISDICTION]** GBAC shall consist of the following nine (9) members: (1), (2), (3), (4), (5), (6), (7), (8), and (9).

(1) Members of the GBAC who are not ex officio members shall have expertise in building construction, development, engineering, natural resources conservation, energy conservation, green building practices, environmental protection, environmental law, or other similar green building expertise.

(2) The Chairperson of the GBAC shall be the Director of the Department of the Environment.

(3) All members of the GBAC shall either work in, or be residents of the **[JURISDICTION]**, and shall serve without compensation.

(4) The members shall serve a 2-year term.

(5) A member appointed to fill a vacancy or after a term has begun, shall serve only for the remainder of the term or until a successor is appointed.

(6) The GBAC shall advise the **[GOVERNING BODY]** on:

i. The development, adoption, and revisions of this ordinance, including suggestions for additional incentives to promote green building practices;

ii. The evaluation of the effectiveness of the **[JURISDICTION]**'s green building policies and their impact on the **[JURISDICTION]**'s environmental health, including the relation of the development of the **[JURISDICTION]**'s green building policies to the specific environmental challenges facing the **[JURISDICTION]**;

iii. The green building practices to be included in the triennial revisions of the Construction Codes; and

iv. The promotion of green building education, including educating relevant **[JURISDICTION]** employees, the building community, and the public regarding the benefits and techniques of high-performance building standards.

v. The GBAC shall meet at least six (6) times each year.

(7) The GBAC shall issue an annual report of its recommendations to the **[GOVERNING BODY]**. The report shall include recommended updates of green building standards, building systems monitoring and data compiled from **[JURISDICTION]**-owned or **[JURISDICTION]** instrumentality-owned and operated buildings, and an analysis of the building projects exempted by the **[GOVERNING BODY]** under section 11. The report shall be distributed to all members of the Council and the **[GOVERNING BODY]** and made available to the general public within 30 days after its issuance.

Section. 11. Exemptions and extensions. The **[GOVERNING BODY]** may, in unusual circumstances and only upon a showing of good cause, grant an exemption from any of the requirements of this ordinance based on:

(a) Substantial evidence of a practical infeasibility or hardship of meeting a required green building standard;

(b) A determination that the public interest would not be served by complying with such requirements; or

(c) Other compelling circumstances as determined by the **[GOVERNING BODY]** by rulemaking.

(1) The burden shall be on the applicant to show circumstances to establish hardship or infeasibility under this section.

(2) If the **[GOVERNING BODY]** determines that the required verification requirement is not practicable for a project, the **[GOVERNING BODY]** shall determine if another green building standard is practicable before exempting the project from all green building requirements.

(3) The **[GOVERNING BODY]** shall promulgate rules to establish requirements for the exemption process within 180 days of the effective date of this ordinance.

(d) Notwithstanding any other provision of this ordinance, construction encompassed by building construction permits applied for within 180 days (6 months) of the effective date of this ordinance shall be exempt from the verification requirements of this ordinance.

(e) Notwithstanding any other provision of this ordinance, the **[GOVERNING BODY]**, upon a finding of reasonable grounds, may extend the period for green building verifications required in Section 6(g) sub-parts (1) and (2), for up to three (3) successive 120-day (4-month) periods.

Section 12. Rulemaking. Within 180 days of the effective date of this ordinance, the **[GOVERNING BODY]** shall promulgate rules to implement this ordinance. The proposed rules shall be submitted to the **[GOVERNING BODY]** for a 45-day period of review, excluding Saturdays, Sundays, legal holidays, and days of **[GOVERNING BODY]** recess. If the **[GOVERNING BODY]** does not approve or disapprove the proposed rules, in whole or in part, by resolution within this 45-day review period, the proposed rules shall be deemed approved.

Section13. That the **[JURISDICTION'S KEEPER OF RECORDS]** is hereby ordered and directed to cause this ordinance to be published. (An additional provision may be required to direct the number of times the ordinance is to be published and to specify that it is to be in a newspaper in general circulation. Posting may also be required.)

Section 14. That this ordinance and the rules, regulations, provisions, requirements, orders and matters established and adopted hereby shall take effect and be in full force and effect **[TIME PERIOD]** from and after the date of its final passage and adoption.

APPENDIX D
ENFORCEMENT PROCEDURES

The provisions contained in this appendix are not mandatory unless specifically referenced in the adopting ordinance.

SECTION D101 GENERAL

D101.1 Scope. The provisions of this appendix shall supplement the provisions of Chapter 1 and provide procedures to enforce continued compliance of buildings, structures and building sites constructed and protected under the provisions of this code.

D101.2 Intent. This appendix shall be construed to secure its expressed intent, which is to ensure public health, safety and welfare and protection of the environment insofar as they are affected by the continued occupancy and maintenance of buildings and building sites. Existing buildings, structures and building site improvements that do not comply with these provisions shall be altered or repaired to restore compliance with this code.

SECTION D102 APPLICABILITY

D102.1 General. Equipment, systems, devices, safeguards and protections required by this code or a previous code under which the buildings, structures and building site was constructed, altered or repaired; or under which portions of the building site were protected; shall be maintained.

D102.2 Owner responsibility. Except as otherwise specified in this code, the *owner* or the *owner's* designated agent shall be responsible for the maintenance of buildings, structures and building site. No *owner*, *operator*, or occupant shall cause any service, facility, equipment or utility that is required under this code to be removed or shut off from or discontinued.

D102.3 Existing remedies. The provisions of this chapter shall not be construed to abolish or impair existing remedies of the jurisdiction or its officers or agencies relating to the removal or demolition of any structure or building site improvement that is dangerous, unsafe or causing irreparable harm to environmental systems.

SECTION D103 DEFINITIONS

D103.1 Definitions. The following words and terms shall, for the purposes of this appendix, have the meanings shown herein. Refer to Chapter 2 of this code for general definitions.

OPERATOR. Any person who has charge, care or control of a building, structure and building site that is let or offered for occupancy.

OWNER. Any person, agent, operator, firm or corporation having a legal or equitable interest in a building site; or recorded in the official records of the state, county or municipality as holding title to the building site; or otherwise having control of the building site, including the guardian of an estate of any person, and the executor or administrator of the estate of such person if ordered to take possession of real property by a court.

STRICT LIABILITY OFFENSE. An offense in which the prosecution in a legal proceeding is not required to prove criminal intent as a part of its case. It is enough to prove that the defendant either did an act that is prohibited, or failed to do an act that the defendant was legally required to do.

SECTION D104 DUTIES AND POWERS OF THE CODE OFFICIAL

D104.1 General. The *code official* is hereby authorized and directed to enforce the provisions of this appendix.

D104.2 Inspections. The *code official* is authorized to make all inspections necessary for administration of this appendix, to ensure compliance with maintenance requirements of this code and to resolve identified violations.

D104.3 Right of entry. Where it is necessary to make an inspection to enforce the provisions of this code, or whenever the *code official* has reasonable cause to believe there exists in a building or structure or on a building site a condition in violation of the code, the *code official* is authorized to enter the building site, and if needed to enter the building or structure at reasonable times to inspect or perform the duties imposed by this code. Where the building site or building is occupied, the *code official* shall present identification credentials to the occupant and request entry. If the building site or building is unoccupied, the *code official* shall first make a reasonable effort to locate the owner or other person having charge or control of the building site or building and request entry. If entry is refused, the *code official* shall have recourse to the remedies provided by law to secure entry.

D104.4 Identification. The *code official* shall carry proper identification credentials when inspecting a building or building site in the performance of duties under this code.

D104.5 Notices and orders. The *code official* shall issue all necessary notices or orders to ensure compliance with this code.

SECTION D105 VIOLATIONS

D105.1 Unlawful acts. It shall be unlawful for an owner or an owner's designated agent to be in conflict with, or violation of, any of the provisions of this code.

D105.2 Notice of violation. The *code official* shall serve notice of the violation or issue an order in accordance with Section D106.

D105.3 Prosecution of violation. Any person failing to comply with a notice of violation or order served in accordance with Section D106 shall be deemed guilty of a misdemeanor or civil infraction as determined by the jurisdiction, and the violation shall be deemed a *strict liability offense*. If the notice of violation or order is not complied with, the *code official* shall institute the appropriate preceding at law or in equity to restrain, correct or abate the violation. The expenses incurred by the jurisdiction during action taken by the jurisdiction on the building site or in the building shall be charged against the real estate of the building site and shall be a lien upon such real estate.

D105.4 Violation penalties. Any person who violates the provisions of this code, or fails to comply with the provisions of this code, shall be prosecuted within the limits provided by the laws of the state and jurisdiction. Each day that a violation continues after notice has been served in accordance with Section D106 shall be deemed a separate violation and offense.

D105.5 Abatement of violation. The imposition of penalties under the provisions of this code shall not preclude the legal officer of the jurisdiction from instituting appropriate action to restrain, correct or abate a violation, or to prevent illegal occupancy of a building, structure or building site, or to stop an illegal act, conduct, business or utilization of a building, structure or building site.

SECTION D106 NOTICES AND ORDERS

D106.1 Notice of violation. Whenever the *code official* determines that there has been a violation of this code or has grounds to believe that a violation has occurred, notice shall be given in the form and manner prescribed in Sections D106.2 and D106.3.

D106.2 Form of notice. A notice of violation prescribed in Section D106.1 shall be in accordance with the following:

1. Be in writing;
2. Include a real estate description of the building site sufficient for identification;
3. Include a statement of the violation or violations and why the notice is being issued;
4. Include a correction order allowing a reasonable time to make repairs and improvements required to bring the building, structure and building site into compliance with the provisions of this code;
5. Inform the property *owner* and those receiving the notice and order of the right to appeal; and
6. Include a statement of the right of the jurisdiction to file a lien in accordance with Section D105.3.

D106.3 Service. A notice of violation and order to comply shall be served in accordance with Sections D106.3.1 and D106.3.2.

D106.3.1 Recipient of notice. The notice of violation and order to comply shall be served on the person responsible for the violation of the code. When the person responsible for the violation is someone other than the *owner* of the building and building site, a copy of the notice shall also be served on the property *owner*.

D106.3.2 Method of service. Such notice and order to comply shall be deemed to be properly served if a copy is:

1. Delivered personally; or
2. Sent by certified or first-class mail addressed to the last known address.
3. If a notice served by mail is returned showing that the letter was not delivered, a copy of the notice and order shall be posted in a conspicuous place in or about the building, structure or building site affected by the notice.

D106.4 Unauthorized tampering. Notices, orders, signs, tags or seals posted or affixed by the *code official* shall not be mutilated, destroyed or tampered with, or removed without authorization from the *code official*.

D106.5 Penalties. Penalties for noncompliance with notices and orders shall be as set forth in Section D105.4.

D106.6 Transfer of ownership. It shall be unlawful of the *owner* of any building, structure or building site who has received a compliance order or upon whom a notice of violation has been served, to sell, transfer, mortgage, lease or otherwise dispose of the building, structure or building site to another until the provisions of the compliance order or notice of violation have been complied with, or until such *owner* shall first furnish the grantee, transferee, mortgagee or lessee a true copy of any compliance order or notice of violation issued by the *code official* and shall furnish to the *code official* a signed and notarized statement from the grantee, transferee, mortgagee or lessee, acknowledging the receipt of such compliance order or notice of violation and fully accepting the responsibility without condition for making corrections or repairs required by such compliance order or notice of violation.

SECTION D107 EMERGENCY MEASURES AND ABATEMENT

D107.1 Imminent hazard. When, in the opinion of the *code official*, there is an imminent hazard to the building site or to surrounding public and private property resulting from the failure of a building or building site system, including but not limited to: stormwater management systems; erosion control measures; gray water or rainwater collection systems; or dry vegetation used for vegetative roofs or hardscape shading; which endangers life or which will cause irreparable harm to environmental systems on, or adjacent to, the building site, the *code official* is hereby authorized and empowered to order immediate repair of these systems and measures to restore proper operation.

D107.2 Temporary safeguards. Notwithstanding other provisions of this code, whenever, in the opinion of the *code official*, there is an imminent hazard due to the failures of systems and measures, the *code official* shall order the necessary work done, whether or not the legal procedures specified in this chapter has been instituted; and shall cause such other action to be taken as the *code official* deems necessary to resolve the hazard.

D107.3 Closing streets. When necessary for public safety, the *code official* shall temporarily close or order the authority having jurisdiction to close sidewalks, streets, public ways and bicycle pathways, adjacent to the hazardous location.

D107.4 Emergency repairs. For the purposes of this section, the *code official* shall employ the necessary labor and materials to perform the required work as expeditiously as possible.

D107.5 Costs of emergency repairs. Costs incurred in the performance of the emergency work shall be paid by the jurisdiction. The legal counsel of the jurisdiction shall institute appropriate action against the *owner* of the building site for the recovery of the costs.

D107.6 Hearing. Any person ordered to take emergency measures shall comply with such order forthwith. Any person affected thereafter, upon application to the board of violation appeals shall be afforded a hearing as described in this code.

SECTION D108 MEANS OF APPEAL

D108.1 General. In order to hear and decide appeals of notices of violation and orders of compliance issued by the *code official* pursuant to this chapter, there shall be and is hereby created a board of violation appeals. Where the board of appeals established under Section 108 is in compliance with the provisions of Sections D108.1 through D108.6, the board of appeals shall be permitted to serve as the board of violation appeals.

D108.2 Board of violation appeals. The board of violation appeals shall be appointed by the applicable governing authority and shall hold office at its pleasure.

D108.2.1 Membership of the board. The board of violation appeals shall consist of a minimum of three members who are qualified by experience and training to pass on matters regulated by this code. Members shall be appointed to serve staggered and overlapping terms. Members of the board shall not be employees of the jurisdiction. The *code official* shall be an ex-officio member but shall have no vote on any matter before the board.

D108.2.2 Alternate members. The governing authority shall appoint two or more alternate members who shall be called by the chair of the board to hear appeals during the absence or disqualification of a member. Alternate members shall possess the qualifications required for board membership.

D108.2.3 Board chair. The board shall annually select one of its members to serve as chair.

D108.2.4 Disqualification of member. A member shall not hear an appeal in which that member has a personal, professional or financial interest.

D108.2.5 Secretary. The *code official* shall designate a qualified person to serve as secretary to the board. The secretary shall file a detailed record of all proceedings in the office of the *code official*.

D108.3 Application for appeal. Any person receiving a notice of violation issued by the *code official* pursuant to this chapter, shall have the right to appeal to the board of violation appeals. The application for appeal shall be in writing and filed within 20 days after the day the notice of violation was served. An application for appeal shall be based on a claim that the requirements of this code are adequately satisfied.

D108.4 Stays of enforcement. Other than notices of imminent hazard, appeals of notices and orders shall stay the enforcement of the notice and order until the appeal is heard by the board of violation appeals.

D108.5 Hearing. Hearings of appeals shall be in accordance with Sections D108.5.1 through D108.5.5.

D108.5.1 Notice. The board shall meet upon notice from the chair, within 20 days of the filing of an appeal, or at stated periodic meetings. Notice of the board meeting shall be published in the newspaper of record for the jurisdiction. Written notice shall be provided to the *owner* of the building site subject to the notice of violation as well as any person cited in the notice of violation.

D108.5.2 Open hearing. All hearings before the board shall be open to the public. The appellant, the appellant's representative, the *code official* and any person whose interests are affected shall be given an opportunity to be heard.

D108.5.3 Quorum. A quorum shall consist of not less than two-thirds of the board membership. A quorum shall be present in order for a hearing to proceed.

D108.5.4 Hearing procedures. The board shall adopt and make available to the public through the secretary procedures under which a hearing will be conducted. The procedures shall not require compliance under strict rules of evidence, but shall mandate that only relevant information be received.

D108.5.5 Postponement of hearing. When the full board is not present to hear an appeal, either the appellant or the appellant's representative shall have the right to request postponement of the hearing.

D108.6 Board decision. The board of violations appeal shall either uphold, modify or reverse the decision of the *code official*. The board shall modify or reverse the decision of the *code official* only by a concurring vote of a majority of the total number of appointed board members. The decision of the board to uphold, modify or reverse the decision of the *code official* shall be in writing and shall direct actions appropriate to implement the decision.

D108.6.1 Records and copies. The decision and directive of the board shall be recorded. Copies shall be furnished to the appellant, the *owner* of the building site and to the *code official*.

D108.6.2 Implementation. The *code official* shall take immediate action in accordance with the decision of the board.

D108.7 Court review. Any person, whether or not a previous party of the appeal, shall have the right to apply to the appropriate court for a writ of certiorari to correct errors of law. Application for review shall be made in the manner and time required by law.

INDEX

D

E

F

G

H

I

J

K

L

M

N

O

P

R

T

U

V

W

Z

ANSI/ASHRAE/USGBC/IES Standard 189.1-2011
(Supersedes ANSI/ASHRAE/USGBC/IES Standard 189.1-2009)

Standard for the Design of High-Performance Green Buildings

Except Low-Rise Residential Buildings

A Compliance Option of the International Green Construction Code™

See Appendix I for approval dates by the ASHRAE Standards Committee, the ASHRAE Board of Directors, the U.S. Green Building Council, the Illuminating Engineering Society of North America, and the American National Standards Institute.

This standard is under continuous maintenance by a Standing Standard Project Committee (SSPC) for which the Standards Committee has established a documented program for regular publication of addenda or revisions, including procedures for timely, documented, consensus action on requests for change to any part of the standard. The change submittal form, instructions, and deadlines may be obtained in electronic form from the ASHRAE Web site (www.ashrae.org), or in paper form from the ASHRAE Manager of Standards.

The latest edition of an ASHRAE Standard may be purchased on the ASHRAE Web site (www.ashrae.org) or from ASHRAE Customer Service, 1791 Tullie Circle, NE, Atlanta, GA 30329-2305, telephone: 404-636-8400 (worldwide), or toll free 1-800-527-4723 (for orders in the United States and Canada), or e-mail: orders@ashrae.org. For reprint permission, go to www.ashrae.org/permissions.

 ISSN 1041-2336

ASHRAE Standing Standard Project Committee 189.1
Cognizant TC: TC 2.8, Building Environmental Impacts and Sustainability
SPLS Liaison: Rita M. Harrold/Allan B. Fraser
ASHRAE Staff Liaison: Bert E. Etheredge
IES Liaison: Rita M. Harrold
USGBC Liaison: Brendan Owens

Name	Affiliation
Dennis Stanke, Chair*	Trane Commercial Systems, Ingersoll-Rand
Richard Heinisch, Vice Chair *	Acuity Brands Lighting
Dan Nall, Vice Chair*	WSP Flack+Kurtz
Andrew Persily, Vice Chair *	NIST
Leon Alevantis*	California Department of Public Health
Jim Bowman*	American Forest & Paper Association, Inc.
Harvey Bryan*	Arizona State University
Ron Burton (BOMA)*	BOMA International
Dimitri Contoyannis*	IES
Dru Crawley*	Bentley Systems
John Cross*	American Institute of Steel Construction
Lance DeLaura*	Southern California Gas Company
Charles Eley (AIA)*	Architectural Energy Corporation
Anthony Floyd*	City of Scottsdale
Susan Gitlin*	U.S. Environmental Protection Agency
Gregg Gress*	International Code Council
Donald Horn*	U.S. General Services Administration
Roy Hubbard*	Johnson Controls Inc.
John Koeller*	Koeller and Company
Michael Jouaneh*	Lutron
Tom Lawrence*	University of Georgia
Neil Leslie*	Gas Technology Institute
Bing Liu*	Pacific Northwest National Laboratory
Richard Lord*	UT Carrier Corp
Merle McBride*	Owens Corning
Jim McClendon*	Walmart Stores
Molly McGuire*	Taylor Engineering
Jonathan McHugh*	McHugh Energy Consultants
Teresa Rainey*	Skidmore Owing Merrill
Steve Rosenstock (EEI)*	Edison Electric Institute
Jeff Ross-Bain*	Ross-Bain Green Building
Lawrence Schoen*	Schoen Engineering Inc.
Boggarm Setty*	Setty & Associates
Wayne Stoppelmoor*	Schneider Electric
Martha VanGeem*	CTLGroup
David Viola*	IAPMO
Susan Anderson	Osram Sylvania
Ernie Conrad	Landmark Facilities
Julia Beabout	Simulated Solutions
James Benya	Benya Lighting Design
Lee Burgett	Trane Commercial Systems, Ingersoll-Rand
Paula Cino	National Multi Housing Council
Steven Clark	Aquatherm
Daryn Cline	Evapco Inc.
Peyton Collie	Sheet Metal and Air Conditioning Contractor's National Association
Peter Dahl	Sebesta Blomberg
Michael DeWein	Building Codes Assistance Project
William Dillard	Mechanical Services of Florida
Nicola Ferzacca	Architecture Engineers
Katherine Hammack	Ernst and Young
Josh Jacobs	UL Environment
Stephen Kennedy	Georgia Power
Carl Lawson	Hanson Professional Services
Mark MacCracken	Calmac Manufacturing Corp
Thomas Marseille	WSP F + K
Kent Peterson	P2S Engineering Inc.
John Pulley	HOK
Jeffery Rutt	U.S. Department of Defense
Harvey Sachs	American Council for an Energy-Efficient Economy
Joshua Saunders	Underwriters Laboratories
Charles Seyffer	Camfil Farr
Melanie Shepherdson	Natural Resource Defense Council
Swati Ogale	Ecoways Consulting, Ltd., UK
Jeffrey Stone	American Forest & Paper Association
Christian Taber	Big Ass Fans
Robert Thompson	U.S. Environmental Protection Agency
Robert Timmerman	AtSite Real Estate
Timothy Wentz	University of Nebraska
David Williams	LHB Inc.
Steven Winkel	The Preview Group

** Denotes members of voting status when the document was approved for publication*

SPECIAL NOTE

This American National Standard (ANS) is a national voluntary consensus standard developed under the auspices of the American Society of Heating, Refrigerating and Air-Conditioning Engineers (ASHRAE). *Consensus* is defined by the American National Standards Institute (ANSI), of which ASHRAE is a member and which has approved this standard as an ANS, as "substantial agreement reached by directly and materially affected interest categories. This signifies the concurrence of more than a simple majority, but not necessarily unanimity. Consensus requires that all views and objections be considered, and that an effort be made toward their resolution." Compliance with this standard is voluntary until and unless a legal jurisdiction makes compliance mandatory through legislation.

ASHRAE obtains consensus through participation of its national and international members, associated societies, and public review.

ASHRAE Standards are prepared by a Project Committee appointed specifically for the purpose of writing the Standard. The Project Committee Chair and Vice-Chair must be members of ASHRAE; while other committee members may or may not be ASHRAE members, all must be technically qualified in the subject area of the Standard. Every effort is made to balance the concerned interests on all Project Committees.

The Manager of Standards of ASHRAE should be contacted for:

a. interpretation of the contents of this Standard,
b. participation in the next review of the Standard,
c. offering constructive criticism for improving the Standard, or
d. permission to reprint portions of the Standard.

DISCLAIMER

ASHRAE uses its best efforts to promulgate Standards and Guidelines for the benefit of the public in light of available information and accepted industry practices. However, ASHRAE does not guarantee, certify, or assure the safety or performance of any products, components, or systems tested, installed, or operated in accordance with ASHRAE's Standards or Guidelines or that any tests conducted under its Standards or Guidelines will be nonhazardous or free from risk.

ASHRAE INDUSTRIAL ADVERTISING POLICY ON STANDARDS

ASHRAE Standards and Guidelines are established to assist industry and the public by offering a uniform method of testing for rating purposes, by suggesting safe practices in designing and installing equipment, by providing proper definitions of this equipment, and by providing other information that may serve to guide the industry. The creation of ASHRAE Standards and Guidelines is determined by the need for them, and conformance to them is completely voluntary.

In referring to this Standard or Guideline and in marking of equipment and in advertising, no claim shall be made, either stated or implied, that the product has been approved by ASHRAE.

CONTENTS

ANSI/ASHRAE/USGBC/IES Standard 189.1-2011, Standard for the Design of High-Performance Green Buildings Except Low-Rise Residential Buildings

SECTION **PAGE**

NOTE

Approved addenda, errata, or interpretations for this standard can be downloaded free of charge from the ASHRAE Web site at www.ashrae.org/technology.

1791 Tullie Circle NE
Atlanta, GA 30329
www.ashrae.org

(This foreword is not part of this standard. It is merely informative and does not contain requirements necessary for conformance to the standard. It has not been processed according to the ANSI requirements for a standard and may contain material that has not been subject to public review or a consensus process. Unresolved objectors on informative material are not offered the right to appeal at ASHRAE or ANSI.)

FOREWORD

ANSI/ASHRAE/USGBC/IES Standard 189.1 was created through a collaborative effort involving ASHRAE, the U.S. Green Building Council, and the Illuminating Engineering Society of North America. Like its 2009 predecessor, the 2011 version of the standard is written in code-intended (mandatory and enforceable) language so that it may be readily referenced or adopted by enforcement authorities to provide the minimum acceptable level of design criteria specifically for high-performance green buildings within their jurisdiction. States and local jurisdictions within the United States that wish to adopt Standard 189.1 into law may want to review applicable federal laws regarding preemption and related waivers that are available from the U.S. Department of Energy (www1.eere.energy.gov/buildings/appliance_standards/ state_petitions.html).

Building projects in general, including their design, construction, and operation, result in potentially significant energy and environmental impacts. Development frequently converts land from biologically diverse natural habitat to impervious hardscape with greatly reduced biodiversity. The U.S. Green Building Council has reported that buildings in the United States produce 39% of U.S. carbon dioxide (CO2) emissions, are responsible for 40% of U.S. energy consumption, account for 13% of U.S. water consumption, and contribute 15% to GDP per year.

While buildings increase the national energy and environmental footprint, they also contribute significantly to the national economy and offer great potential for reducing energy use, greenhouse gas emissions, water use, heat island and light pollution effects, and impacts on the atmosphere, materials, and resources.

The far-reaching influence of buildings leads to calls for action to reduce their energy and environmental impact. To help meet its ongoing responsibility to support such actions, ASHRAE Standing Standard Project Committee (SSPC) 189.1 uses the ASHRAE continuous maintenance process to update this standard in response to input from all segments of the building community, the public at large, and project committee members. Compliance with these updated provisions will further reduce negative energy and environmental impacts through high-performance building design, construction, and operation.

The project committee members and consultants considered a variety of factors to develop the provisions of this standard, including published research, justification for proposals received from outside the committee, and the committee members' professional judgment. However, new provisions within the standard were not uniformly subjected to economic assessment. Cost-benefit assessment, while an important consideration in general, was not a necessary criterion for acceptance of any given proposed change to the standard. The development of an economic threshold value associated with the energy or environmental benefit of each provision falls outside the scope of this standard.

Standard 189.1 addresses site sustainability, water use efficiency, energy use efficiency, indoor environmental quality (IEQ), and the building's impact on the atmosphere, materials, and resources. The standard devotes a section to each of these subject areas, as well as a separate section related to plans for construction and high-performance operation.

Many of the provisions of this 2011 version of the standard differ from those of the 2009 version. New provisions include the following:

- *Since Standard 189.1 adopts by reference many requirements from other ASHRAE standards (particularly Standards 62.1 and 90.1), this version updates requirements to reflect the most current version of each referenced standard. Most importantly, it refers to Standards 90.1-2010 and 62.1-2010 rather than the 2007 version of each.*
- *The standard limits the requirement for condensate recovery from mechanical cooling equipment to regions where significant amounts of condensate can be expected based on climate conditions.*
- *The standard replaces the Standard 90.1 across-the-board approach to reduction in interior lighting power density (LPD) with an LPD reduction based on specific building and space types.*
- *For lighted signs visible during daytime hours, automatic controls are now required to reduce the lighting power to 35% of full power. For other signs, automatic controls must now turn off lighting during daytime hours and reduce the lighting power to 70% of full power after midnight.*
- *For hotel guest rooms, automatic controls are now required to turn off power for lighting, television, and switched outlets and to reset HVAC setpoints within 30 minutes after the guest room becomes unoccupied.*
- *As an alternative to permanent projections for shading, building projects may now employ automatically controlled building façade systems, such as dynamic glazing and shading systems, which modify solar heat gain factor (SHGF) in response to daylight levels or solar intensity.*
- *Additional federal minimum efficiency requirements for commercial refrigeration equipment, effective January 1, 2012, have been incorporated into Table C-16.*
- *Prescribed on-site renewable energy must be based on roof area rather than conditioned space area, and the renewable energy requirement for multiple-story buildings now exceeds the requirement for single-story buildings.*
- *Invasive plants-those not indigenous to the building site-must be removed from the building site and destroyed.*
- *Open-graded (uniformed size) aggregate and porous pavers (e.g., open-grid pavers) qualify as a hardscape*

surface for heat island mitigation with no further testing. Permeable pavement and permeable pavers must meet a minimum percolation rate rather than a minimum solar reflectance index (SRI).

- *Roofs must meet both an initial and an aged (three year) minimum solar reflectance index.*
- *Pedestrian walkways must be provided to connect transit stops to primary building entrances.*

As was the case in the 2009 version of the standard, each section follows a similar format:

x.1 General. *This subsection includes a statement of scope and addresses other broad issues for the section.*

x.2 Compliance Paths. *This subsection indicates the compliance options available within a given section.*

x.3 Mandatory Provisions. *This subsection contains mandatory provisions that apply to all projects (i.e., provisions that must be met and may not be ignored in favor of equal or more stringent provisions found in other subsections).*

x.4 Prescriptive Option. *This subsection-an alternative to the Performance Option-contains prescribed provisions that must be met in addition to all mandatory provisions. Prescribed provisions offer a simple compliance approach that involves minimal calculations.*

x.5 Performance Option. *This subsection-an alternative to the Prescriptive Option-contains performance-based provisions that must be met in addition to all mandatory provisions. Performance provisions offer a more complex alternate compliance approach that typically involves simulation or other calculations.*

SSPC 189.1 considers and administers changes to this continuous maintenance standard and provides interpretations as requested. Proposed changes to the standard may originate within or outside of the committee. The committee welcomes proposals for improving the standard using the ANSI-approved ASHRAE continuous maintenance procedure. A continuous maintenance proposal (CMP) form can be found online at http://www.ashrae.org/technology/page/97 as well as in the back of this standard, and may be completed and submitted at any time. The committee takes formal action on every proposal received, which often results in changes to the published standard. ASHRAE posts approved addenda in publication notices on the ASHRAE Web site. To receive notice of all public reviews, approved and published addenda, errata, and interpretations as well as meeting notices, ASHRAE encourages interested parties to sign up for the free ASHRAE Internet Listserv for this standard (http://www.ashrae.org/publications/detail/14931).

1. PURPOSE

The purpose of this standard is to provide minimum requirements for the siting, design, construction, and plan for operation of high-performance green buildings to:

a. balance environmental responsibility, resource efficiency, occupant comfort and well being, and community sensitivity, and
b. support the goal of development that meets the needs of the present without compromising the ability of future generations to meet their own needs.

2. SCOPE

2.1 This standard provides minimum criteria that:

a. apply to the following elements of *building projects*:
   1. new buildings and their systems
   2. new portions of buildings and their systems
   3. new systems and equipment in existing buildings
b. address *site* sustainability, water use efficiency, energy efficiency, indoor environmental quality (IEQ), and the building's impact on the atmosphere, materials, and resources.

2.2 The provisions of this standard do not apply to:

a. single-family houses, multi-family structures of three stories or fewer above grade, manufactured houses (mobile homes) and manufactured houses (modular), and
b. buildings that use none of the following: electricity, fossil fuel, or water.

2.3 This standard shall not be used to circumvent any safety, health, or environmental requirements.

3. DEFINITIONS, ABBREVIATIONS, AND ACRONYMS

3.1 General. Certain terms, abbreviations, and acronyms are defined in this section for the purposes of this standard. These definitions are applicable to all sections of this standard.

Terms that are not defined herein, but that are defined in standards that are referenced herein (e.g., ANSI/ASHRAE/IES Standard 90.1), shall have the meanings as defined in those standards.

Other terms that are not defined shall have their ordinarily accepted meanings within the context in which they are used. Ordinarily accepted meanings shall be based upon American standard English language usage, as documented in an unabridged dictionary accepted by the *authority having jurisdiction*.

3.2 Definitions

acceptance representative: An entity identified by the *owner* who leads, plans, schedules, and coordinates the activities needed to implement the building acceptance testing activities. The acceptance representative may be a qualified employee or consultant of the *owner*. The individual serving as the acceptance representative shall be independent of the project design and construction management, though this individual may be an employee of a firms providing those services.

adapted plants: see *plants, adapted.*

adequate transit service: at least two buses (including bus rapid transit), streetcars, or light rail trains per hour on weekdays, operating between 6:00 a.m. and 9:00 a.m., and between 3:00 p.m. and 6:00 p.m., or at least five heavy passenger rail or ferries operating between 6:00 a.m. and 9:00 a.m., and between 3:00 p.m. and 6:00 p.m.

agricultural land: land that is, or was within ten years prior to the date of the building permit application for the *building project*, primarily devoted to the commercial production of horticultural, viticultural, floricultural, dairy, apiary, vegetable, or animal products or of berries, grain, hay, straw, turf, seed, finfish in upland hatcheries, or livestock, and that has long-term commercial significance for agricultural production. Land that meets this definition is *agricultural land* regardless of how the land is zoned by the local government with zoning jurisdiction over that land.

air, outdoor: see ANSI/ASHRAE Standard 62.1

airflow rate, minimum outdoor: the rate of outdoor airflow provided by a ventilation system when running when all *densely occupied spaces* with *demand control ventilation* are unoccupied.

alternate on-site sources of water: see *water, alternate on-site sources of.*

annual load factor: the calculated annual electric consumption, in kWh, divided by the product of the calculated annual peak electric demand, in kW, and 8760 hours.

attic and other roofs: see ANSI/ASHRAE/IES Standard 90.1.

authority having jurisdiction (AHJ): the agency or agent responsible for enforcing this standard.

baseline building design: see ANSI/ASHRAE/IES Standard 90.1.

baseline building performance: see ANSI/ASHRAE/IES Standard 90.1.

basis of design (BOD): a document that records the concepts, calculations, decisions, and product selections used to meet the *owner's project requirements* and to satisfy applicable regulatory requirements, standards, and guidelines. The document includes both narrative descriptions and lists of individual items that support the design process. (See *owner's project requirements.*)

biobased product: a commercial or industrial product (other than food or feed) that is composed, in whole or in significant part, of biological products or renewable agricultural materials (including plant, animal, and marine materials) or forestry materials.

bio-diverse plantings: nonhomogeneous, multiple-species plantings.

breathing zone: see ANSI/ASHRAE Standard 62.1.

brownfield site: a *site* documented as contaminated by means of an ASTM E1903 Phase II Environmental Site Assessment or a *site* classified as a brownfield by a local, state, or federal government agency.

building entrance: see ANSI/ASHRAE/IES Standard 90.1.

building envelope: see ANSI/ASHRAE/IES Standard 90.1.

building project: a building, or group of buildings, and *site* that utilize a single submittal for a construction permit or that are within the boundary of contiguous properties under single ownership or effective control (see *owner*).

carbon dioxide equivalent (CO_2e): a measure used to compare the impact of various greenhouse gases based on their global warming potential (GWP). CO_2e approximates the time-integrated warming effect of a unit mass of a given greenhouse gas, relative to that of carbon dioxide (CO_2). GWP is an index for estimating the relative global warming contribution of atmospheric emissions of 1 kg of a particular greenhouse gas compared to emissions of 1 kg of CO_2. The following GWP values are used based on a 100-year time horizon: 1 for CO_2, 25 for methane (CH_4), and 298 for nitrous oxide (N_2O).

classroom: a space primarily used for scheduled instructional activities.

climate zone: see Section 5.1.4 of ANSI/ASHRAE/IES Standard 90.1.

cognizant authority: see ANSI/ASHRAE Standard 62.1.

commissioning authority (CxA): An entity identified by the *owner* who leads, plans, schedules, and coordinates the commissioning team to implement the building commissioning process. (*See commissioning process*)

commissioning plan: A document that outlines the organization, schedule, allocation of resources, and documentation requirements of the building commissioning process. (See *commissioning process*.)

commissioning process: A quality-focused process for enhancing the delivery of a project. The process focuses upon verifying and documenting that the facility and all of its systems and assemblies are planned, designed, installed, tested, operated, and maintained to meet the *owner's project requirements*. (See *owner's project requirements*.)

complete operational cycle: a period of time as long as one year so as to account for climactic variations affecting outdoor water consumption.

conditioned space: see ANSI/ASHRAE/IES Standard 90.1.

construction checklist: a form used by the contractor to verify that appropriate components are onsite, ready for installation, correctly installed, and functional.

continuous air barrier: the combination of interconnected materials, assemblies, and flexible sealed joints and components of the *building envelope* that provide airtightness to a specified permeability. (See *building envelope*.)

continuous daylight dimming: method of automatic lighting control using daylight photosensors where the lights are dimmed continuously or use at least four preset levels with at least a five-second fade between levels, and where the control turns the lights off when sufficient daylight is available.

cycles of concentration: the ratio of makeup rate to the sum of the blowdown and drift rates.

daylight area:

a. ***primary sidelighted area*** **(See Figure 3.1):** The total *primary sidelighted area* is the combined *primary sidelighted area* without double-counting overlapping areas. The floor area for each *primary sidelighted area* is directly adjacent to *vertical fenestration* in exterior wall with an area equal to the product of the *primary sidelighted area* width and the *primary sidelighted area* depth. The *primary sidelighted area* width is the width of the window plus, on each side, the smallest of

   1. 2 ft (0.6 m) or
   2. the distance to any 60 in. (1.5 m) or higher vertical obstruction.

   The *primary sidelighted area* depth is the horizontal distance perpendicular to the glazing which is the smallest of:

   1. the distance from the floor to the top of the glazing or
   2. the distance to any 60 in (1.5 m) or higher vertical obstruction.

b. ***under skylights*** **(see Figure 3.2):** The total *daylight area* under *skylights* is the combined *daylight area* without double-counting overlapping areas. The *daylight area* under *skylights* is bounded by the *skylight* opening, plus horizontally in each direction, the smallest of

   1. 70% of the ceiling height [0.7 × CH] or
   2. the distance to any *daylight area* under *roof monitors* or
   3. the distance to the front face of any vertical obstruction where any part of the obstruction is farther away from the nearest edge of the *skylight* opening than 70% of the distance between the top of the obstruction and the ceiling [0.7 × (CH – OH)].

   where

   CH ≡ the height of the ceiling at the lowest edge of the *skylight*

   OH ≡ the height to the top of the obstruction

c. ***under roof monitor*** **(see Figure 3.3):** The total *daylight area* under *roof monitors* is the combined *daylight area* without double-counting overlapping areas. The *daylight area* under *roof monitors* is equal to the product of the width of the *vertical fenestration* above the ceiling level and the smallest of

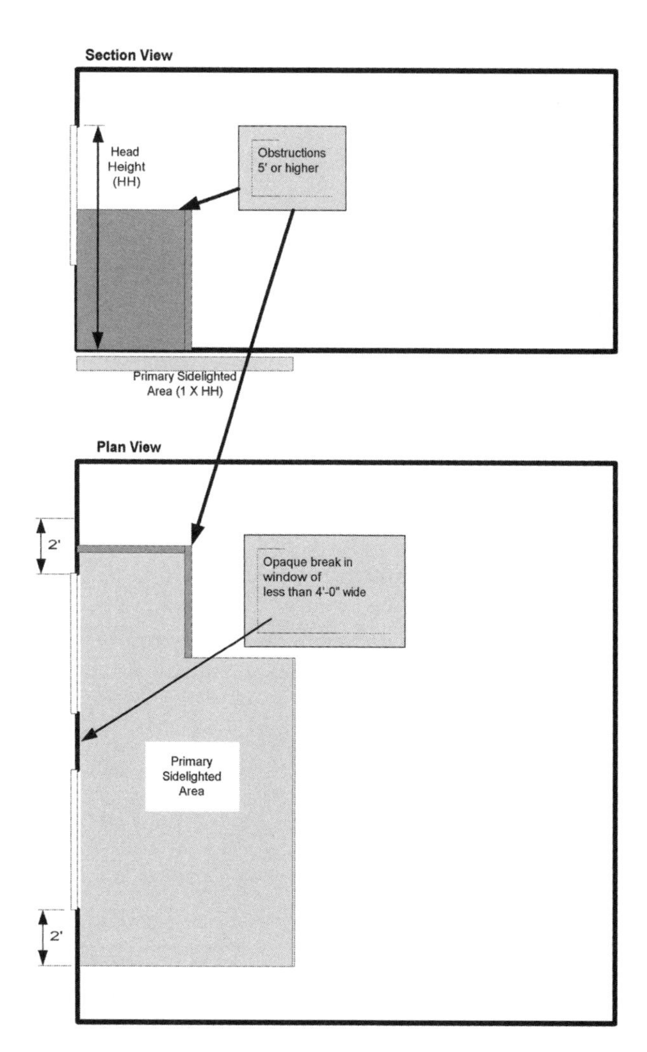

Figure 3.1 Section and plan views of primary sidelighted area.

the following horizontal distances inward from the bottom edge of the glazing:

1. the vertical distance from the floor to the bottom edge of the monitor glazing or
2. the distance to the edge of any *primary sidelighting area* or
3. the distance to the front face of any vertical obstruction where any part of the obstruction is farther away than the difference between obstruction height and the monitor sill height (MSH –OH).

daylight hours: the period from 30 minutes after sunrise to 30 minutes before sunset.

demand control ventilation (DCV): see ANSI/ASHRAE/IES Standard 90.1.

densely occupied space: those spaces with a design occupant density greater than or equal to 25 people per 1000 ft^2 (100 m^2).

designated park land: federal-, state-, or local-government-owned land that is formally designated and set aside as park land or wildlife preserve.

development footprint: the total land area of a project *site* that will be developed with impervious surfaces, constructed as part of the project such as buildings, streets, other areas that have been graded so as to be effectively impervious, and parking areas.

dwelling unit: see ANSI/ASHRAE/IES Standard 90.1.

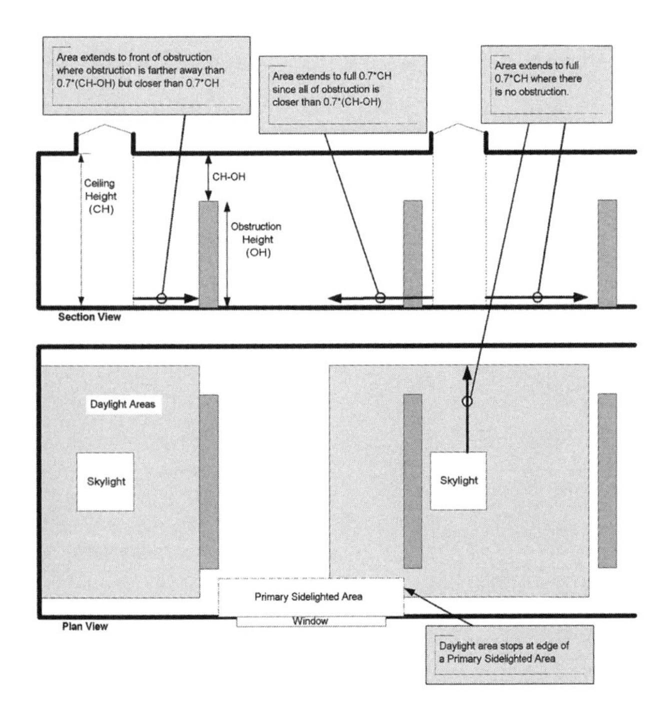

Figure 3.2 Section and plan views of daylight area under skylight.

emergency ride home: access to transportation home in the case of a personal emergency, or unscheduled overtime for employees who commute via transit, carpool, or vanpool.

evapotranspiration (ET): the sum of evaporation and plant transpiration. Evaporation accounts for the movement of water to the air from sources such as the soil, canopy interception, and water bodies. Transpiration accounts for the movement of water within a plant and the subsequent loss of water as vapor through stomata in its leaves.

> ***ET_c:*** *Evapotranspiration* of the plant material derived by multiplying ET_0 by the appropriate plant coefficient.
>
> ***ET_0:*** Maximum *evapotranspiration* as defined by the standardized Penman-Monteith equation or from the National Weather Service, where available.

expressway: a divided highway with a minimum of four lanes, which has controlled access for a minimum of 10 miles (16 kilometers) and a posted minimum speed of at least 45 mph (70 km/h).

fenestration: see ANSI/ASHRAE/IES Standard 90.1.

fenestration area: see ANSI/ASHRAE/IES Standard 90.1.

fish and wildlife habitat conservation area: areas with which state or federally designated endangered, threatened, or sensitive species have a primary association.

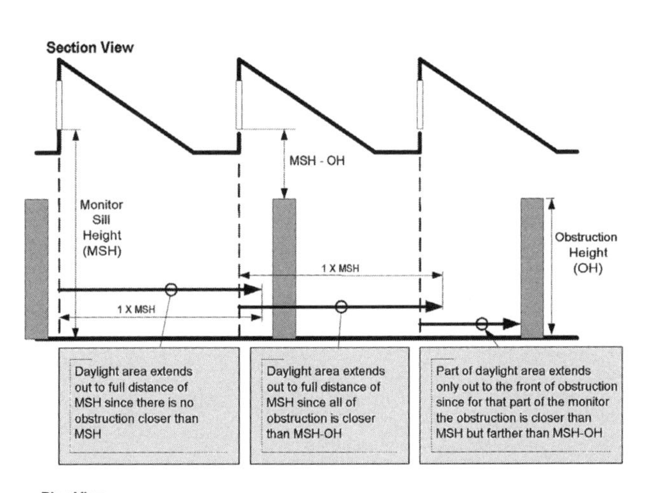

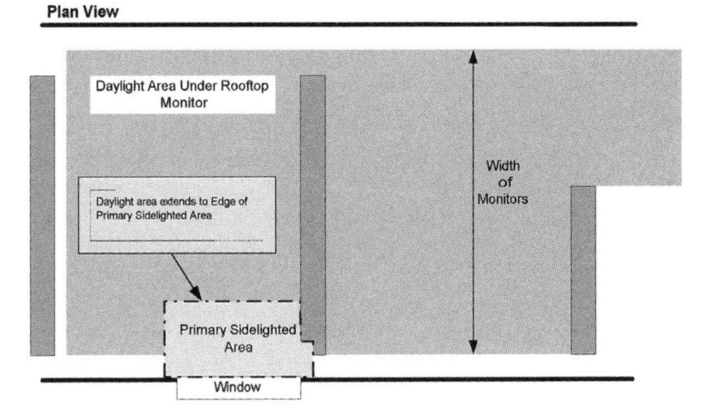

Figure 3.3 Section and plan views of daylight area under roof monitor.

forest land: all designated state forests, national forests, and all land that is, or was within 10 years prior to the date of the building permit for the *building project*, primarily devoted to growing trees for long-term commercial timber production.

generally accepted engineering standard: see ANSI/ ASHRAE/IES Standard 90.1.

geothermal energy: heat extracted from the Earth's interior and used to produce electricity, mechanical power, or provide thermal energy for heating buildings or processes. Geothermal energy does not include systems that use energy independent of the geothermal source to raise the temperature of the extracted heat, such as heat pumps.

greenfield site: a *site* of which 20% or less has been previously developed with impervious surfaces.

greyfield site: a *site* of which more than 20% is already developed with impervious surfaces.

gross roof area: see ANSI/ASHRAE/IES Standard 90.1.

gross wall area: see ANSI/ASHRAE/IES Standard 90.1.

hardscape: *site* paved areas including roads, driveways, parking lots, walkways, courtyards, and plazas.

heat island effect: the tendency of urban areas to be at a warmer temperature than surrounding rural areas.

high-performance green building: a building designed, constructed, and capable of being operated in a manner that increases environmental performance and economic value over time, seeks to establish an indoor environment that supports the health of occupants, and enhances satisfaction and productivity of occupants through integration of environmentally preferable building materials and water-efficient and energy-efficient systems.

hydrozoning: to divide the landscape irrigation system into sections in order to regulate each zone's water needs based on plant materials, soil, and other factors.

improved landscape: any disturbed area of the *site* where new plant and/or grass materials are to be used, including green *roofs*, plantings for stormwater controls, planting boxes, and similar vegetative use. *Improved landscape* shall not include *hardscape* areas such as sidewalks, driveways, or other paved areas, and swimming pools or decking.

integrated design process: a design process utilizing early collaboration amongst representatives of each stakeholder and participating consultant on the project. Unlike the conventional or linear design process, integrated design requires broad stakeholder/consultant participation.

integral heat recovery: a refrigeration system that allows heating and cooling to be transferred between air-conditioned zones in a refrigeration system. In the case of a VRF system, this is typically done using a three-pipe system with compression discharge gas, liquid refrigerant, and suction gas piping being piped to each conditioned zone.

***integrated project delivery*:** see *integrated design process.*

interior projection factor: see *projection factor, interior.*

irrigation adequacy: a representation of how well irrigation meets the needs of the plant material. This reflects the percentage of required water for turf or plant material supplied by rainfall and controller-scheduled irrigations.

irrigation excess: a representation of the amount of irrigation water applied beyond the needs of the plant material. This reflects the percentage of water applied in excess of 100% of required water.

landscape establishment period: a time period, beginning on the date of completion of permanent plantings and not exceeding 18 months, intended to allow the permanent landscape to become sufficiently established to remain viable.

life cycle assessment (LCA): a compilation and evaluation of the inputs, outputs, and the potential environmental impacts of a building system throughout its life cycle. LCA addresses the environmental aspects and potential environmental impacts (e.g., use of resources and environmental consequences of releases) throughout a building's life cycle, from raw material acquisition through manufacturing, construction, use, operation, end-of-life treatment, recycling, and final disposal (end of life). The purpose is to identify opportunities to improve the environmental performance of buildings throughout their life cycles.

light rail: a streetcar-type vehicle that has step entry or level boarding entry and is operated on city streets, semi-exclusive rights-of-way, or exclusive rights-of-way.

lighting power allowance: see ANSI/ASHRAE/IES Standard 90.1.

lighting zone (LZ): An area defining limitations for outdoor lighting.

> ***LZ0:*** Undeveloped areas within national parks, state parks, forest land, rural areas, and other undeveloped areas as defined by the *AHJ*.
>
> ***LZ1:*** Developed areas of national parks, state parks, forest land, and rural areas.
>
> ***LZ2:*** Areas predominantly consisting of residential zoning, neighborhood business districts, light industrial with limited nighttime use, and residential mixed-use areas.
>
> ***LZ3:*** All areas not included in LZ0, LZ1, LZ2, or LZ4.
>
> ***LZ4:*** High activity commercial districts in major metropolitan areas as designated by the local jurisdiction.

liner system (Ls): An insulation system for a metal building *roof* that includes the following components. A continuous membrane is installed below the purlins and uninterrupted by framing members. Uncompressed, unfaced insulation rests on top of the membrane between the purlins. For multilayer installations, the last rated R-value of insulation is for unfaced insulation draped over purlins and then compressed when the metal *roof* panels are attached. A minimum R-3 (R-0.5) thermal spacer block between the purlins and the metal *roof* panels is required, unless compliance is shown by the overall assembly U-factor, or otherwise noted.

low-impact trail: erosion-stabilized pathway or track that utilizes natural groundcover or installed system greater than 50% pervious. The pathway or track is designed and used only for pedestrian and nonmotorized vehicles (excluding power-assisted conveyances for individuals with disabilities).

maintenance plan: see *maintenance program* in ANSI/ASHRAE/ACCA Standard 180.

minimum outdoor airflow rate: see *airflow rate, minimum outdoor.*

native plants: see *plants, native.*

non-potable water: see *water, non-potable.*

nonresidential: see ANSI/ASHRAE/IES Standard 90.1.

north-oriented: facing within 45 degrees of true north within the northern hemisphere (however, facing within 45 degrees of true south in the southern hemisphere).

occupiable space: see ANSI/ASHRAE Standard 62.1.

office furniture system: either a panel-based workstation comprised of modular interconnecting panels, hang-on components, and drawer/filing components, or a freestanding grouping of furniture items and their components that have been designed to work in concert.

on-site renewable energy system: photovoltaic, solar thermal, *geothermal energy,* and wind systems used to generate energy and located on the *building project.*

once-through cooling: The use of water as a cooling medium where the water is passed through a heat exchanger one time and is then discharged to the drainage system. This also includes the use of water to reduce the temperature of condensate or process water before discharging it to the drainage system.

open-graded (uniform-sized) aggregate: materials such as crushed stone or decomposed granite that provide 30%–40% void spaces.

open plan workstation: see ANSI/BIFMA M7.1.

outdoor air: see *air, outdoor.*

***owner*:** The party in responsible control of development, construction, or operation of a project at any given time.

owner's project requirements (OPR): a written document that details the functional requirements of a project and the expectations of how it will be used and operated. These include project goals, measurable performance criteria, cost considerations, benchmarks, success criteria, and supporting information.

permanently installed: see ANSI/ASHRAE/IES Standard 90.1.

permeable pavement: pervious concrete or porous asphalt that allows the movement of water and air through the paving material, and primarily used as paving for roads, parking lots, and walkways. Permeable paving materials have an open-graded coarse aggregate with interconnected voids.

permeable pavers: units that present a solid surface but allow natural drainage and migration of water into the base below by permitting water to drain through the spaces between the pavers.

plants:

a. ***adapted plants:*** *plants* that reliably grow well in a given habitat with minimal attention from humans in the form of winter protection, pest protection, water irrigation, or fertilization once root systems are established in the soil. Adapted *plants* are considered to be low maintenance but not invasive.
b. ***invasive plants:*** Species of *plants* that are not native to the *building project site* and that cause or are likely to cause environmental harm. At a minimum, the list of invasive species for a *building project site* includes *plants* included in city, county, and regional lists and State and Federal Noxious Weeds laws.
c. ***native plants:*** *plants* that adapted to a given area during a defined time period and are not invasive. In America, the term often refers to *plants* growing in a region prior to the time of settlement by people of European descent.

porous pavers (open-grid pavers): units where at least 40% of the surface area consists of holes or openings that are filled with sand, gravel, other porous material, or vegetation.

post-consumer recycled content: proportion of *recycled material* in a product generated by households or by commercial, industrial, and institutional facilities in their role as end-users of the product, which can no longer be used for its intended purpose. This includes returns of material from the distribution chain. (See *recycled material.*)

potable water: see *water, potable.*

pre-consumer recycled content: proportion of *recycled material* in a product diverted from the waste stream during the manufacturing process. Content that shall not be considered pre-consumer recycled includes the re-utilization of materials such as rework, regrind, or scrap generated in a process and capable of being reclaimed within the same process that generated it. (See *recycled material.*)

private office workstation: see ANSI/BIFMA M7.1.

projection factor (PF): see ANSI/ASHRAE/IES Standard 90.1.

projection factor, interior: the ratio of the horizontal depth of the interior shading projection divided by the sum of the height of the *fenestration* above the interior shading projection and, if the interior projection is below the bottom of the *fenestration*, the vertical distance from the bottom of the *fenestration* to the top of the farthest point of the interior shading projection, in consistent units.

proposed building performance: see ANSI/ASHRAE/IES Standard 90.1.

proposed design: see ANSI/ASHRAE/IES Standard 90.1.

public way: A street, alley, transit right of way, or other parcel of land open to the outdoors leading to a street or transit right of way that has been deeded, dedicated, or otherwise permanently appropriated to the public for public use and that has a clear width and height of not less than 10 ft (3 m).

recovered material: material that would have otherwise been disposed of as waste or used for energy recovery (e.g., incinerated for power generation), but has instead been collected and recovered as a material input, in lieu of new primary material, for a recycling or a manufacturing process.

recycled content: proportion, by mass, of *recycled material* in a product or packaging. Only pre-consumer and post-consumer materials shall be considered as *recycled content*. (See *recycled material.*)

recycled material: material that has been reprocessed from *recovered* (reclaimed) *material* by means of a manufacturing process and made into a final product or into a component for incorporation into a product. (See *recovered material.*)

residential: see ANSI/ASHRAE/IES Standard 90.1.

roof: see ANSI/ASHRAE/IES Standard 90.1.

roof area, gross: see ANSI/ASHRAE/IES Standard 90.1.

roof monitor: a raised central portion of a *roof* having *vertical fenestration.*

seating: task and guest chairs used with office furniture systems.

semiheated space: see ANSI/ASHRAE/IES Standard 90.1.

service water heating: see ANSI/ASHRAE/IES Standard 90.1.

sidelighting: daylighting provided by *vertical fenestration* mounted below the ceiling plane.

sidelighting effective aperture: the relationship of daylight transmitted through windows to the *primary sidelighted areas*. The *sidelighting effective aperture* is calculated according to the following formula:

$$\text{Sidelighting Effective Aperture} = \frac{\sum \text{Window Area} \times \text{Window VLT}}{\text{Area of Primary Sidelighted Area}}$$

where Window VLT is the visible light transmittance of windows as determined in accordance with Section 5.8.2.6 of ASHRAE/IESNA Standard 90.1.

single-rafter roof: see ANSI/ASHRAE/IES Standard 90.1.

skylight: see ANSI/ASHRAE/IES Standard 90.1.

site: a contiguous area of land that is under the ownership or control of one entity.

smart controller (weather-based irrigation controller): a device that estimates or measures depletion of water from the soil moisture reservoir and operates an irrigation system to replenish water as needed while minimizing excess.

soil gas retarder system: a combination of measures that retard vapors in the soil from entering the occupied space.

solar energy system: any device or combination of devices or elements that rely upon direct sunlight as an energy source, including but not limited to any substance or device that collects sunlight for use in:

a. the heating or cooling of a structure or building;
b. the heating or pumping of water;
c. industrial, commercial, or agricultural processes; or
d. the generation of electricity.

solar heat gain coefficient (SHGC): see ANSI/ASHRAE/IES Standard 90.1.

solar reflectance index (SRI): a measure of a constructed surface's ability to reflect solar heat, as shown by a small temperature rise. A standard black surface (reflectance 0.05, emittance 0.90) is 0 and a standard white surface (reflectance 0.80, emittance 0.90) is 100.

SWAT: smart water application technology as defined by the Irrigation Association.

toplighting: lighting building interiors with daylight admitted through *fenestration* located on the *roof* such as *skylights* and *roof monitors*.

tubular daylighting device: a means to capture sunlight from a rooftop. Sunlight is then redirected down from a highly reflective shaft and diffused throughout interior space.

turfgrass: grasses that are regularly mowed and, as a consequence, form a dense growth of leaf blades, shoots, and roots.

vendor: a company that furnishes products to project contractors and/or subcontractors for on-site installation.

variable air volume (VAV) system: see ANSI/ASHRAE/IES Standard 90.1.

verification: the process by which specific documents, components, equipment, assemblies, systems, and interfaces among systems are confirmed to comply with the criteria described in the *owner's project requirements*. (See *owner's project requirements.*)

vertical fenestration: see ANSI/ASHRAE/IES Standard 90.1.

wall: see ANSI/ASHRAE/IES Standard 90.1.

wall area, gross: see ANSI/ASHRAE/IES Standard 90.1.

water, alternate on-site sources of: *alternate on-site sources of water* include, but are not limited to:

a. rainwater or stormwater harvesting,
b. air conditioner condensate,
c. gray water from interior applications and treated as required,
d. swimming pool filter backwash water,

e. cooling tower blowdown water,
f. foundation drain water,
g. industrial process water, or
h. on-site wastewater treatment plant effluent.

water, non-potable: water that is not *potable water.* (See *water, potable.*)

water, potable: water from public drinking water systems or from natural freshwater sources such as lakes, streams, and aquifers where water from such natural sources would or could meet drinking water standards.

water factor (WF):

a. ***clothes washer (residential and commercial):*** the quantity of water in gal (L) used to wash each ft^3 (m^3) of machine capacity.
b. ***residential dishwasher:*** the quantity of water use in gal (L) per full machine wash and rinse cycle.

weatherproofing system: a group of components including associated adhesives and primers that when installed create a protective envelope against water and wind.

wetlands: those areas that are inundated or saturated by surface or groundwater at a frequency and duration sufficient to support, and that under normal circumstances do support, a prevalence of vegetation adapted for life in saturated soil conditions. This definition incorporates all areas that would meet the definition of "wetlands" under applicable federal or state guidance whether or not they are officially designated, delineated, or mapped, including man-made areas that are designed, constructed, or restored to include the ecological functions of natural wetlands.

yearly average day-night average sound levels: level of the time-mean-square A-weighted sound pressure averaged over a one-year period with ten dB added to sound levels occurring in each night-time period from 2200 hours to 0700 hours, expressed in dB.

3.3 Abbreviations and Acronyms

AC	alternating current
AHJ	*authority having jurisdiction*
AHRI	Air-Conditioning, Heating, and Refrigeration Institute
ANSI	American National Standards Institute
ASHP	air-source heat pump
ASHRAE	American Society of Heating, Refrigerating and Air-Conditioning Engineers, Inc.
ASME	American Society of Mechanical Engineers
ASTM	American Society for Testing and Materials International
BIFMA	The Business and Institutional Furniture Manufacturer's Association
BMS	Building Management System
BOD	*basis of design*
Btu	British thermal unit
Btu/h	British thermal unit per hour
CDPH	California Department of Public Health
CFC	chlorofluorocarbon
cfm	ft^3/min
ci	continuous insulation
CIE	Commission Internationale de L'Eclairage (International Commission on Illumination)
CITES	Convention on International Trade in Endangered Species of Wild Fauna and Flora
cm	centimeter
CO_2	carbon dioxide
CO_2e	*carbon dioxide equivalent*
CSA	Canadian Standards Association
CxA	commissioning authority
dB	decibel
DB	dry bulb
DC	direct current
DCV	*demand control ventilation*
DX	direct expansion
EA_{vf}	effective aperture for *vertical fenestration*
EISA	Energy Independence and Security Act
EMS	Energy Management System
EPAct	U.S. Energy Policy Act
ESC	erosion and sedimentation control
ET_c	*evapotranspiration*
ET_o	*maximum evapotranspiration*
ETS	environmental tobacco smoke
fc	footcandle
FF&E	furniture, fixtures, and equipment
ft	foot
gal	gallon
gpm	gallons per minute
GWP	global warming potential
h	hour
ha	hectare
HCFC	hydrochlorofluorocarbon
HVAC	heating, ventilation, and air conditioning
HVAC&R	heating, ventilation, air conditioning, and refrigeration
IAPMO	International Association of Plumbing and Mechanical Officials
I-P	inch-pound
IA	Irrigation Association
IAQ	indoor air quality
IEQ	indoor environmental quality
IES	Illuminating Engineering Society of North America
in.	inch
kg	kilogram
kL	kiloliter
km	kilometer
kVA	kilovolt-ampere

kW	kilowatt
kWh	kilowatt-hour
L	liter
lb	pound
LCA	*life-cycle assessment*
LID	low impact development
lm	lumen
LPD	lighting power density
Ls	*liner system*
LZ	*lighting zone*
m	meter
M&V	measurement and verification
µg	microgram
mg	milligram
MCWB	maximum coincident wet bulb
MDF	medium density fiberboard
MERV	minimum efficiency reporting value
mi	mile
min	minute
MJ	megaJoule
mm	millimeter
mph	miles per hour
NA	not applicable
NAECA	National Appliance Energy Conservation Act
NC	noise criterion
NR	not required
O&M	operation and maintenance
OITC	outdoor-indoor transmission class
OPR	*owner's project requirements*
Pa	Pascal
PF	*projection factor*
ppb	parts per billion
ppm	parts per million
s	second
SCAQMD	South Coast Air Quality Management District
SHGC	*solar heat gain coefficient*
SMACNA	Sheet Metal and Air Conditioning Contractors National Association
SRI	*solar reflectance index*
STC	sound transmission class
TMP	transportation management plan
UL	Underwriters Laboratory
USDA	United States Department of Agriculture
USDOE	United States Department of Energy
USEPA	United States Environmental Protection Agency
USFEMA	United States Federal Emergency Management Agency
USGBC	United States Green Building Council
USGSA	United States General Services Administration
VAV	*variable air volume*
VOC	volatile organic compound
VRF	variable refrigerant flow system
WB	wet bulb
WF	water factor
yr	year

4. ADMINISTRATION AND ENFORCEMENT

4.1 General. *Building projects* shall comply with Sections 4 through 11. Within each of those sections, *building projects* shall comply with all Mandatory Provisions (x.3); and, where offered, either

a. Prescriptive Option (x.4) or

b. Performance Option (x.5).

4.1.1 Normative Appendices. The normative appendices to this standard are considered to be integral parts of the mandatory requirements of this standard, which for reasons of convenience, are placed apart from all other normative elements.

4.1.2 Informative Appendices. The informative appendices to this standard and informative notes located within this standard contain additional information and are not mandatory or part of this standard.

5. SITE SUSTAINABILITY

5.1 Scope. This section addresses requirements for *building projects* that pertain to *site* selection, *site* development, mitigation of *heat island effect*, and light pollution reduction.

5.2 Compliance. The *site* shall comply with Section 5.3, "Mandatory Provisions," and either

a. Section 5.4, "Prescriptive Option," or
b. Section 5.5, "Performance Option."

5.3 Mandatory Provisions

5.3.1 Site Selection. The *building project* shall comply with 5.3.1.1 and 5.3.1.2.

5.3.1.1 Allowable Sites. The *building project* shall take place on one of the following:

a. in an existing *building envelope.*
b. on a *brownfield site.*
c. on a *greyfield site.*
d. on a *greenfield site* that is within 1/2 mi (800 m) of *residential* land that is developed, or that has one or more buildings under construction, with an average density of 10 *dwelling units* per acre (4 units per ha) unless that *site* is *agricultural land* or *forest land.* Proximity is determined by drawing a circle with a 1/2 mi (800 m) radius around the center of the proposed *site.*
e. on a *greenfield site* that is within 1/2 mi (800 m) of not less than ten basic services and that has pedestrian access between the building and the services unless that *site* is *agricultural land* or *forest land.* Basic services include, but are not limited to: (1) financial institutions, (2) places of worship, (3) convenience or grocery stores, (4) day care facilities, (5) dry cleaners, (6) fire stations, (7) beauty shops, (8) hardware stores, (9) laundry facilities, (10) libraries, (11) medical/dental offices, (12) senior care facilities, (13) parks, (14) pharmacies, (15) post offices, (16) restaurants, (17) schools, (18) supermarkets, (19) theaters, (20) community centers, (21) fitness centers, (22) museums, and (23) local government facilities. Proximity is determined by drawing a circle with a 1/2 mi (800 m) radius around the center of the proposed *site.*
f. on a *greenfield site* that is either within 1/2 mi (800 m) of an existing, or planned and funded, commuter rail, *light rail* or subway station or within 1/4 mi (400 m) of *adequate transit service* usable by building occupants unless that *site* is *agricultural land* or *forest land.* Proximity is determined by drawing a circle with a 1/2 mi (800 m) radius around the center of the proposed *site.*
g. on a *greenfield site* that is *agricultural land* and the building's purpose is related to the agricultural use of the land.
h. on a *greenfield site* that is *forest land* and the building's purpose is related to the forestry use of the land.
i. on a *greenfield site* that is *designated park land* and the building's purpose is related to the use of the land as a park.

5.3.1.2 Prohibited Development Activity. There shall be no *site* disturbance or development of the following:

a. previously undeveloped land having an elevation lower than 5 ft (1.5 m) above the elevation of the 100 year flood as defined by USFEMA.

   Exception to 5.3.1.2a: In alluvial "AO" designated flood zones, development is allowed when provided with engineered floodproofing for building structures up to an elevation that is at least as high as the minimum lowest floor elevation determined by the *AHJ.* Drainage paths shall be constructed to guide floodwaters around and away from the structures.

b. within 150 ft (50 m) of any *fish and wildlife habitat conservation area* unless the *site* disturbance or development involves plantings or habitat enhancement of the functions and values of the area.
c. within 100 ft (35 m) of any *wetland* unless the *site* disturbance or development involves plantings or habitat enhancement of the functions and values of the *wetland.*

Exception to 5.3.1.2: Development of a *low-impact trail* is allowed within 15 ft (4.5 m) of a *fish and wildlife habitat conservation area* or *wetland.*

5.3.2 Mitigation of Heat Island Effect

5.3.2.1 Site Hardscape. For the purposes of this section, the *site hardscape* includes roads, sidewalks, courtyards, and parking lots but not the constructed building surfaces and not any portion of the *site hardscape* covered by photovoltaic panels generating electricity or other *solar energy systems* used for space heating or water heating. At least 50% of the *site hardscape* shall be provided with one or any combination of the following:

a. existing trees and vegetation or new *bio-diverse plantings* of *native plants* and *adapted plants* located to provide shade within ten years of issuance of the final certificate of occupancy. The effective shade coverage on the *hardscape* shall be the arithmetic mean of the shade coverage calculated at 10 a.m., noon, and 3 p.m. on the summer solstice.
b. paving materials with a minimum initial *SRI* of 29. A default *SRI* value of 35 for new concrete without added color pigment is allowed to be used instead of measurements.
c. *open-graded (uniform-sized) aggregate, permeable pavement, permeable pavers,* and *porous pavers (open-grid pavers). Permeable pavement* and *permeable pavers* shall have a percolation rate of not less than 2 gal/min·ft^2 (100 L/min·m^2).
d. shading through the use of structures, provided that the top surface of the shading structure complies with the provisions of Section 5.3.2.3.
e. parking under a building, provided that the *roof* of the building complies with the provisions of Section 5.3.2.3.
f. buildings or structures that provide shade to the *site hardscape.* The effective shade coverage on the *hardscape* shall be the arithmetic mean of the shade coverage

calculated at 10 a.m., noon, and 3 p.m. on the summer solstice.

Exception: Section 5.3.2.1 shall not apply to *building projects* in *climate zones* 6, 7, and 8.

5.3.2.2 Walls. Above-grade building *walls* and retaining *walls* shall be shaded in accordance with this section. The building is allowed to be rotated up to 45 degrees to the nearest cardinal orientation for purposes of calculations and showing compliance. Compliance with this section shall be achieved through the use of shade-providing *plants*, manmade structures, existing buildings, hillsides, permanent building projections, *on-site renewable energy systems* or a combination of these, using the following criteria:

a. shade shall be provided on at least 30% of the east and west above-grade *walls* and retaining *walls* from grade level to a height of 20 ft (6 m) above grade or the top of the exterior *wall*, whichever is less, within five years of issuance of the final certificate of occupancy. Shade coverage shall be calculated at 10 a.m. for the east *walls* and 3 p.m. for the west *walls* on the summer solstice.
b. where shading is provided by vegetation, such vegetation (including trees) shall be existing trees and vegetation or new *bio-diverse plantings* of *native plants* and *adapted plants* and appropriately sized, selected, planted, and maintained so that they do not interfere with overhead or underground utilities. Such trees shall be placed a minimum of 5 ft (1.5 m) from and within 50 ft (15 m) of the building or retaining *wall*.

Exceptions:

1. The requirements of this section are satisfied if 75% or more of the opaque *wall* surfaces on the east and west have a minimum *SRI* of 29. Each *wall* is allowed to be considered separately for this exception.
2. East *wall* shading is not required for buildings located in *climate zones* 5, 6, 7, and 8. West *wall* shading is not required for buildings located in *climate zones* 7 and 8.

5.3.2.3 Roofs. This section applies to the building and covered parking *roof* surfaces for *building projects* in *climate zones* 1, 2, and 3. A minimum of 75% of the entire *roof* surface not used for *roof* penetrations and associated equipment, *on-site renewable energy systems* such as photovoltaics or solar thermal energy collectors including necessary space between rows of panels or collectors, portions of the roof used to capture heat for building energy technologies, rooftop decks or walkways, or vegetated (green) roofing systems shall be covered with products that comply with one or more of the following:

a. have a minimum initial *SRI* of 78 for a low-sloped *roof* (a slope less than or equal to 2:12) and a minimum initial *SRI* of 29 for a steep-sloped *roof* (a slope of more than 2:12).
b. comply with the criteria for the USEPA's ENERGY STAR® Program Requirements for Roof Products—Eligibility Criteria.

Exceptions:

1. *Building projects* where an annual energy analysis simulation demonstrates that the total annual building energy cost and total annual CO_2e, as calculated in accordance with Sections 7.5.2 and 7.5.3, are both a minimum of 2% less for the proposed *roof* than for a *roof* material complying with the requirements of Section 5.3.2.3(a), or
2. *Roofs* used to shade or cover parking and *roofs* over semi-heated spaces provided that they have a minimum initial *SRI* of 29. A default *SRI* value of 35 for new concrete without added color pigment is allowed to be used instead of measurements.

5.3.2.4 Solar Reflectance Index. The *SRI* shall be calculated in accordance with ASTM E1980 for medium-speed wind conditions. The *SRI* shall be based upon solar reflectance as measured in accordance with ASTM E1918 or ASTM C1549, and the thermal emittance as measured in accordance with ASTM E408 or ASTM C1371. For roofing products, the values for solar reflectance and thermal emittance shall be determined by a laboratory accredited by a nationally recognized accreditation organization, and shall be certified by the manufacturer. For building materials other than roofing products, the values for solar reflectance and thermal emittance shall be determined by an independent third party.

5.3.3 Reduction of Light Pollution

5.3.3.1 General. Exterior lighting systems shall comply with Section 9 of ANSI/ASHRAE/IES Standard 90.1 and with Sections 5.3.3.2 and 5.3.3.3 of this standard.

5.3.3.2 Backlight and Glare

a. All building-mounted luminaires located less than two mounting heights from any property line shall meet the maximum allowable Glare Ratings in Table 5.3.3.2B.
b. All other luminares shall meet the maximum allowable Backlight and Glare Ratings in Table 5.3.3.2A.

5.3.3.3 Uplight. All exterior lighting shall meet one of the following Uplight requirements:

a. Exterior luminares shall meet the maximum allowable Uplight Ratings of Table 5.3.3.2A or
b. Exterior lighting shall meet the Uplight requirements of Table 5.3.3.3.

Exceptions:

1. Lighting in *lighting zones* 3 and 4, solely for uplighting structures, building facades, or landscaping.
2. Lighting in *lighting zones* 1 and 2, solely for uplighting structures, building facades, or landscaping provided the applicable lighting power densities do not exceed 50% of the *lighting power allowances* in ANSI/ASHRAE/IES Standard 90.1, Table 9.4.3B.

TABLE 5.3.3.2A Maximum Allowable Backlight, Uplight, and Glare (BUG) Ratings[1, 2, 3, 4]

	LZ0	LZ1	LZ2	LZ3	LZ4
Allowed Backlight Rating					
>2 mounting heights from property line	B0	B1	B2	B3	B4
1 to 2 mounting heights from property line	B0	B1	B2	B3	B3
0.5 to 1 mounting height to property line	B0	B0	B1	B2	B2
<0.5 mounting height to property line	B0	B0	B0	B1	B2
Allowed Uplight Rating	U0	U1	U2	U3	U4
Allowed Glare Rating	G0	G1	G2	G3	G4

Notes to Table 5.3.3.2A:

1. Fixtures mounted two mounting heights or less from a property line shall have backlight towards the property line, except when mounted on buildings.
2. For property lines that abut public walkways, bikeways, plazas, and parking lots, the property line may be considered to be 5 feet (1.5 m) beyond the actual property line for purpose of determining compliance with this section. For property lines that abut public roadways and public transit corridors, the property line may be considered to be the centerline of the public roadway or public transit corridor for the purpose of determining compliance with this section.
3. If the luminaire is installed in other than the intended manner, or is an adjustable luminaire for which the aiming is specified, the rating shall be determined by the actual photometric geometry in the aimed orientation.
4. Backlight, Uplight, and Glare ratings are defined based on specific lumen limits per IES TM-15 Addendum A.

TABLE 5.3.3.2B Maximum Allowable Glare Ratings for Building Mounted Luminaires Within Two Mounting Heights of Any Property Line

	LZ0	LZ1	LZ2	LZ3	LZ4
Glare	G0	G0	G1	G1	G2

Notes to Table 5.3.3.2B:

1. For property lines that abut public walkways, bikeways, plazas, and parking lots, the property line may be considered to be 5 feet (1.5 m) beyond the actual property line for purpose of determining compliance with this section. For property lines that abut public roadways and public transit corridors, the property line may be considered to be the centerline of the public roadway or public transit corridor for the purpose of determining compliance with this section
2. Backlight, Uplight, and Glare ratings are defined based on specific lumen limits per IES TM-15 Addendum A.

TABLE 5.3.3.3 Maximum Allowable Percentage of Uplight

	LZ0	LZ1	LZ2	LZ3	LZ4
Percentage of total exterior fixture lumens allowed to be emitted above 90 degrees or higher from nadir (straight down)	0%	0%	1%	2%	5%

Exceptions to Sections 5.3.3.2 and 5.3.3.3:

1. Specialized signal, directional, and marker lighting associated with transportation.
2. Advertising signage or directional signage.
3. Lighting integral to equipment or instrumentation and installed by its manufacturer.
4. Lighting for theatrical purposes, including performance, stage, film production, and video production.
5. Lighting for athletic playing areas.
6. Lighting that is in use for no more than 60 continuous days and is not re-installed any sooner than 60 days after being uninstalled.
7. Lighting for industrial production, material handling, transportation sites, and associated storage areas.
8. Theme elements in theme/amusement parks.
9. Roadway lighting required by governmental authorities.
10. Lighting classified for and used in hazardous locations as specified in NFPA 70.
11. Lighting for swimming pools and water features.

5.3.4 Plants

5.3.4.1 Invasive Plants. *Invasive plants* shall be removed from the *building project site* and destroyed or disposed of in a land fill. *Invasive plants* shall not be planted on the *building project site.*

5.3.5 Mitigation of Transportation Impacts

5.3.5.1 Pedestrian and Transit Connectivity

5.3.5.1.1 Walkways. Each *primary building entrance* shall be provided with a pedestrian walkway that extents to either a *public way* or a transit stop. Walkways across parking lots shall be clearly delineated.

5.4 Prescriptive Option

5.4.1 Site Development. *Building projects* shall comply with Sections 5.4.1.1 and 5.4.1.2.

5.4.1.1 Effective Pervious Area for All Sites. A minimum of 40% of the entire *site* shall incorporate one or any combination of the following:

a. shall be vegetated with a minimum depth of growing medium of 12 in. (300 mm). Such vegetated areas include bioretention facilities, rain gardens, filter strips, grass swales, vegetated level spreaders, constructed *wetlands*, planters, and open space with plantings. At least 60% of the vegetated area shall consist of *biodiverse planting* of *native plants* and/or *adapted plants* other than turfgrass.
b. shall have a vegetated (green) *roof* with a minimum depth of growing medium of 3 in. (75 mm).
c. shall have *porous pavers (open grid pavers)*.
d. shall have *permeable pavement, permeable pavers,* or *open graded (uniform-sized) aggregate* with a minimum percolation rate of 2 gal/min·ft^2 (100 L/min·m^2) and a minimum of 6 in. (150 mm) of open-graded base below.

Exceptions:

1. The effective pervious surface is allowed to be reduced to a minimum of 20% of the entire *site* if 10% of the average annual rainfall for the entire *development footprint* is captured on *site* and reused for *site* or building water use.
2. The effective pervious surface is not required if 50% of the average annual rainfall for the entire *development footprint* is captured on *site* and reused for *site* or building water use.
3. Locations with less than 10 in. (250 mm) of average annual rainfall.
4. Areas of *building projects* on a *brownfield site* where contamination has been left in place.

5.4.1.2 Greenfield Sites. On a *greenfield site*:

a. where more than 20% of the area of the predevelopment *site* has existing *native plants* or *adapted plants*, a minimum of 20% of the area of *native plants* or *adapted plants* shall be retained.
b. where 20% or less of the area of the predevelopment *site* has existing *native plants* or *adapted plants*, a minimum of 20% of the *site* shall be developed or retained as vegetated area. Such vegetated areas include bioretention facilities, rain gardens, filter strips, grass swales, vegetated level spreaders, constructed *wetlands*, planters, and open space with plantings. A minimum of 60% of such vegetated area shall consist of *biodiverse planting* of *native plants* and/or *adapted plants* other than turfgrass.

Exception to 5.4.1.2(b): Locations with less than 10 in. (250 mm) of average annual rainfall.

5.5 Performance Option

5.5.1 Site Development. *Building projects* shall comply with the following:

a. If the project is in an existing *building envelope*, a minimum of 20% of the average annual rainfall on the *development footprint* shall be managed through infiltration, reuse, or *ET*.
b. If the project is not in an existing *building envelope*, but is on a *greyfield site* or a *brownfield site*, a minimum of 40% of the average annual rainfall on the *development footprint* shall be managed through infiltration, reuse, or *ET*.
c. For all other *sites*, a minimum of 50% of the average annual rainfall on the *development footprint* shall be managed through infiltration, reuse, or *ET*.

6. WATER USE EFFICIENCY

6.1 Scope. This section specifies requirements for *potable water* and *non-potable water* use efficiency, both for the *site* and for the building, and water monitoring.

6.2 Compliance. The water systems shall comply with Section 6.3, "Mandatory Provisions," and either

a. Section 6.4, "Prescriptive Option," or
b. Section 6.5, "Performance Option."

Site water use and building water use are not required to use the same option, i.e., prescriptive or performance, for demonstrating compliance.

6.3 Mandatory Provisions

6.3.1 Site Water Use Reduction

6.3.1.1 Landscape Design. A minimum of 60% of the area of the *improved landscape* shall be in *bio-diverse planting* of *native plants* and *adapted plants* other than *turfgrass*.

Exception: The area of dedicated athletic fields, golf courses, and driving ranges shall be excluded from the calculation of the *improved landscape* for schools, *residential* common areas, or public recreational facilities.

6.3.1.2 Irrigation System Design. *Hydrozoning* of automatic irrigation systems to water different plant materials such as *turfgrass* versus shrubs is required. Landscaping sprinklers shall not be permitted to spray water directly on a building and within 3 ft (1 m) of a building.

6.3.1.3 Controls. Any irrigation system for the project *site* shall be controlled by a qualifying *smart controller* that uses *ET* and weather data to adjust irrigation schedules and that complies with the minimum requirements or an on-site rain or moisture sensor that automatically shuts the system off after a predetermined amount of rainfall or sensed moisture in the soil. Qualifying *smart controllers* shall meet the minimum requirements as listed below when tested in accordance with IA *SWAT* Climatological Based Controllers 8th Draft Testing Protocol. *Smart controllers* that use *ET* shall use the following inputs for calculating appropriate irrigation amounts:

a. *Irrigation adequacy*—80% minimum ET_c.
b. *Irrigation excess*—not to exceed 10%.

Exception: A temporary irrigation system used exclusively for the establishment of new landscape shall be exempt from this requirement. Temporary irrigation systems shall be removed or permanently disabled at such time as the *landscape establishment period* has expired.

6.3.2 Building Water Use Reduction

6.3.2.1 Plumbing Fixtures and Fittings. Plumbing fixtures (water closets and urinals) and fittings (faucets and showerheads) shall comply with the following requirements:

a. Water closets (toilets)—flushometer valve type: For single flush, maximum flush volume shall be determined in accordance with ASME A112.19.2/CSA B45.1 and shall be 1.28 gal (4.8 L). For dual-flush, the effective flush volume shall be determined in accordance with ASME A112.19.14 and shall be 1.28 gal (4.8 L).
b. Water closets (toilets)—tank-type: Tank-type water closets shall be certified to the performance criteria of the U.S. EPA WaterSense Tank-Type High-Efficiency Toilet Specification and shall have a maximum flush volume of 1.28 gal (4.8 L).
c. Urinals: Maximum flush volume when determined in accordance with ASME A112.19.2/CSA B45.1—0.5 gal (1.9 L). Non-water urinals shall comply with ASME A112.19.19 (vitreous china) or IAPMO Z124.9 (plastic) as appropriate.
d. Public lavatory faucets: Maximum flow rate—0.5 gpm (1.9 L/min) when tested in accordance with ASME A112.18.1/CSA B125.1.
e. Public metering self-closing faucet: Maximum water use—0.25 gal (1.0 L) per metering cycle when tested in accordance with ASME A112.18.1/CSA B125.1.
f. *Residential* bathroom lavatory sink faucets: Maximum flow rate—1.5 gpm (5.7 L/min) when tested in accordance with ASME A112.18.1/CSA B125.1. *Residential* bathroom lavatory sink faucets shall comply with the performance criteria of the USEPA WaterSense High-Efficiency Lavatory Faucet Specification.
g. *Residential* kitchen faucets: Maximum flow rate—2.2 gpm (8.3 L/min) when tested in accordance with ASME A112.18.1/CSA B125.1.
h. *Residential* showerheads: Maximum flow rate—2.0 gpm (7.6 L/min) when tested in accordance with ASME A112.18.1/CSA B125.1.
i. *Residential* shower compartment (stall) in *dwelling units* and guest rooms: The allowable flow rate from all shower outlets (including rain systems, waterfalls, bodysprays, and jets) that can operate simultaneously shall be limited to a total of 2.0 gpm (7.6 L/min).

   Exception: Where the area of a shower compartment exceeds 2600 in.^{2} (1.7 m^2), an additional flow of 2.0 gpm (7.6 L/min) shall be permitted for each multiple of 2600 in.^{2} (1.7 m^2) of floor area or fraction thereof.

6.3.2.2 Appliances

a. Clothes washers and dishwashers installed within *dwelling units* shall comply with the ENERGY STAR Program Requirements for Clothes Washers and ENERGY STAR Program Requirements for Dishwashers. Maximum water use shall be as follows:
   1. Clothes Washers—maximum *Water Factor* of 6.0 gal/ft^3 of drum capacity (800 L/m^3 of drum capacity).
   2. Dishwashers—maximum *Water Factor* of 5.8 gal/full operating cycle (22 L/full operating cycle).

   (See also the energy efficiency requirements in Section 7.4.7.3.)

b. Clothes washers installed in publicly accessible spaces (e.g., multifamily and hotel common areas) and coin- and card-operated clothes washers of any size used in laundromats shall have a maximum *Water Factor* of

TABLE 6.3.2.1 Plumbing Fixtures and Fittings Requirements

Plumbing Fixture	Maximum
Water closets (toilets)—flushometer valve type	Single flush volume of 1.28 gal (4.8 L)
Water closets (toilets)—flushometer valve type	Effective dual flush volume of 1.28 gal (4.8 L)
Water closets (toilets)—tank-type	Single flush volume of 1.28 gal (4.8 L)
Water closets (toilets)—tank-type	Effective dual flush volume of 1.28 gal (4.8 L)
Urinals	Flush volume 0.5 gal (1.9 L)
Public lavatory faucets	Flow rate—0.5 gpm (1.9 L/min)
Public metering self-closing faucet	0.25 gal (1.0 L) per metering cycle
Residential bathroom lavatory sink faucets	Flow rate—1.5 gpm (5.7 L/min)
Residential kitchen faucets	Flow rate— 2.2 gpm (8.3 L/min)
Residential showerheads	Flow rate—2.0 gpm (7.6 L/min)
Residential shower compartment (stall) in *dwelling units* and guest rooms	Flow rate from all shower outlets total of 2.0 gpm (7.6 L/min)

7.5 gal/ft^3 of drum capacity-normal cycle (1.0 kL/m^{3}of drum capacity-normal cycle). (See also the energy efficiency requirements in Sections 7.4.7.3 and 7.4.7.4.)

6.3.2.3 HVAC Systems and Equipment

a. *Once-through cooling* with *potable water* is prohibited.
b. Cooling towers and evaporative coolers shall be equipped with makeup and blowdown meters, conductivity controllers, and overflow alarms in accordance with the thresholds listed in Table 6.3.3B. Cooling towers shall be equipped with efficient drift eliminators that achieve drift reduction to a maximum of 0.002% of the recirculated water volume for counterflow towers and 0.005% of the recirculated water flow for cross-flow towers.
c. *Building projects* located in regions where the ambient mean coincident wet-bulb temperature at 1% design cooling conditions is greater than or equal to 72°F (22°C) shall have a system for collecting condensate from air-conditioning units with a capacity greater than 65,000 Btu/h (19 kW), and the condensate shall be recovered for re-use.

6.3.2.4 Roofs

a. The use of *potable water* for *roof* spray systems to thermally condition the *roof* is prohibited.
b. The use of *potable water* for irrigation of vegetated (green) *roofs* is prohibited once plant material has been established. After the *landscape establishment period* is completed, the *potable water* irrigation system shall be removed or permanently disconnected.

6.3.3 Water Consumption Measurement

6.3.3.1 Consumption Management. Measurement devices with remote communication capability shall be provided to collect water consumption data for the domestic water supply to the building. Both potable and reclaimed water entering the *building project* shall be monitored or submetered. In addition, for individual leased, rented, or other tenant or sub-tenant space within any building totaling in excess of 50,000 ft^2 (5000 m^2), separate submeters shall be provided. For subsystems with multiple similar units, such as multi-cell cooling towers, only one measurement device is required for the subsystem. Any project or building, or tenant or sub-tenant space within a project or building, such as a commercial car wash or aquarium, shall be submetered where consumption is projected to exceed 1000 gal/day (3800 L/day).

TABLE 6.3.3A Water Supply Source Measurement Thresholds

Water Source	Main Measurement Threshold
Potable water	1000 gal/day (3800 L/day)
Municipally reclaimed water	1000 gal/day (3800 L/day)
Alternate sources of water	500 gal/day (1900 L/day)

Measurement devices with remote capability shall be provided to collect water use data for each water supply source (e.g., *potable water,* reclaimed water, rainwater) to the *building project* that exceeds the thresholds listed in Table 6.3.3A. Utility company service entrance/interval meters are allowed to be used.

Provide sub-metering with remote communication measurement to collect water use data for each of the building subsystems, if such subsystems are sized above the threshold levels listed in Table 6.3.3B.

6.3.3.2 Consumption Data Collection. All building measurement devices, monitoring systems, and sub-meters installed to comply with the thresholds limits in Section 6.3.3.1 shall be configured to communicate water consumption data to a meter data management system. At a minimum, meters shall provide daily data and shall record hourly consumption of water.

TABLE 6.3.3B Subsystem Water Measurement Thresholds

Subsystem	Sub-Metering Threshold
Cooling towers (meter on makeup water and blowdown)	Cooling tower flow through tower >500 gpm (30 L/s)
Evaporative coolers	Makeup water >0.6 gpm (0.04 L/s)
Steam and hot-water boilers	>500,000 Btu/h (50 kW) input
Total Irrigated landscape area with controllers	>25,000 ft^2 (2500 m^2)
Separate campus or project buildings	Consumption >1000 gal/day (3800 L/day)
Separately leased or rental space	Consumption >1000 gal/day (3800 L/day)
Any large water using process	Consumption >1000 gal/day (3800 L/day)

6.3.3.3 Data Storage and Retrieval. The meter data management system shall be capable of electronically storing water meter, monitoring systems, and submeter data and creating user reports showing calculated hourly, daily, monthly, and annual water consumption for each measurement device and submeter and provide alarming notification capabilities as needed to support the requirements of the Water User Efficiency Plan for Operation in Section 10.3.2.1.2.

6.4 Prescriptive Option

6.4.1 Site Water Use Reduction. For golf courses and driving ranges, only municipally reclaimed water and/or *alternate on-site sources of water* shall be used to irrigate the landscape. For other landscaped areas, a maximum of one-third of *improved landscape* area is allowed to be irrigated with *potable water*. The area of dedicated athletic fields shall be excluded from the calculation of the *improved landscape* for schools, *residential* common areas, or public recreational facilities. All other irrigation shall be provided from *alternate on-site sources of water* or municipally reclaimed water.

Exception: *Potable water* is allowed to be temporarily used on such newly installed landscape for the *landscape establishment period*. The amount of *potable water* that may be applied to the newly planted areas during the temporary *landscape establishment period* shall not exceed 70% of ET_o for *turfgrass* and 55% of ET_o for other plantings. If municipally-reclaimed water is available at a water main within 200 ft (60 m) of the project *site*, it shall be used in lieu of *potable water* during the *landscape establishment period*. After the *landscape establishment period* has expired, all irrigation water use shall comply with the requirements established elsewhere in this standard.

6.4.2 Building Water Use Reduction

6.4.2.1 Cooling Towers. The water being discharged from cooling towers for air conditioning systems such as chilled-water systems shall be limited in accordance with method (a) or (b):

a. For makeup waters having less than 200 ppm (200 mg/L) of total hardness expressed as calcium carbonate, by achieving a minimum of five *cycles of concentration*.

b. For makeup waters with more than 200 ppm (200 mg/L) of total hardness expressed as calcium carbonate, by achieving a minimum of 3.5 *cycles of concentration*.

Exception: Where the total dissolved solids concentration of the discharge water exceeds 1500 mg (1500 ppm/L), or the silica exceeds 150 ppm (150 mg/L) measured as silicon dioxide before the above cycles of concentration are reached.

6.4.2.2 Commercial Food Service Operations. Commercial food service operations (e.g., restaurants, cafeterias, food preparation kitchens, caterers, etc.):

a. shall use high-efficiency pre-rinse spray valves (i.e., valves which function at 1.3 gpm (4.9 L/min) or less and comply with a 26-second performance requirement when tested in accordance with ASTM F2324),
b. shall use dishwashers that comply with the requirements of the ENERGY STAR Program for Commercial Dishwashers,
c. shall use boilerless/connectionless food steamers that consume no more than 2.0 gal/hour (7.5 L/hour) in the full operational mode,
d. shall use combination ovens that consume not more than 10 gal/hour (38 L/hour) in the full operational mode,
e. shall use air-cooled ice machines that comply with the requirements of the ENERGY STAR Program for Commercial Ice Machines, and
f. shall be equipped with hands-free faucet controllers (foot controllers, sensor-activated, or other) for all faucet fittings within the food preparation area of the kitchen and the dish room, including pot sinks and washing sinks.

6.4.2.3 Medical and Laboratory Facilities. Medical and laboratory facilities, including clinics, hospitals, medical centers, physician and dental offices, and medical and non-medical laboratories of all types shall:

a. use only water-efficient steam sterilizers equipped with (1) water-tempering devices that allow water to flow only when the discharge of condensate or hot water from the sterilizer exceeds 140°F (60°C) and (2) mechanical vacuum equipment in place of venturi-type vacuum systems for vacuum sterilizers.

b. use film processor water recycling units where large frame x-ray films of more than 6 in. (150 mm) in either length or width are processed. Small dental x-ray equipment is exempt from this requirement.
c. use digital imaging and radiography systems where the digital networks are installed.
d. use a dry-hood scrubber system or, if the applicant determines that a wet-hood scrubber system is required, the scrubber shall be equipped with a water recirculation system. For perchlorate hoods and other applications where a hood wash-down system is required, the hood shall be equipped with self-closing valves on those wash-down systems.
e. use only dry vacuum pumps, unless fire and safety codes for explosive, corrosive or oxidative gasses require a liquid ring pump.
f. use only efficient water treatment systems that comply with the following criteria:
   1. For all filtration processes, pressure gauges shall determine and display when to backwash or change cartridges.
   2. For all ion exchange and softening processes, recharge cycles shall be set by volume of water treated or based upon conductivity or hardness.
   3. For reverse osmosis and nanofiltration equipment, with capacity greater than 27 gal/h (100 L/h), reject water shall not exceed 60% of the feed water and shall be used as scrubber feed water or for other beneficial uses on the project *site*.
   4. Simple distillation is not acceptable as a means of water purification.
g. Food service operations within medical facilities shall comply with Section 6.4.2.2.

6.4.3 Special Water Features. Water use shall comply with the following:

a. Ornamental fountains and other ornamental water features shall be supplied either by *alternate on-site sources of water* or by municipally reclaimed water delivered by the local water utility acceptable to the *AHJ*. Fountains and other features shall be equipped with: (1) makeup water meters (2) leak detection devices that shut off water flow if a leak of more than 1.0 gal/h (3.8 L/h) is detected, and (3) equipment to recirculate, filter, and treat all water for reuse within the system.

   Exception: Where *alternate on-site sources of water* or municipally reclaimed water are not available within 500 ft (150 m) of the *building project site*, *potable water* is allowed to be used for water features with less than 10,000 gallon (38,000 L) capacity.

b. Pools and spas:
   1. Backwash water: Recover filter backwash water for reuse on landscaping or other applications, or treat and reuse backwash water within the system.
   2. Filtration: For filters with removable cartridges, only reusable cartridges and systems shall be used. For filters with backwash capability, use only pool filter equipment that includes a pressure drop gauge to determine when the filter needs to be backwashed and a sight glass enabling the operator to determine when to stop the backwash cycle.
   3. Pool splash troughs, if provided, shall drain back into the pool system.

6.5 Performance Option. Calculations shall be done in accordance with *generally accepted engineering standards* and handbooks acceptable to the *AHJ*.

6.5.1 Site Water Use Reduction. *Potable water* (and municipally reclaimed water, where used) intended to irrigate *improved landscape* shall be limited to 35% of the water demand for that landscape. The water demand shall be based upon *ET* for that climatic area and shall not exceed 70% of ET_o for *turfgrass* areas and 55% of ET_o for all other plant material after adjustment for rainfall.

6.5.2 Building Water Use Reduction. The *building project* shall be designed to have a total annual interior water use less than or equal to that achieved by compliance with Sections 6.3.2, 6.4.2, and 6.4.3.

7. ENERGY EFFICIENCY

7.1 Scope. This section specifies requirements for energy efficiency for buildings and appliances, for *on-site renewable energy systems*, and for energy measuring.

7.2 Compliance. The energy systems shall comply with Section 7.3, "Mandatory Provisions," and either

a. Section 7.4, "Prescriptive Option," or
b. Section 7.5, "Performance Option."

7.3 Mandatory Provisions

7.3.1 General. *Building projects* shall be designed to comply with Sections 5.4, 6.4, 7.4, 8.4, 9.4, and 10.4 of ANSI/ASHRAE/IES Standard 90.1.

7.3.2 On-Site Renewable Energy Systems. *Building project* design shall show allocated space and pathways for future installation of *on-site renewable energy systems* and associated infrastructure that provide the annual energy production equivalent of not less than 6.0 kBtu/ft^2 (20 kWh/m^2) for single-story buildings and not less than 10.0 kBtu/ft^2 (32 kWh/m^2) multiplied by the total *roof* area in ft^2 (m^2) for all other buildings.

Exceptions:

1. *Building projects* that have an annual daily average incident solar radiation available to a flat plate collector oriented due south at an angle from horizontal equal to the latitude of the collector location less than 1.2 kBtu/ft^2·day (4.0 kWh/m^2·day), accounting for existing buildings, permanent infrastructure that is not part of the *building project*, topography, or trees.
2. *Building projects* that comply with Section 7.4.1.1.

7.3.3 Energy Consumption Management

7.3.3.1 Consumption Management. Measurement devices with remote communication capability shall be provided to collect energy consumption data for each energy supply source to the building, including gas, electricity, and district energy, that exceeds the thresholds listed in Table 7.3.3.1A. The measurement devices shall have the capability to automatically communicate the energy consumption data to a data acquisition system.

For all buildings that exceed the threshold in Table 7.3.3.1A, subsystem measurement devices with remote capability (including current sensors or flowmeters) shall be provided to measure energy consumption data of each subsystem for each use category that exceeds the thresholds listed in Table 7.3.3.1B.

The energy consumption data from the subsystem measurement devices shall be automatically communicated to the data acquisition system.

7.3.3.2 Energy Consumption Data Collection. All building measurement devices shall be configured to automatically communicate the energy data to the data acquisition system. At a minimum, measurement devices shall provide daily data and shall record hourly energy profiles. Such hourly energy profiles shall be capable of being used to assess building performance at least monthly.

7.3.3.3 Data Storage and Retrieval. The data acquisition system shall be capable of electronically storing the data from the measurement devices and other sensing devices, for a minimum of 36 months, and creating user reports showing hourly, daily, monthly, and annual energy consumption.

Exception: Portions of buildings used as *residential.*

7.4 Prescriptive Option

7.4.1 General Comprehensive Prescriptive Requirements. When a requirement is provided below, it supersedes the requirement in ANSI/ASHRAE/IES Standard 90.1. For all other criteria, the *building project* shall comply with the requirements of ANSI/ASHRAE/IES Standard 90.1.

7.4.1.1 On-Site Renewable Energy Systems. *Building projects* shall contain *on-site renewable energy systems* that provide the annual energy production equivalent of not less than 6.0 kBtu/ft^2 (20 kWh/m^2) multiplied by the total *roof* area in ft^2 (m^2) for single-story buildings and not less than 10.0 kBtu/ft^2 (32 kWh/m^2) multiplied by the total *roof* area in ft^2 (m^2) for all other buildings. The annual energy production

TABLE 7.3.3.1A Energy Source Thresholds

Energy Source	Threshold
Electrical service	>200 kVA
On-site renewable electric power	All systems > 1 kVA (peak)
Gas and district services	>1,000,000 Btu/h (300 kW)
Geothermal energy	>1,000,000 Btu/h (300 kW) heating
On-site renewable thermal energy	>100,000 Btu/h (30 kW)

TABLE 7.3.3.1B System Energy Use Thresholds

Use (Total of All Loads)	Subsystem Threshold
HVAC system	Connected electric load > 100kVA
HVAC system	Connected gas or district services load > 500,000 Btu/h (150 kW)
People moving	Sum of all feeders > 50 kVA
Lighting	Connected load > 50 kVA
Process and plug process	Connected load > 50 kVA Connected gas or district services load > 250,000 Btu/h (75 kW)

shall be the combined sum of all *on-site renewable energy systems*.

Exception: Buildings that demonstrate compliance with both of the following are not required to contain *on-site renewable energy systems*:

1. An annual daily average incident solar radiation available to a flat plate collector oriented due south at an angle from horizontal equal to the latitude of the collector location less than 4.0 kWh/m^2·day, accounting for existing buildings, permanent infrastructure that is not part of the *building project*, topography, and trees.
2. A commitment to purchase renewable electricity products complying with the Green-e Energy National Standard for Renewable Electricity Products of at least 7 kWh/ft^2 (75 kWh/m^2) of conditioned space each year until the cumulative purchase totals 70 kWh/ft^2 (750 kWh/m^2) of conditioned space.

7.4.2 Building Envelope. The *building envelope* shall comply with Section 5 of ANSI/ASHRAE/IES Standard 90.1 with the following modifications and additions:

7.4.2.1 Building Envelope Requirements. The *building envelope* shall comply with the requirements in Tables A-1 to A-8 in Normative Appendix A. These requirements supersede the requirements in Tables 5.5-1 to 5.5-8 of ANSI/ASHRAE/IES Standard 90.1.

Exception: Buildings that comply with Section 8.3.4 regardless of building area are exempt from the *SHGC* criteria for *skylights*.

7.4.2.2 Roof Insulation. *Roofs* shall comply with the provisions of Section 5.3.2.3 and Tables A-1 to A-8 of this standard. Section 5.5.3.1.1 of ANSI/ASHRAE/IES Standard 90.1 and Table 5.5.3.1 of ANSI/ASHRAE/IES Standard 90.1 shall not apply.

7.4.2.3 Single-Rafter Roof Insulation. *Single-rafter roofs* shall comply with the requirements in Table A-9 in Normative Appendix A. These requirements supersede the requirements in Section A2.4.2.4 of ANSI/ASHRAE/IES Standard 90.1. Section A2.4.2.4 and Table A2.4.2 of ANSI/ASHRAE/IES Standard 90.1 shall not apply.

7.4.2.4 Vertical Fenestration Area. The total *vertical fenestration area* shall be less than 40% of the *gross wall area*. This requirement supersedes the requirement in Section 5.5.4.2.1 of ANSI/ASHRAE/IES Standard 90.1.

7.4.2.5 Permanent Projections. For *climate zones* 1–5, the *vertical fenestration* on the west, south, and east shall be shaded by permanent projections that have an area-weighted average *PF* of not less than 0.50. The building is allowed to be rotated up to 45 degrees to the nearest cardinal orientation for purposes of calculations and showing compliance.

Exceptions:

1. *Vertical fenestration* that receives direct solar radiation for fewer than 250 hours per year because of shading by permanent external buildings, existing permanent infrastructure, or topography.
2. *Vertical fenestration* with automatically controlled shading devices capable of modulating in multiple steps the amount of solar gain and light transmitted into the space in response to daylight levels or solar intensity that comply with all of the following:
   a. Exterior shading devices shall be capable of providing at least 90% coverage of the *fenestration* in the closed position.
   b. Interior shading devices shall be capable of providing at least 90% coverage of the *fenestration* in the closed position and have a minimum solar reflectance of 0.50 for the surface facing the *fenestration*.
   c. A manual override located in the same *enclosed space* as the *vertical fenestration* shall override operation of automatic controls no longer than 4 hours.
   d. Acceptance testing and commissioning shall be conducted as required by Section 10 to verify that automatic controls for shading devices respond to changes in illumination or radiation intensity.
3. *Vertical fenestration* with automatically controlled *dynamic glazing* capable of modulating in multiple steps the amount of solar gain and light transmitted into the space in response to daylight levels or solar intensity that comply with all of the following:
   a. *Dynamic glazing* shall have a lower labeled *SHGC* equal to or less than 0.12, lowest labeled *VT* no greater than 0.05, and highest labeled *VT* no less than 0.40.
   b. A manual override located in the same *enclosed space* as the *vertical fenestration* shall override operation of automatic controls no longer than 4 hours.
   c. Acceptance testing and commissioning shall be conducted as required by Section 10 to verify that automatic controls for *dynamic glazing* respond to changes in illumination or radiation intensity.

7.4.2.6 SHGC of Vertical Fenestration. For *SHGC* compliance, the methodology in exception (b) to Section 5.5.4.4.1 of ANSI/ASHRAE/IES Standard 90.1 is allowed, provided that the *SHGC* multipliers in Table 7.4.2.6 are used. This requirement supersedes the requirement in Table 5.5.4.4.1 of ANSI/ASHRAE/IES Standard 90.1. Table 5.5.4.4.1 of ANSI/ASHRAE/IES Standard 90.1 shall not apply. *Vertical fenestration* that is *north-oriented* shall be allowed to have a maximum *SHGC* of 0.10 greater than that specified in Tables A-1 through A-8 in Normative Appendix A. When this exception is utilized, separate calculations shall be performed for these sections of the *building envelope*, and these values shall not be averaged with any others for compliance purposes.

7.4.2.7 Building Envelope Trade-Off Option. The *building envelope* trade-off option in Section 5.6 of ANSI/ASHRAE/IES Standard 90.1 shall not apply unless the procedure incorporates the modifications and additions to ANSI/ASHRAE/IES Standard 90.1 noted in Section 7.4.2.

TABLE 7.4.2.6 SHGC Multipliers for Permanent Projections

PF	*SHGC* Multiplier (All Other Orientations)	*SHGC* Multiplier (North-Oriented)
0–0.60	1.00	1.00
>0.60–0.70	0.92	0.96
>0.70–0.80	0.84	0.94
>0.80–0.90	0.77	0.93
>0.90–1.00	0.72	0.90

7.4.2.8 Fenestration Orientation. To reduce solar gains from the east and west in *climate zones* 1 through 4 and from the west in *climate zones* 5 and 6, the *fenestration area* and *SHGC* shall comply with the following requirements:

a. For *climate zones* 1, 2, 3, and 4:

$$(A_N \times SHGC_N + A_S \times SHGC_S) \geq 1.1 \times (A_E \times SHGC_E + A_W \times SHGC_W)$$

b. For *climate zones* 5 and 6:

$$1/3 \times (A_N \times SHGC_N + A_S \times SHGC_S + A_E \times SHGC_E) \geq 1.1 \times (A_W \times SHGC_W)$$

where

$SHGC_x$ = the *SHGC* for orientation x
A_x = *fenestration area* for orientation x
N = north (oriented less than 45 degrees of true north)
S = south (oriented less than 45 degrees of true south)
E = east (oriented less than or equal to 45 degrees of true east)
W = west (oriented less than or equal to 45 degrees of true west)

Exceptions:

a. *Vertical fenestration* that complies with the exception to Section 5.5.4.4.1 (c) of ANSI/ASHRAE/IES Standard 90.1.
b. Buildings that have an existing building or existing permanent infrastructure within 20 ft (6 m) to the south or north that is at least half as tall as the proposed building.
c. Buildings with shade on 75% of the west- and east-oriented *vertical fenestration areas* from existing buildings, existing permanent infrastructure, or topography at 9 a.m. and 3 p.m. on the summer solstice.
d. Alterations and additions with no increase in vertical fenestration area.

7.4.2.9 Continuous Air Barrier. The *building envelope* shall be designed and constructed with a *continuous air barrier* that complies with Normative Appendix B to control air leakage into, or out of, the *conditioned space*. All air barrier components of each envelope assembly shall be clearly identified on construction documents and the joints, interconnections, and penetrations of the air barrier components shall be detailed.

Exception: *Building envelopes* of *semiheated spaces* provided that the *building envelope* complies with Section 5.4.3.1 of ANSI/ASHRAE/IES Standard 90.1.

7.4.3 Heating, Ventilating, and Air Conditioning. The heating, ventilating, and air conditioning shall comply with Section 6 of ANSI/ASHRAE/IES Standard 90.1 with the following modifications and additions.

7.4.3.1 Minimum Equipment Efficiencies. Projects shall comply with one of the following:

a. **EPAct baseline.** Products shall comply with the minimum efficiencies addressed in the National Appliance Energy Conservation Act (NAECA), Energy Policy Act (EPAct), and the Energy Independence and Security Act (EISA).
b. **Higher Efficiency.** Products shall comply with the greater of the ENERGY STAR requirements in Section 7.4.7.3 and the values in Normative Appendix C. These requirements supersede the requirements in Tables 6.8.1A to 6.8.1G of ANSI/ASHRAE/IES Standard 90.1. The building project shall comply with Sections 7.4.1.1 and 7.4.5.1 with the following modifications:
   1. The *on-site renewable energy systems* required in Section 7.4.1.1 shall provide an annual energy production of not less than 4.0 kBtu/ft^2 (13 kWh/m^2) multiplied by the total *roof* area in ft^2 (m^2) for single-story buildings and not less than 7.0 kBtu/ft^2 (22 kWh/m^2) multiplied by the total *roof* area in ft^2 (m^2) for all other buildings.
   2. The peak load reduction systems required in Section 7.4.5.1 shall be capable of reducing electric peak demand by not less than 5% of the projected peak demand.

7.4.3.2 Ventilation Controls for Densely Occupied Spaces. *DCV* is required for *densely occupied spaces*. This requirement supersedes the occupant density threshold in Section 6.4.3.9 of ANSI/ASHRAE/IES Standard 90.1.

The *DCV* system shall be designed to be in compliance with ANSI/ASHRAE Standard 62.1. Occupancy assumptions shall be shown in the design documents for spaces required to have *DCV*. All CO_2 sensors used as part of a *DCV* system or any other system that dynamically controls outdoor air shall meet the following requirements:

a. Spaces with CO_2 sensors or air sampling probes leading to a central CO_2 monitoring station shall have one sensor or probe for each 10,000 ft^2 (1000 m^2) of floor space and shall be located in the room between 3 and 6 ft (1 and 2 m) above the floor.
b. CO_2 sensors must be accurate to ±50 ppm at 1000 ppm.
c. *Outdoor air* CO_2 concentrations shall be determined by one of the following:

TABLE 7.4.3.3 Minimum System Size for Which an Economizer is Required

Climate Zones	Cooling Capacity for Which an Economizer is Required*
1A, 1B	No economizer requirement
2A, 2B, 3A, 3B, 3C, 4A, 4B, 4C, 5A, 5B, 5C, 6A, 6B, 7, 8	≥33,000 Btu/h (9.7 kW)[a]

* Where economizers are required, the total capacity of all systems without economizers shall not exceed 480,000 Btu/h (140 kW) per building or 20% of the building's air economizer capacity, whichever is greater.

1. *Outdoor air* CO_2 concentrations shall be dynamically measured using a CO_2 sensor located in the path of the *outdoor air* intake.
2. When documented statistical data are available on the local ambient CO_2 concentrations, a fixed value typical of the location where the building is located shall be allowed in lieu of an outdoor sensor.

d. Occupant CO_2 generation rate assumptions shall be shown in the design documents

7.4.3.3 Economizers. Systems shall have economizers meeting the requirements in Section 6.5.1 of ANSI/ASHRAE/IES 90.1 except as noted below.

1. The minimum size requirements for economizers are defined in Table 7.4.3.3 and supersede the requirements in Table 6.5.1 of ANSI/ASHRAE/IES Standard 90.1.
2. Rooftop units with a capacity of less than 60,000 Btu/h (18 kW) shall have two stages of capacity control, with the first stage used for cooling with the economizer and the second stage to add mechanical cooling.
3. For systems that control to a fixed leaving air temperature (i.e., *VAV* systems), the system shall be capable of resetting the supply air temperature up at least 5°F (3°C) during economizer operation.

Exceptions: All the exceptions in Section 6.5.1 of ANSI/ASHRAE/IES Standard 90.1 shall apply except as noted below.

1. The use of exception (i) to Section 6.5.1 of ANSI/ASHRAE/IES Standard 90.1 shall be permitted to eliminate the economizer requirement provided the requirements in Table 6.3.2 of ANSI/ASHRAE/IES Standard 90.1 are applied to the efficiency requirements required by Section 7.4.3.1.
2. For water-cooled units with a capacity less than 54,000 Btu/h (16 kW) that are used in systems where heating and cooling loads are transferred within the building (i.e., water-source heat pump systems), the requirement for an air or water economizer can be eliminated if the condenser-water temperature controls are capable of being set to maintain full load heat rejection capacity down to a 55°F (12°C) condenser-water supply temperature and the HVAC equipment is capable of operating with a 55°F (12°C) condenser-water supply temperature.

7.4.3.4 Zone Controls. The exceptions to Section 6.5.2.1 of ANSI/ASHRAE/IES Standard 90.1 shall be modified as follows:

1. Exception (a) shall not be used.
2. Exception (b)1.ii shall be replaced by the following text: "the design outdoor airflow rate for the zone."

7.4.3.5 Fan System Power Limitation. Systems shall have fan power limitations 10% below limitations specified in Table 6.5.3.1.1A of ANSI/ASHRAE/IES Standard 90.1. This requirement supersedes the requirement in Section 6.5.3.1 and Table 6.5.3.1.1A of ANSI/ASHRAE/IES Standard 90.1. All exceptions in Section 6.5.3.1 of ANSI/ASHRAE/IES Standard 90.1 shall apply.

7.4.3.6 Exhaust Air Energy Recovery. The exhaust air energy recovery requirements defined in Section 6.5.6.1 of ANSI/ASHRAE/IES Standard 90.1 shall be used except that the energy recovery effectiveness shall be 60% and the requirements of Table 7.4.3.6 shall be used instead of those of Table 6.5.6.1 of ANSI/ASHRAE/IES Standard 90.1.

7.4.3.7 Variable-Speed Fan Control for Commercial Kitchen Hoods. In addition to the requirements in Section 6.5.7.1 of ANSI/ASHRAE/IES Standard 90.1, commercial kitchen Type I and Type II hood systems shall have variable-speed control for exhaust and makeup air fans to reduce hood airflow rates at least 50% during those times when cooking is not occurring and the cooking appliances are up to temperature in a standby, ready-to-cook mode. All exceptions in Section 6.5.7.1 of ANSI/ASHRAE/IES Standard 90.1 shall apply.

7.4.3.8 Duct Insulation. Duct insulation shall comply with the minimum requirements in Tables C-9 and C-10 in Normative Appendix C. These requirements supersede the requirements in Tables 6.8.2A and 6.8.2B of ANSI/ASHRAE/IES Standard 90.1.

7.4.3.9 Automatic Control of HVAC and Lights in Hotel/Motel Guest Rooms. In hotels and motels with over 50 guest rooms, the lighting, switched outlets, television, and HVAC equipment serving each guest room shall be automatically controlled such that the power for lighting, switched outlets, and televisions will be turned off within 30 minutes after all occupants leave the guest room and the HVAC setpoint raised by at least 5°F (3°C) in the cooling mode and lowered by at least 5°F (3°C) in the heating mode within 30 minutes after all occupants leave the guest room.

Exception: Guest rooms where the lighting, switched outlets, and televisions are turned off and the HVAC setpoints are raised by at least 5°F (3°C) in the cooling mode and lowered by at least 5°F (3°C) in the heating mode when the occupant removes the card from a captive key system.

TABLE 7.4.3.6 Energy Recovery Requirement (I-P)

	% Outside Air at Full Design Flow							
Climate Zone	≥10% and <20%	≥20% and <30%	≥30% and <40%	≥40% and <50%	≥50% and <60%	≥60% and <70%	≥70% and <80%	≥80%
	Design Supply Fan Flow, cfm							
3B, 3C, 4B, 4C, 5B	NR	NR	NR	NR	NR	NR	≥5000	≥5000
1B, 2B, 5C	NR	NR	NR	NR	≥26,000	≥12,000	≥5000	≥4000
6B	NR	≥22,500	≥11,000	≥5500	≥4500	≥3500	≥2500	≥1500
1A, 2A, 3A, 4A, 5A, 6A	≥30,000	≥13,000	≥5500	≥4500	≥3500	≥2000	≥1000	≥0
7, 8	≥4000	≥3000	≥2500	≥1000	≥0	≥0	≥0	≥0

TABLE 7.4.3.6 Energy Recovery Requirement (SI)

	% Outside Air at Full Design Flow							
Climate Zone	≥10% and <20%	≥20% and <30%	≥30% and <40%	≥40% and <50%	≥50% and <60%	≥60% and <70%	≥70% and <80%	≥80%
	Design Supply Fan Flow, L/s							
3B, 3C, 4B, 4C, 5B	NR	NR	NR	NR	NR	NR	≥2360	≥2360
1B, 2B, 5C	NR	NR	NR	NR	≥12,271	≥5663	≥2360	≥1888
6B	NR	≥10,619	≥5191	≥2596	≥2124	≥1652	≥1180	≥708
1A, 2A, 3A, 4A, 5A, 6A	≥14,158	≥6135	≥2596	≥2124	≥1652	≥944	≥472	>0
7, 8	≥1888	≥1416	≥1180	≥472	>0	>0	>0	>0

7.4.4 Service Water Heating. The *service water heating* shall comply with Section 7 of ANSI/ASHRAE/IES Standard 90.1 with the following modifications and additions.

7.4.4.1 Equipment Efficiency. Equipment shall comply with the minimum efficiencies in Table C-11 in Normative Appendix C. These requirements supersede the requirements in Table 7.8 of ANSI/ASHRAE/IES Standard 90.1.

7.4.4.2 Insulation for Spa Pools. Pools heated to more than 90°F (32°C) shall have side and bottom surfaces insulated on the exterior with a minimum insulation value of R-12 (R-2.1).

7.4.5 Power. The power shall comply with Section 8 of ANSI/ASHRAE/IES Standard 90.1 with the following modifications and additions.

7.4.5.1 Peak Load Reduction. *Building projects* shall contain automatic systems, such as demand limiting or load shifting, that are capable of reducing electric peak demand of the building by not less than 10% of the projected peak demand. Standby power generation shall not be used to achieve the reduction in peak demand.

7.4.6 Lighting. The lighting shall comply with Section 9 of ANSI/ASHRAE/IES Standard 90.1 and the following modifications and additions.

7.4.6.1 Lighting Power Allowance. The interior *lighting power allowance* shall be a maximum of the values determined in accordance with Sections 9.5 and 9.6 of ANSI/ASHRAE/IES Standard 90.1 multiplied by an LPD Factor specified in Table 7.4.6.1A for those areas where the Building Area Method is used and in Table 7.4.6.1B for those areas where the Space-by-Space Method is used. Control factors from Table 9.6.2 in ANSI/ASHRAE/IES Standard 90.1 shall not be used for the control methodologies required in this standard. The exterior *lighting power allowance* shall be a maximum of the values determined in accordance with Sections 9.4.3. of ANSI/ASHRAE/IES Standard 90.1 multiplied by the corresponding factor found in Table 7.4.6.1C. This requirement supersedes the requirements in Sections 9.4.3 of ANSI/ASHRAE/IES Standard 90.1.

7.4.6.2 Occupancy Sensor Controls with Multi-Level Switching or Dimming. The lighting in the following areas shall be controlled by an occupant sensor with multi-level switching or dimming system that reduces lighting power a minimum of 50% when no persons are present:

a. Hallways in multifamily, dormitory, hotel, and motel buildings.
b. Commercial and industrial storage stack areas.
c. Library stack areas.

Exception: Areas lit by HID lighting with a lighting power density of 0.8 W/ft^2 or less.

7.4.6.3 Automatic Controls for Egress and Security Lighting. Lighting in any area within a building that is required to be continuously illuminated for reasons of building security or emergency egress shall not exceed 0.1 W/ft^2

TABLE 7.4.6.1A LPD Factors when Using the Building Area Method

Building Area Type	LPD Factor
Courthouse	0.95
Dining—Cafeteria/Fast Food	0.95
Dining—Family	0.95
Dormitory	0.95
Exercise Center	0.95
Healthcare Clinic	0.95
Hospital	0.95
Library	0.95
Multifamily	0.95
Office	0.95
Penitentiary	0.95
Police Station	0.95
Religious Building	0.95
School/University	0.90
Town Hall	0.95
Transportation	0.95
All Other Building Area Types	1.00

(1 W/m^2). Additional egress and security lighting shall be allowed, provided it is controlled by an automatic control device that turns off the additional lighting.

7.4.6.4 Occupancy Sensors. Occupancy sensors shall have "manual ON", "automatic OFF" controls or shall be controlled to automatically turn the lighting on to not more than 50% power, except in the following spaces where full automatic-on is allowed:

1. occupancy sensor controls required in Section 7.4.6.2,
2. public corridors and stairwells,
3. restrooms,
4. primary building entrance areas and lobbies, and
5. areas where manual-on operation would endanger the safety or security of the room or building occupant(s).

7.4.6.5 Controls for Exterior Sign Lighting. All exterior sign lighting, including internally illuminated signs and lighting on externally illuminated signs, shall comply with the requirements of Sections 7.4.6.5.1 or 7.4.6.5.2.

Exceptions:

a. Sign lighting that is specifically required by a health or life safety statute, ordinance, or regulation.

b. Signs in tunnels.

7.4.6.5.1 All sign lighting that operates more than one hour per day during *daylight hours* shall include controls to automatically reduce the input power to a maximum of 35% of full power for a period from one hour after sunset to one hour before sunrise.

TABLE 7.4.6.1B LPD Factors when Using the Space-by-Space Method

Common Space Type	LPD Factor
Classroom/Lecture/Training	0.85
Conference Meeting/Multipurpose	0.90
Corridor/Transition	0.85
Dining Area	0.90
Dining Area for Family Dining	0.85
Laboratory for Medical/Industrial Research	0.95
Lobby	0.95
Lobby for Elevator	0.85
Lobby for Motion Picture Theater	0.95
Lounge/Recreation	0.85
Office—Enclosed	0.95
Office—Open Plan	0.85
Sales Area	0.95
All Other Common Space Types	1.00
Building-Specific Space Type	**LPD Factor**
Convention Center—Exhibit Space	0.85
Courthouse—Courtroom	0.85
Fitness Center—Fitness Area	0.85
Gymnasium—Audience Seating/Permanent Seating	0.85
Gymnasium—Fitness Area	0.85
Hospital—Emergency	0.95
Hospital—Exam/Treatment	0.85
Hospital—Laundry/Washing	0.95
Hospital—Lounge/Recreation	0.85
Hospital—Medical Supply	0.90
Hospital—Nursery	0.85
Hospital—Nurses' Station	0.90
Hospital—Patient Room	0.90
Hospital—Physical Therapy	0.85
Library—Card File and Cataloguing	0.90
Library—Stacks	0.95
Manufacturing Facility—High Bay	0.85
Manufacturing Facility—Low Bay	0.85
Motel—Dining Area	0.90
Transportation—Air/Train/Bus—Baggage Area	0.90
Transportation—Airport—Concourse	0.90
Transportation—Terminal—Ticket Counter	0.85
Warehouse—Medium/Bulky Material Storage	0.85
All Other Building-Specific Space Types	1.00

TABLE 7.4.6.1C Lighting Power Allowance Factors

	Lighting Zone				
	LZ0	LZ1	LZ2	LZ3	LZ4
For Tradable Areas	1.00	0.90	0.90	0.95	0.95
For Nontradable Areas	1.00	0.95	0.95	0.95	0.95

Exception: Sign lighting using metal halide, high-pressure sodium, induction, cold cathode, or neon lamps that includes controls to automatically reduce the input power to a maximum of 70% of full power for a period from one hour after sunset to one hour before sunrise.

7.4.6.5.2 All other sign lighting shall include:

a. controls to automatically reduce the input power to a maximum of 70% of full power for a period from midnight or within one hour of the end of business operations, whichever is later, until 6:00 am or business opening, whichever is earlier, and
b. controls to automatically turn off during *daylight hours*.

7.4.7 Other Equipment. The other equipment shall comply with Section 10 of ANSI/ASHRAE/IES Standard 90.1 with the following modifications and additions.

7.4.7.1 Electric Motors. Motors shall comply with the minimum requirements in Table C-12 in Normative Appendix C. These requirements supersede the requirements in Section 10.4.1 and Table 10.8 of ANSI/ASHRAE/IES Standard 90.1.

7.4.7.2 Supermarket Heat Recovery. Supermarkets with a floor area of 25,000 ft^2 (2500 m^2) or greater shall recover waste heat from the condenser heat rejection on permanently installed refrigeration equipment meeting one of the following criteria:

1. 25% of the refrigeration system full load total heat rejection.
2. 80% of the space heat, service water heating and dehumidification reheat.

If a recovery system is used that is installed in the refrigeration system, the system shall not increase the saturated condensing temperature at design conditions by more than 5°F (3°C) and shall not impair other head pressure control/energy reduction strategies.

7.4.7.3 ENERGY STAR Equipment. The following equipment within the scope of the applicable ENERGY STAR program shall comply with the equivalent criteria required to achieve the ENERGY STAR label if installed prior to the issuance of the certificate of occupancy:

a. Appliances
   1. Clothes washers: ENERGY STAR Program Requirements for Clothes Washers (see also the water efficiency requirements in Section 6.3.2.2)
   2. Dehumidifiers: ENERGY STAR Program Requirements for Dehumidifiers
   3. Dishwashers: ENERGY STAR Program Requirements Product Specifications for Residential Dishwashers (see also the water efficiency requirements in Section 6.3.2.2)
   4. Refrigerators and freezers: ENERGY STAR Program Requirements for Refrigerators and Freezers
   5. Room air conditioners: ENERGY STAR Program Requirements and Criteria for Room Air Conditioners (see also the energy efficiency requirements in Section 7.4.1)
   6. Room air cleaners: ENERGY STAR Program Requirements for Room Air Cleaners
   7. Water coolers: ENERGY STAR Program Requirements for Water Coolers

b. Heating and Cooling
   1. Residential air-source heat pumps: ENERGY STAR Program Requirements for ASHPs and Central Air Conditioners (see also the energy efficiency requirements in Section 7.4.1)
   2. Residential boilers: ENERGY STAR Program Requirements for Boilers (see also the energy efficiency requirements in Section 7.4.1)
   3. Residential central air conditioners: ENERGY STAR Program Requirements for ASHPs and Central Air Conditioners (see also the energy efficiency requirements in Section 7.4.1)
   4. Residential ceiling fans: ENERGY STAR Program Requirements for Residential Ceiling Fans
   5. Dehumidifiers: ENERGY STAR Program Requirements for Dehumidifiers
   6. Programmable thermostats: ENERGY STAR Program Requirements for Programmable Thermostats
   7. Ventilating fans: ENERGY STAR Program Requirements for Residential Ventilating Fans
   8. Residential warm air furnaces: ENERGY STAR Program Requirements for Furnaces
   9. Residential geothermal heat pumps: ENERGY STAR Program Requirements for Geothermal Heat Pumps

c. Electronics
   1. Cordless phones: ENERGY STAR Program Requirements for Telephony
   2. Audio and video: ENERGY STAR Program Requirements for Audio and Video
   3. Televisions: ENERGY STAR Program Requirements for Televisions
   4. Set-top boxes: ENERGY STAR Program Requirements for Set-Top Boxes

d. Office Equipment
   1. Computers: ENERGY STAR Program Requirements for Computers
   2. Copiers: ENERGY STAR Program Requirements for Imaging Equipment
   3. Fax machines: ENERGY STAR Program Requirements for Imaging Equipment
   4. Laptops: ENERGY STAR Program Requirements for Computers
   5. Mailing machines: ENERGY STAR Program Requirements for Imaging Equipment
   6. Monitors: ENERGY STAR Program Requirements for Displays
   7. Multifunction devices (printer/fax/scanner): Program Requirements for Imaging Equipment
   8. Printers: ENERGY STAR Program Requirements for Imaging Equipment
   9. Scanners: ENERGY STAR Program Requirements for Imaging Equipment

10. Computer servers: ENERGY Star Program Requirements for Computer Servers

e. Water Heaters: ENERGY STAR Program Requirements for Residential Water Heaters

f. Lighting
   1. Compact fluorescent light bulbs (CFLs): ENERGY STAR Program Requirements for CFLs
   2. Residential light fixtures: ENERGY STAR Program Requirements for Residential Light Fixtures
   3. Integral LED lamps: ENERGY STAR Program Requirements for Integral LED Lamps

g. Commercial Food Service
   1. Commercial fryers: ENERGY STAR Program Requirements for Commercial Fryers
   2. Commercial hot food holding cabinets: ENERGY STAR Program Requirements for Hot Food Holding Cabinets
   3. Commercial refrigerators and freezers: ENERGY STAR Program Requirements for Commercial Refrigerators and Freezers
   4. Commercial steam cookers: ENERGY STAR Program Requirements for Commercial Steam Cookers (see also water efficiency requirements in Section 6.4.2.2)
   5. Commercial ice machines: ENERGY STAR Program Requirements for Commercial Ice Machines
   6. Commercial dishwashers: ENERGY STAR Program Requirements for Commercial Dishwashers
   7. Commercial griddles: ENERGY STAR Program Requirements for Commercial Griddles
   8. Commercial ovens: ENERGY STAR Program Requirements for Commercial Ovens

h. Other Products
   1. Battery charging systems: ENERGY STAR Program Requirements for Products with Battery Charger Systems (BCSs)
   2. External power adapters: ENERGY STAR Program Requirements for Single-Voltage AC-DC and AC-AC Power Supplies
   3. Vending machines: ENERGY STAR Program Requirements for Refrigerated Beverage Vending Machines

Exception: Products with minimum efficiencies addressed in the Energy Policy Act (EPAct) and the Energy Independence and Security Act (EISA) when complying with Section 7.4.3.1a.

7.4.7.4 Commercial Refrigerators, Freezers, and Clothes Washers

a. Commercial refrigerators and freezers shall comply with the minimum efficiencies in Table C-13 in Normative Appendix C. Open refrigerated display cases not covered by strips or curtains are prohibited. Lighting loads, including all power supplies or ballasts, for commercial reach-in refrigerator/freezer display cases shall not exceed 42 watts per door for case doors up to 5 ft (1.5 m) in height and 46 watts per door for case doors greater than 5 ft (1.5 m) in height.

b. Commercial clothes washers shall comply with the minimum efficiencies in Table C-14 in Normative Appendix C.

7.4.8 Energy Cost Budget. The Energy Cost Budget option in Section 11 of ANSI/ASHRAE/IES Standard 90.1 shall not be used.

7.5 Performance Option

7.5.1 General Comprehensive Performance Requirements. Projects shall comply with Sections 7.5.2, 7.5.3, and 7.5.4.

7.5.2 Annual Energy Cost. The *building project* shall have an annual energy cost less than or equal to that achieved by compliance with Sections 7.3 and 7.4, and Sections 5.3.2.2, 5.3.2.3, 6.3.2, 6.4.2, 8.3.1, 8.3.4, and 8.4.1. Comparisons shall be made using Normative Appendix D.

7.5.3 Annual Carbon Dioxide Equivalent (CO_2e). The *building project* shall have an annual CO_2e less than or equal to that achieved by compliance with Sections 7.3 and 7.4, and Sections 5.3.2.2, 5.3.2.3, 6.3.2, 6.4.2, 8.3.1, 8.3.4, and 8.4.1. Comparisons shall be made using Normative Appendix D provided that the baseline building design is calculated in accordance with Section 7.5.2. To determine the CO_2e value for each energy source supplied to the *building project*, multiply the energy consumption by the emissions factor. CO_2e emission factors shall be taken from Table 7.5.3.

7.5.4 Annual Load Factor/Peak Electric Demand. The *building project* shall have the same or less peak electric demand than achieved by compliance with Sections 7.3 and 7.4, and Sections 5.3.2.2, 5.3.2.3, 6.3.2, 6.4.2, 8.3.1, 8.3.4, and 8.4.1. Comparisons shall be made using Normative Appendix D provided that the baseline building design is calculated in accordance with Section 7.5.2. In addition, the *building project* shall have a minimum electrical *annual load factor* of 0.25.

TABLE 7.5.3 CO_2e Emission Factors

Building Project Energy Source	CO_2e lb/kWh (kg/kWh)
Grid delivered electricity and other fuels not specified in this table	1.670 (0.758)
LPG or propane	0.602 (0.274)
Fuel oil (residual)	0.686 (0.312)
Fuel oil (distillate)	0.614 (0.279)
Coal (except lignite)	0.822 (0.373)
Coal (lignite)	1.287 (0.583)
Gasoline	0.681 (0.309)
Natural gas	0.510 (0.232)

8. INDOOR ENVIRONMENTAL QUALITY (IEQ)

8.1 Scope. This section specifies requirements for indoor environmental quality, including indoor air quality, environmental tobacco smoke control, *outdoor air* delivery monitoring, thermal comfort, *building entrances*, acoustic control, daylighting, and low emitting materials.

8.2 Compliance. The indoor environmental quality shall comply with Section 8.3, "Mandatory Provisions," and either:

a. Section 8.4, "Prescriptive Option," or
b. Section 8.5, "Performance Option."

Daylighting and low-emitting materials are not required to use the same option, i.e., prescriptive or performance, for demonstrating compliance.

8.3 Mandatory Provisions

8.3.1 Indoor Air Quality. The building shall comply with Sections 4 through 7 of ANSI/ASHRAE Standard 62.1 with the following modifications and additions. When a requirement is provided below, this supersedes the requirements in ANSI/ASHRAE Standard 62.1.

8.3.1.1 Minimum Ventilation Rates. The Ventilation Rate Procedure of ANSI/ASHRAE Standard 62.1 shall be used.

8.3.1.2 Outdoor Air Delivery Monitoring

8.3.1.2.1 Spaces Ventilated by Mechanical Systems. A permanently mounted, direct total outdoor airflow measurement device shall be provided that is capable of measuring the system *minimum outdoor airflow rate*. The device shall be capable of measuring flow within an accuracy of ±15% of the *minimum outdoor airflow rate*. The device shall also be capable of being used to alarm the building operator or for sending a signal to a building central monitoring system when flow rates are not in compliance.

Exception: Constant volume air supply systems that use a damper position feedback system are not required to have a direct total outdoor airflow measurement device.

8.3.1.3 Filtration and Air Cleaner Requirements

a. **Particulate Matter**
   1. **Wetted Surfaces.** Particulate matter filters or air cleaners provided upstream of wetted surfaces in accordance with Section 5.8 of ANSI/ASHRAE Standard 62.1 shall have a MERV of not less than 8.
   2. **Particulate Matter Smaller than 10 Micrometers (PM10).** Particulate matter filters or air cleaners provided to reduce PM10 in outdoor intake in accordance with 6.2.1.1 of ANSI/ASHRAE Standard 62.1 shall have a MERV of not less than 8.
   3. **Particulate Matter Smaller than 2.5 Micrometers (PM2.5).** Particulate matter filters or air cleaners provided to reduce PM2.5 in outdoor intake air in accordance with Section 6.2.1.2 of ANSI/ASHRAE Standard 62.1 shall have a MERV of not less than 13.
b. **Ozone.** In addition to Section 6.2.1.3 of ANSI/ASHRAE Standard 62.1, when the building is located in an area that is designated "non-attainment" with the National Ambient Air Quality Standards for ozone as determined by the *AHJ*, air-cleaning devices having a removal efficiency of no less than the efficiency specified in Section 6.2.1.3 of ANSI/ASHRAE Standard 62.1 shall be provided to clean outdoor air prior to its introduction to occupied spaces.
c. **Bypass Pathways.** All filter frames, air cleaner racks, access doors, and air cleaner cartridges shall be sealed.

8.3.1.4 Environmental Tobacco Smoke

a. Smoking shall not be allowed inside the building. Signage stating such shall be posted within 10 ft (3 m) of each building entrance.
b. Any exterior designated smoking areas shall be located a minimum of 25 ft (7.5 m) away from *building entrances*, *outdoor air* intakes, and operable windows.

8.3.1.5 Building Entrances. All *building entrances* shall employ an entry mat system that shall have a scraper surface, an absorption surface, and a finishing surface. Each surface shall be a minimum of the width of the entry opening, and the minimum length is measured in the primary direction of travel.

Exceptions:
1. Entrances to individual *dwelling units*.
2. Length of entry mat surfaces is allowed to be reduced due to a barrier, such as a counter, partition, or wall, or local regulations prohibiting the use of scraper surfaces outside the entry. In this case entry mat surfaces shall have a minimum length of 3 ft (1 m) of indoor surface, with a minimum combined length of 6 ft (2 m).

8.3.1.5.1 Scraper Surface. The scraper surface shall comply with the following:

a. Shall be the first surface stepped on when entering the building.
b. Shall be either immediately outside or inside the entry.
c. Shall be a minimum of 3 ft (1 m) long.
d. Shall be either permanently mounted grates or removable mats with knobby or squeegee-like projections.

8.3.1.5.2 Absorption Surface. The absorption surface shall comply with the following:

a. Shall be the second surface stepped on when entering the building.
b. Shall be a minimum of 3 ft (1 m) long, and made from materials that can perform both a scraping action and a moisture wicking action.

8.3.1.5.3 Finishing Surface. The finishing surface shall comply with the following:

a. Shall be the third surface stepped on when entering the building.
b. Shall be a minimum of 4 ft (1.2 m) long, and made from material that will both capture and hold any remaining particles or moisture.

TABLE 8.3.4.1 Minimum Toplighting Area

Lighting Power Density or *Lighting Power Allowances* in Daylight Area, W/ft^2 (W/m^2)	Minimum Toplighting Area to Daylight Area Ratio
1.4 W/ft^2 (14 W/m^2) < LPD	3.6%
1.0 W/ft^2 (10 W/m^2) < LPD < 1.4 W/ft^2 (14 W/m^2)	3.3%
0.5 W/ft^2 (5 W/m^2) < LPD < 1.0 W/ft^2 (10 W/m^2)	3.0%

8.3.2 Thermal Environmental Conditions for Human Occupancy. The building shall be designed in compliance with ANSI/ASHRAE Standard 55, Sections 6.1, "Design," and 6.2, "Documentation."

Exception: Spaces with special requirements for processes, activities, or contents that require a thermal environment outside that which humans find thermally acceptable, such as food storage, natatoriums, shower rooms, saunas, and drying rooms.

8.3.3 Acoustical Control

8.3.3.1 Exterior Sound. *Wall* and roof-ceiling assemblies that are part of the *building envelope* shall have a composite OITC rating of 40 or greater or a composite STC rating of 50 or greater, and *fenestration* that is part of the *building envelope* shall have an OITC or STC rating of 30 or greater for any of the following conditions:

a. Buildings within 1000 ft (300 m) of *expressways*.
b. Buildings within 5 mi (8 km) of airports serving more than 10,000 commercial jets per year.
c. Where *yearly average day-night average sound levels* at the property line exceed 65 decibels.

Exception: Buildings that may have to adhere to functional and operational requirements such as factories, stadiums, storage, enclosed parking structure, and utility buildings.

8.3.3.2 Interior Sound. Interior *wall* and floor/ceiling assemblies separating interior rooms and spaces shall be designed in accordance with all of the following:

a. *Wall* and floor-ceiling assemblies separating adjacent *dwelling units*, *dwelling units* and public spaces, adjacent tenant spaces, tenant spaces and public places, and adjacent *classrooms* shall have a composite STC rating of 50 or greater.
b. *Wall* and floor-ceiling assemblies separating hotel rooms, motel rooms, and patient rooms in nursing homes and hospitals shall have a composite STC rating of 45 or greater.
c. *Wall* and floor-ceiling assemblies separating *classrooms* from rest rooms and showers shall have a composite STC rating of 53 or greater.
d. *Wall* and floor-ceiling assemblies separating *classrooms* from music rooms, mechanical rooms, cafeteria, gymnasiums, and indoor swimming pools shall have a composite STC rating of 60 or greater.

8.3.3.3 Outdoor-Indoor Transmission Class and Sound Transmission Class. OITC values for assemblies and components shall be determined in accordance with ASTM E1332. STC values for assemblies and components shall be determined in accordance with ASTM E90 and ASTM E413.

8.3.4 Daylighting by Toplighting. There shall be a minimum *fenestration area* providing daylighting by *toplighting* for large enclosed spaces. In buildings three stories and less above grade, conditioned or unconditioned enclosed spaces that are greater than 20,000 ft^2 (2000 m^2) directly under a *roof* with finished ceiling heights greater than 15 ft (4 m) and that have a *lighting power allowance* for general lighting equal to or greater than 0.5 W/ft^2 (5.5 W/m^2) shall comply with the following.

Exceptions:
1. Buildings in *climate zones* 7 or 8.
2. Auditoria, theaters, museums, places of worship, and refrigerated warehouses.

8.3.4.1 Minimum Daylight Area by Toplighting. A minimum of 50% of the floor area directly under a *roof* in spaces with a lighting power density or *lighting power allowance* greater than 0.5 W/ft^2 (5 W/m^2) shall be in the *daylight area*. Areas that are daylit shall have a minimum *toplighting* area to *daylight area* ratio as shown in Table 8.3.4.1. For purposes of compliance with Table 8.3.4.1, the greater of the space lighting power density and the space *lighting power allowance* shall be used.

8.3.4.2 Skylight Characteristics. *Skylights* used to comply with Section 8.3.4.1 shall have a glazing material or diffuser that has a measured haze value greater than 90%, tested according to ASTM D1003 (notwithstanding its scope) or other test method approved by the *AHJ*.

Exceptions:
1. *Skylights* with a measured haze value less than or equal to 90% whose combined area does not exceed 5% of the total *skylight* area.
2. *Tubular daylighting devices* having a diffuser.
3. *Skylights* that are capable of preventing direct sunlight from entering the occupied space below the well during occupied hours. This shall be accomplished using one or more of the following:
   a. orientation
   b. automated shading or diffusing devices
   c. diffusers
   d. fixed internal or external baffles
4. Airline terminals, convention centers, and shopping malls.

TABLE 8.4.1.1 Minimum Sidelighting Effective Aperture

Climate Zone	Minimum Sidelighting Effective Aperture
1, 2, 3A, 3B	0.10
3C, 4, 5, 6, 7, 8	0.15

8.3.5 Isolation of the Building from Pollutants in Soil. Building projects that include construction or expansion of a ground-level foundation and which are located on *brownfield sites* or in "Zone 1" counties identified to have a significant probability of radon concentrations higher than 4 picocuries/liter on the USEPA map of radon zones, shall have a *soil gas retarding system* installed between the newly constructed space and the soil.

8.4 Prescriptive Option

8.4.1 Daylighting by Sidelighting

8.4.1.1 Minimum Sidelighting Effective Aperture. Office spaces and *classrooms* shall comply with the following criteria:

a. All north-, south-, and east-facing facades for those spaces shall have a minimum *sidelighting effective aperture* as prescribed in Table 8.4.1.1.
b. The combined width of the *primary sidelighted areas* shall be at least 75% of the length of the façade wall.
c. Opaque interior surfaces in *daylight areas* shall have visible light reflectances greater than or equal to 80% for ceilings and 70% for partitions higher than 60 in. (1.8 m) in *daylight areas*.

Exceptions:
1. Spaces with programming that requires dark conditions (e.g., photographic processing).
2. Spaces with *toplighting* in compliance with Section 8.3.4.
3. *Daylight areas* where the height of existing adjacent structures above the window is at least twice the distance between the window and the adjacent structures, measured from the top of the glazing.

8.4.1.2 Office Space Shading. Each west-, south-, and east-facing facade, shall be designed with a shading *PF*. The *PF* shall be not less than 0.5. Shading is allowed to be external or internal using the *interior PF*. The building is allowed to be rotated up to 45 degrees for purposes of calculations and showing compliance. The following shading devices are allowed to be used:

a. Louvers, sun shades, light shelves, and any other permanent device. Any *vertical fenestration* that employs a combination of interior and external shading is allowed to be separated into multiple segments for compliance purposes. Each segment shall comply with the requirements for either external or *interior projection factor*.
b. Building self-shading through *roof* overhangs or recessed windows.

Exceptions:
1. Translucent panels and glazing systems with a measured haze value greater than 90%, tested according to ASTM D1003 (notwithstanding its scope) or other test method approved by the *AHJ*, and that are entirely 8 ft (2.5 m) above the floor, do not require external shading devices.
2. *Vertical fenestration* that receives direct solar radiation for less than 250 hours per year because of shading by permanent external buildings, existing permanent infrastructure, or topography.
3. *Vertical fenestration* with automatically controlled shading devices in compliance with Exception 2 of Section 7.4.2.5.
4. *Vertical fenestration* with automatically controlled *dynamic glazing* in compliance with Exception 3 of Section 7.4.2.5.

8.4.2 Materials. Reported emissions or VOC contents specified below shall be from a representative product sample and conducted with each product reformulation or at a minimum of every three years. Products certified under third-party certification programs as meeting the specific emission or VOC content requirements listed below are exempted from this three-year testing requirement but shall meet all the other requirements as listed below.

8.4.2.1 Adhesives and Sealants. Products in this category include carpet, resilient, and wood flooring adhesives; base cove adhesives; ceramic tile adhesives; drywall and panel adhesives; aerosol adhesives; adhesive primers; acoustical sealants; firestop sealants; HVAC air duct sealants, sealant primers; and caulks. All adhesives and sealants used on the interior of the building (defined as inside of the *weatherproofing system* and applied on-site) shall comply with the requirements of either Section 8.4.2.1.1 or 8.4.2.1.2:

8.4.2.1.1 Emissions Requirements. Emissions shall be determined according to CDPH/EHLB/Standard Method V1.1 (commonly referred to as *California Section 01350*) and shall comply with the limit requirements for either office or *classroom* spaces regardless of the space type.

8.4.2.1.2 VOC Content Requirements. VOC content shall comply with and shall be determined according to the following limit requirements:

a. Adhesives, sealants and sealant primers: SCAQMD Rule 1168. HVAC duct sealants shall be classified as "Other" category within the SCAQMD Rule 1168 sealants table.
b. Aerosol adhesives: Green Seal Standard GS-36.

Exceptions: The following solvent welding and sealant products are not required to meet the emissions or the VOC content requirements listed above.
1. Cleaners, solvent cements, and primers used with plastic piping and conduit in plumbing, fire suppression, and electrical systems.
2. HVAC air duct sealants when the air temperature of the space in which they are applied is less than 40°F (4.5°C).

8.4.2.2 Paints and Coatings. Products in this category include sealers, stains, clear wood finishes, floor sealers and coatings, waterproofing sealers, primers, flat paints and coatings, non-flat paints and coatings, and rust-preventative coatings. Paints and coatings used on the interior of the building (defined as inside of the *weatherproofing system* and applied on-site) shall comply with either Section 8.4.2.2.1 or 8.4.2.2.2.

8.4.2.2.1 Emissions Requirements. Emissions shall be determined according to CDPH/EHLB/Standard Method V1.1 (commonly referred to as California Section 01350) and shall comply with the limit requirements for either office or *classroom* spaces regardless of the space type.

8.4.2.2.2 VOC Content Requirements. VOC content shall comply with and be determined according to the following limit requirements:

a. Architectural paints, coatings and primers applied to interior surfaces: Green Seal Standard GS-11.
b. Clear wood finishes, floor coatings, stains, sealers, and shellacs: SCAQMD Rule 1113.

8.4.2.3 Floor Covering Materials. Floor covering materials installed in the building interior shall comply with the following:

a. Carpet: Carpet shall be tested in accordance with and shown to be compliant with the requirements of CDPH/EHLB/Standard Method V1.1 (commonly referred to as California Section 01350). Products that have been verified and labeled to be in compliance with Section 9 of CDPH/EHLB/Standard Method V1.1 (commonly referred to as California Section 01350) comply with this requirement.
b. Hard surface flooring in office spaces and *classrooms*: Materials shall be tested in accordance with and shown to be compliant with the requirements of CDPH/EHLB/Standard Method V1.1 (commonly referred to as California Section 01350).

8.4.2.4 Composite Wood, Wood Structural Panel and Agrifiber Products. Composite wood, wood structural panel, and agrifiber products used on the interior of the building (defined as inside of the *weatherproofing system*) shall contain no added urea-formaldehyde resins. Laminating adhesives used to fabricate on-site and shop-applied composite wood and agrifiber assemblies shall contain no added urea-formaldehyde resins. Composite wood and agrifiber products are defined as: particleboard, medium density fiberboard (MDF), wheatboard, strawboard, panel substrates, and door cores. Materials considered furniture, fixtures and equipment (FF&E) are not considered base building elements and are not included in this requirement. Emissions for products covered by this section shall be determined according to and shall comply with one of the following:

a. Third-party certification shall be submitted indicating compliance with the California Air Resource Board's (CARB) regulation, *Airborne Toxic Control Measure to Reduce Formaldehyde Emissions from Composite Wood Products*. Third-party certifier shall be approved by CARB.
b. CDPH/EHLB/Standard Method V1.1 (commonly referred to as California Section 01350) and shall comply with the limit requirements for either office or classroom spaces regardless of the space type.

Exception: Structural panel components such as plywood, particle board, wafer board, and oriented strand board identified as "EXPOSURE 1," "EXTERIOR," or "HUD-APPROVED" are considered acceptable for interior use.

8.4.2.5 Office Furniture Systems and Seating. All office furniture systems and seating installed prior to occupancy shall be tested according to ANSI/BIFMA M7.1 and shall not exceed the limit requirements listed in Normative Appendix E of this standard.

8.4.2.6 Ceiling and Wall Systems. These systems include ceiling and wall insulation, acoustical ceiling panels, tackable wall panels, gypsum wall board and panels, and wall coverings. Emissions for these products shall be determined according to CDPH/EHLB/Standard Method V1.1 (commonly referred to as California Section 01350) and shall comply with the limit requirements for either office or classroom spaces regardless of the space type.

8.5 Performance Option

8.5.1 Daylighting Simulation

8.5.1.1 Usable Illuminance in Office Spaces and Classrooms. The design for the *building project* shall demonstrate an illuminance of at least 30 fc (300 lux) on a plane 2.5 ft (0.8 m) above the floor, within 75% of the area of the *daylight area.* The simulation shall be made at noon on the equinox using an accurate physical model or computer daylighting model.

a. Computer models shall be built using daylight simulation software based on the ray-tracing or radiosity methodology.
b. Simulation shall be done using either the CIE Overcast Sky Model or the CIE Clear Sky Model.

Exception: Where the simulation demonstrates that existing adjacent structures preclude meeting the illuminance requirements.

8.5.1.2 Direct Sun Limitation on Worksurfaces in Offices. It shall be demonstrated that direct sun does not strike anywhere on a worksurface in any daylit space for more than 20% of the occupied hours during an equinox day in regularly occupied office spaces. If the worksurface height is not defined, a height of 2.5 ft (0.75 m) above the floor shall be used.

8.5.2 Materials. The emissions of all the materials listed below and used within the building (defined as inside of the *weatherproofing system* and applied onsite) shall be modeled for individual VOC concentrations. The sum of each individual VOC concentration from the materials listed below shall be shown to be in compliance with the limits as listed in Section 4.3 of the CDPH/EHLB/Standard Method

V1.1 (commonly referred to as California Section 01350) and shall be compared to 100% of its corresponding listed limit. In addition, the modeling for the building shall include at a minimum the criteria listed in Normative Appendix F. Emissions of materials used for modeling VOC concentrations shall be obtained in accordance with the testing procedures of CDPH/EHLB/Standard Method V1.1 (commonly referred to as California Section 01350) unless otherwise noted below.

a. Tile, strip, panel, and plank products, including vinyl composition tile, resilient floor tile, linoleum tile, wood floor strips, parquet flooring, laminated flooring, and modular carpet tile.

b. Sheet and roll goods, including broadloom carpet, sheet vinyl, sheet linoleum, carpet cushion, wallcovering, and other fabric.

c. Rigid panel products, including gypsum board, other wall paneling, insulation board, oriented strand board, medium density fiber board, wood structural panel, acoustical ceiling tiles, and particleboard.

d. Insulation products.

e. Containerized products, including adhesives, sealants, paints, other coatings, primers, and other "wet" products.

f. Cabinets, shelves, and worksurfaces that are permanently attached to the building before occupancy. Emissions of these items shall be obtained in accordance with the ANSI/BIFMA M7.1.

g. Office furniture systems and seating installed prior to initial occupancy. Emissions of these items shall be obtained in accordance with the ANSI/BIFMA M7.1.

Exception: Salvaged materials that have not been refurbished or refinished within one year prior to installation.

9. THE BUILDING'S IMPACT ON THE ATMOSPHERE, MATERIALS, AND RESOURCES

9.1 Scope. This section specifies requirements for the building's impact on the atmosphere, materials, and resources, including construction waste management, refrigerants, storage and collection of recyclables, and reduced impact materials.

9.2 Compliance. The building materials shall comply with Section 9.3, "Mandatory Provisions," and either

a. Section 9.4, "Prescriptive Option," or
b. Section 9.5, "Performance Option."

9.3 Mandatory Provisions

9.3.1 Construction Waste Management

9.3.1.1 Diversion. A minimum of 50% of nonhazardous construction and demolition waste material generated prior to the issuance of the final certificate of occupancy shall be diverted from disposal in landfills and incinerators by recycling and/or reuse. Reuse includes donation of materials to charitable organizations, salvage of existing materials onsite, and packaging materials returned to the manufacturer, shipper, or other source that will reuse the packaging in future shipments. Excavated soil and land-clearing debris shall not be included in the calculation. Calculations are allowed to be done by either weight or volume, but shall be consistent throughout. Specific area(s) on the construction *site* shall be designated for collection of recyclable and reusable materials. Off-site storage and sorting of materials shall be allowed. Diversion efforts shall be tracked throughout the construction process.

9.3.1.2 Total Waste. For new *building projects* on *sites* with less than 5% existing buildings, structures or *hardscape*, the total amount of construction waste generated prior to the issuance of the final certificate of occupancy on the project shall not exceed 42 yd^3 or 12,000 lbs per 10,000 ft^2 (35 m^3 or 6000 kg per 1000 m^2) of new building floor area. This shall apply to all waste whether diverted, landfilled, incinerated, or otherwise disposed of. Excavated soil and land-clearing debris shall not be included in the calculation. The amount of waste shall be tracked throughout the construction process.

9.3.2 Extracting, Harvesting, and/or Manufacturing. This section applies to all materials, products, and/or assemblies installed prior to the issuance of the final certificate of occupancy.

Materials shall be harvested and/or extracted and products and/or assemblies shall be manufactured according to the laws and regulations of the country of origin.

Wood products in the project, other than recovered or reused wood, shall not contain wood from endangered wood species unless the trade of such wood conforms with the requirements of the Convention on International Trade in Endangered Species of Wild Fauna and Flora (CITES).

9.3.3 Refrigerants. CFC-based refrigerants in HVAC&R systems shall not be used. Fire suppression systems shall not contain ozone-depleting substances (CFCs, HCFCs, or Halons).

9.3.4 Storage and Collection of Recyclables and Discarded Goods

9.3.4.1 Recyclables. There shall be an area that serves the entire building and is dedicated to the collection and storage of non-hazardous materials for recycling, including paper, corrugated cardboard, glass, plastics, and metals. The size and functionality of the recycling areas shall be coordinated with the anticipated collection services to maximize the effectiveness of the dedicated areas.

9.3.4.2 Reusable goods. For *building projects* with *residential* spaces, there shall be an area that serves the entire building and is designed for the collection and storage of discarded but clean items in good condition. Charitable organizations or others to arrange for periodic pickups shall be identified and posted.

9.3.4.3 Fluorescent and HID Lamps and Ballasts. An area shall be provided that serves the entire building and is designed for the collection and storage of fluorescent and HID lamps and ballasts and facilitates proper disposal and/or recycling according to state and local hazardous waste requirements.

9.4 Prescriptive Option

9.4.1 Reduced Impact Materials. The *building project* shall contain materials that comply with Section 9.4.1.1, 9.4.1.2, or 9.4.1.3. Components of mechanical, electrical, plumbing, fire safety systems, and transportation devices shall not be included in the calculations except for piping, plumbing fixtures, ductwork, conduit, wiring, cabling, and elevator and escalator framing. Calculations shall only include materials *permanently installed* in the project. A value of 45% of the total construction cost is allowed to be used in lieu of the actual total cost of materials.

9.4.1.1 Recycled Content. The sum of *post-consumer recycled content* plus one-half of the *pre-consumer recycled content* shall constitute a minimum of 10%, based on cost, of the total materials in the *building project*. The *recycled content* of a material shall be determined by weight. The recycled fraction of the material in an assembly shall then be multiplied by the cost of assembly to determine its contribution to the 10% requirement.

The annual average industry values, by country of production, for the *recycled content* of steel products manufactured in basic oxygen furnaces and electric arc furnaces are allowed to be used as the *recycled content* of the steel. For the purpose of calculating the *recycled content* contribution of concrete, the constituent materials in concrete (e.g., the cementitious materials, aggregates, and water) are allowed to be treated as separate components and calculated separately.

9.4.1.2 Regional Materials. A minimum of 15% of building materials or products used, based on cost, shall be regionally extracted/harvested/recovered or manufactured within a radius of 500 mi (800 km) of the project *site*. If only a fraction of a product or material is extracted/harvested/recovered or manufactured locally, then only that percentage (by weight) shall contribute to the regional value.

Exception: For building materials or products shipped in part by rail or water, the total distance to the project shall be determined by weighted average, whereby that portion of the distance shipped by rail or water shall be multiplied by 0.25 and added to that portion not shipped by rail or water, provided that the total does not exceed 500 mi (800 km).

9.4.1.3 Biobased Products. A minimum of 5% of building materials used, based on cost, shall be *biobased products*. *Biobased products* shall comply with the minimum biobased contents of the USDA's Designation of Biobased Items for Federal Procurement, contain the "USDA Certified *Biobased Product*" label, or be composed of solid wood, engineered wood, bamboo, wool, cotton, cork, agricultural fibers, or other biobased materials with at least 50% biobased content.

9.4.1.3.1 Wood Building Components. Wood building components including, but not limited to, structural framing, sheathing, flooring, sub-flooring, wood window sash and frames, doors, and architectural millwork used to comply with this requirement shall contain not less than 60% certified wood content tracked through a chain of custody process either by physical separation or percentage-based approaches. Acceptable certified wood content documentation shall be provided by sources certified through a forest certification system with principles, criteria, and standards developed using ISO/IEC Guide 59, or the WTO Technical Barriers to Trade. Wood building components from a *vendor* are allowed to comply when the annual average amount of certified wood products purchased by the *vendor*, for which they have chain of custody *verification* not older than two years, is 60% or greater of their total annual wood products purchased.

9.5 Performance Option

9.5.1 Life-Cycle Assessment. A *LCA* shall be performed in accordance with ISO Standard 14044 for a minimum of two building alternatives, considering at least those material components included for consideration in Section 9.4.1, both of which shall conform to the *OPR*. Each building alternative shall consist of a common design, construction, and materials for the locale, including building size and use, as commonly approved by the *AHJ*. Each building alternative shall comply with Sections 6, 7, and 8. The service life of the buildings shall be not less than that determined using Table 10.3.2.3, except that the design life of long-life buildings shall be no less than 75 years.

9.5.1.1 LCA Performance Metric. The building alternative chosen for the project shall have a 5% improvement over the other building alternative assessed in the *LCA* in a minimum of two of the impact categories. The impact categories are: land use (or habitat alteration), resource use, climate change, ozone layer depletion, human health effects, ecotoxicity, smog, acidification, and eutrophication.

9.5.1.2 Procedure. The *LCA* shall include the following three steps:

Step 1: Perform a life-cycle inventory (LCI). The LCI accounts for all the individual environmental flows to and from the material components in a building throughout its life cycle.

1. The LCI shall include the materials and energy consumed and the emissions to air, land, and water for each of the following stages:
   a. Extracting and harvesting materials and fuel sources from nature.
   b. Processing building materials and manufacturing building components.
   c. Transporting materials and components.
   d. Assembly and construction.
   e. Maintenance, repair, and replacement during the design life with or without operational energy consumption.
   f. Demolition, disposal, recycling, and reuse of the building at the end of its life cycle.
2. The LCI shall account for emissions to air for the following:
   a. The six principal pollutants for which the USEPA has set National Ambient Air Quality Standards as required by the Clean Air Act and its amendments: carbon monoxide, nitrogen dioxide, lead, sulfur oxides, particulate matter (PM_{10} and $PM_{2.5}$), and ozone.
   b. Greenhouse gases (not including water vapor and ozone) as described in the Inventory of U.S. Greenhouse Gas Emissions and Sinks: carbon dioxide, methane, nitrous oxide, chlorofluorocarbons, hydrochlorofluorocarbons, bromofluorocarbons, hydrofluorocarbons, perfluorocarbons, sulfur hexafluoride, sulfur dioxide, and VOCs.
   c. Hazardous air pollutants listed in the Clean Air Act and its amendments.

Step 2: Compare the two building alternatives using a published third-party impact indicator method that includes, at a minimum the impact categories listed in Section 9.5.1.1. An *LCA* report shall be prepared that meets the requirements for third-party reporting in ISO Standard 14044 and also includes:

1. A description of the two building alternatives, including:
   a. a description of the system boundary used,
   b. the design life of each building, and
   c. the physical differences between buildings.
2. The impact indicator method and impact categories used.
3. The results of the *LCA* indicating a minimum of 5% improvement in the proposed building compared to the other building alternative for a minimum of two impact categories, including an explanation of the rationale for the weighting and averaging of the impacts.

Step 3: Conduct a critical review by an external expert independent of those performing the *LCA*.

9.5.1.3 Reporting. The following shall be submitted to the *AHJ*:

a. The *LCA* report.
b. The documentation of critical peer review by a third party including the results from the review and the reviewer's name and contact information.

10. CONSTRUCTION AND PLANS FOR OPERATION

10.1 Scope. This section specifies requirements for construction and plans for operation, including the *commissioning process*, building acceptance testing, measurement and *verification*, energy use reporting, durability, transportation management, erosion and sediment control, construction, and indoor air quality during construction.

10.2 Compliance. All of the provisions of Section 10 are mandatory provisions.

10.3 Mandatory Provisions

10.3.1 Construction

10.3.1.1 Building Acceptance Testing. Acceptance testing shall be performed on all buildings in accordance with this section using *generally accepted engineering standards* and handbooks acceptable to the *AHJ*.

An acceptance testing process shall be incorporated into the design and construction of the *building project* that verifies systems specified in this section perform in accordance with construction documents.

10.3.1.1.1 Activities Prior to Building Permit. Complete the following:

a. Designate a project *Acceptance Representative* to lead, review, and oversee completion of acceptance testing activities.
b. Construction documents shall indicate who is to perform acceptance tests and the details of the tests to be performed.
c. *Acceptance representative* shall review construction documents to verify relevant sensor locations, devices, and control sequences are properly documented.

10.3.1.1.2 Activities Prior to Building Occupancy. Complete the following:

a. Verify proper installation and start-up of the systems.
b. Perform acceptance tests. For each acceptance test, complete test form and include a signature and license number, as appropriate, for the party who has performed the test.
c. Verify a system manual has been prepared that includes O&M documentation and full warranty information, and provides operating staff the information needed to understand and optimally operate building systems.

10.3.1.1.3 Systems. The following systems, if included in the *building project*, shall have acceptance testing:

a. Mechanical systems: heating, ventilating, air conditioning, IAQ, and refrigeration systems (mechanical and/or passive) and associated controls.
b. Lighting systems: automatic daylighting controls, manual daylighting controls, occupancy sensing devices, and, automatic shut-off controls
c. Fenestration Control Systems: Automatic controls for shading devices and dynamic glazing.
d. Renewable energy systems.
e. Water measurement devices, as required in Section 6.3.3.
f. Energy measurement devices, as required in Section 7.3.3.

10.3.1.1.4 Documentation. The *owner* shall retain completed acceptance test forms.

10.3.1.2 Building Project Commissioning. For buildings that exceed 5000 ft^2 (500 m^2) of gross floor area, commissioning shall be performed in accordance with this section using *generally accepted engineering standards* and handbooks acceptable to the *AHJ*. Buildings undergoing the *commissioning process* will be deemed to comply with the requirements of Section 10.3.1.1, "Building Acceptance Testing."

A *commissioning process* shall be incorporated into the predesign, design, construction, and first year occupancy of the *building project* that verifies that the delivered building and its components, assemblies, and systems comply with the documented *OPR*. Procedures, documentation, tools, and training shall be provided to the building operating staff to sustain features of the building assemblies and systems for the service life of the building. This material shall be assembled and organized into a systems manual that provides necessary information to the building operating staff to operate and maintain all commissioned systems identified within the building project.

10.3.1.2.1 Activities Prior to Building Permit. The following activities shall be completed:

a. Designate a project *commissioning authority* (CxA) to lead, review, and oversee completion of the *commissioning process* activities prior to completion of schematic design.
b. The *owner*, in conjunction with the design team as necessary, shall develop the *OPR* during predesign and updated during the design phase by the design team as necessary, in conjunction with the *owner* and the commissioning team. The *OPR* will be distributed to all parties participating in project programming, design, construction, and operations, and the commissioning team members.
c. The design team shall develop the *BOD*. The *BOD* document shall include all the information required in Section 6.2, "Documentation," of ANSI/ASHRAE Standard 55.
d. The CxA shall review both the *OPR* and *BOD* to ensure that no conflicting requirements or goals exist and that the *OPR* and *BOD*, based on the professional judgment and experience of the CxA, are sufficiently detailed for the project being undertaken.
e. Construction phase commissioning requirements shall be incorporated into project specifications and other construction documents developed by the design team.
f. The CxA shall conduct two focused *OPR* reviews of the construction documents: the first at near 50% design completion and the second of the final construction documents prior to delivery to the contractor. The purpose of these reviews is to verify that the documents achieve

the construction phase *OPR* and the *BOD* document fully supports the *OPR*, with sufficient details.

g. Develop and implement a *commissioning plan* containing all required forms and procedures for the complete testing of all equipment, systems, and controls included in Section 10.3.1.2.4.

10.3.1.2.2 Activities Prior to Building Occupancy. The following activities shall be completed:

a. Verify the installation and performance of the systems to be commissioned, including completion of the *construction checklist* and *verification*.

Exception to 10.3.1.2.2(a): Systems that, because their operation is seasonally dependent, cannot be fully commissioned in accordance with the *commissioning plan* at time of occupancy. These systems shall be commissioned at the earliest time after occupancy when operation of systems is allowed to be fully demonstrated as determined by CxA.

b. It shall be verified that the owner requirements for the training of operating personnel and building occupants is completed. Where systems cannot be fully commissioned at the time of occupancy because of seasonal dependence, the training of personnel and building occupants shall be completed when the systems' operation can be fully demonstrated by the CxA.
c. Complete preliminary commissioning report.
d. Verify a system manual has been prepared that includes O&M documentation, full warranty information, and provides operating staff the information needed to understand and operate the commissioned systems as designed.

10.3.1.2.3 Post-Occupancy Activities. Complete the following:

a. Complete any commissioning activities called out in the *commissioning plan* for systems whose commissioning can only be completed subsequent to building occupancy, including trend logging and off-season testing.
b. Verify the *owner* requirements for training operating personnel and building occupants are completed for those systems whose seasonal operational dependence mean they were unable to be fully commissioned prior to building occupancy.
c. Complete a final commissioning report.

10.3.1.2.4 Systems. The following systems, if included in the *building project*, shall be commissioned:

a. Heating, ventilating, air-conditioning, IAQ, and refrigeration systems (mechanical and/or passive) and associated controls. Control sequences to be verified for compliance with construction documentation as part of *verification*.
b. *Building envelope* systems, components, and assemblies to verify the thermal and moisture integrity.
c. *Building envelope* pressurization to confirm air-tightness if included in *BOD* requirements.
d. Lighting systems.
e. Fenestration control systems: Automatic controls for shading devices and *dynamic glazing*.
f. Irrigation.
g. Plumbing.
h. Domestic and process water pumping and mixing systems.
i. *Service water heating* systems.
j. Renewable energy systems.
k. Water measurement devices, as required in Section 6.3.3.
l. Energy measurement devices, as required in Section 7.3.3.

10.3.1.2.5 Documentation. *Owner* shall retain the System Manual and Final Commissioning Report.

10.3.1.3 Erosion and Sediment Control (ESC). Develop and implement an erosion and sediment control (ESC) plan for all construction activities. The ESC plan shall conform to the erosion and sedimentation control requirements of the most current version of the USEPA NPDES General Permit for Stormwater Discharges From Construction Activities or local erosion and sedimentation control standards and codes, whichever is more stringent and regardless of size of project.

10.3.1.4 Indoor Air Quality (IAQ) Construction Management. Develop and implement an indoor air quality (IAQ) construction management plan to include the following:

a. Air conveyance materials shall be stored and covered so that they remain clean. All filters and controls shall be in place and operational when HVAC systems are operated during building "flush-out" or baseline IAQ monitoring. Except for system startup, testing, balancing, and commissioning, permanent HVAC systems shall not be used during construction.
b. After construction ends, prior to occupancy and with all interior finishes installed, a post-construction, pre-occupancy building flush-out as described under Section 10.3.1.4 (b) 1, or post-construction, pre-occupancy baseline IAQ monitoring as described under Section 10.3.1.4 (b) 2 shall be performed:
    1. Post-Construction, Pre-Occupancy Flush-Out: A total air volume of *outdoor air* in total air changes as defined by Equation 10.3.1.4 shall be supplied while maintaining an internal temperature of a minimum of 60°F (15°C) and relative humidity no higher than 60%. For buildings located in non-attainment areas, filtration and/or air cleaning as described in Section 8.3.1.3 shall be supplied when the Air Quality Index forecast exceeds 100 (category orange, red, purple, or maroon). One of the following options shall be followed:
        (a) Continuous Post-Construction, Pre-Occupancy Flush-Out: The flush-out shall be continuous and supplied at an outdoor airflow rate no less than that determined in Section 8.3.1.1.

TABLE 10.3.1.4 Maximum Concentration of Air Pollutants Relevant to IAQ

Contaminant	Maximum Concentration, µg/m³ (Unless Otherwise Noted)
Nonvolatile Organic Compounds	
Carbon monoxide (CO)	9 ppm and no greater than 2 ppm above outdoor levels
Ozone	0.075 ppm (8-hr)
Particulates ($PM_{2.5}$)	35 (24-hr)
Particulates (PM_{10})	150 (24-hr)
Volatile Organic Compounds	
Acetaldehyde	140
Acrylonitrile	5
Benzene	60
1,3-Butadiene	20
t-Butyl methyl ether (Methyl-t-butyl ether)	8000
Carbon disulfide	800
Caprolactam *	100
Carbon tetrachloride	40
Chlorobenzene	1000
Chloroform	300
1,4-Dichlorobenzene	800
Dichloromethane (Methylene chloride)	400
1,4-Dioxane	3000
Ethylbenzene	2000
Ethylene glycol	400
Formaldehyde	33
2-Ethylhexanoic acid*	25
n-Hexane	7000
1-Methyl-2-pyrrolidinone*	160
Naphthalene	9
Nonanal*	13
Octanal*	7.2
Phenol	200
4-Phenylcyclohexene (4-PCH)*	2.5
2-Propanol (Isopropanol)	7000
Styrene	900
Tetrachloroethene (Tetrachloroethylene, Perchloroethylene)	35
Toluene	300
1,1,1-Trichloroethane (Methyl chloroform)	1000
Trichloroethene (Trichloroethylene)	600
Xylene isomers	700
Total Volatile Organic Compounds (TVOC)	–†

* This test is only required if carpets and fabrics with styrene butadiene rubber (SBR) latex backing material are installed as part of the base building systems.

† TVOC reporting shall be in accordance with CA/DHS/EHLB/R-174 and shall be in conjunction with the individual VOCs listed above.

(b) Continuous Post-Construction, Pre-Occupancy/Post-Occupancy Flush-Out: If occupancy is desired prior to completion of the flush-out, the space is allowed to be occupied following delivery of half of the total air changes calculated from Equation 10.3.1.4 to the space. The space shall be ventilated at a minimum rate of 0.30 cfm per ft² (1.5 L/s per m²) of *outdoor air* or the outdoor airflow rate determined in Section 8.3.1.1, whichever is greater. These conditions shall be maintained until the total air changes calculated according to Equation 10.3.1.4 have been delivered to the space. The flush out shall be continuous.

Equation 10.3.1.4:

$$TAC = V_{ot} \times 1/A \times 1/H \times 60 \text{ min/h} \times 24 \text{ h/day} \times 14 \text{ days (I-P)}$$

$$TAC = V_{ot} \times 1 \text{ m}^3/1000\ L \times 1/A \times 1/H \times 3600 \text{ s/h} \times 24 \text{ h/day} \times 14 \text{ days (SI)}$$

where

TAC = total air changes

V_{ot} = system design *outdoor air* intake flow cfm (L/s) (according to Equation 6-8 of ANSI/ASHRAE Standard 62.1)

A = floor area ft² (m²)

H = ceiling height, ft (m)

2. Post-Construction, Pre-Occupancy Baseline IAQ Monitoring: Baseline IAQ testing shall be conducted after construction ends and prior to occupancy. The ventilation system shall be operated continuously within ±10% of the outdoor airflow rate provided by the ventilation system at design occupancy for a minimum of 24 hours prior to IAQ monitoring. Testing shall be done using protocols consistent with the USEPA Compendium of Methods for the Determination of Toxic Organic Pollutants in Ambient Air, TO-1, TO-11, TO-17 and ASTM Standard Method D 5197. The testing shall demonstrate that the contaminant maximum concentrations listed in Table 10.3.1.4 are not exceeded in the return airstreams of the HVAC systems that serve the space intended to be occupied. If the return airstream of the HVAC system serving the space intended to be occupied cannot be separated from other spaces either already occupied or not occupied at all, for each portion of the building served by a separate ventilation system, the testing shall demonstrate that the contaminant maximum concentrations at *breathing zone* listed in Table 10.3.1.4 are not exceeded in the larger of the following number of locations: (a) no less than one location per 25,000 ft² (2500 m²) or (b) in each contiguous floor area. For each sampling point where the maximum concentration limits are exceeded conduct additional flush-out with outside air and retest the specific parameter(s) exceeded to

demonstrate the requirements are achieved. Repeat procedure until all requirements have been met. When retesting non-complying building areas, take samples from the same locations as in the first test.

10.3.1.5 Moisture Control. The following items to control moisture shall be implemented during construction:

a. Materials stored onsite or materials installed that are absorptive shall be protected from moisture damage.
b. Building construction materials that show visual evidence of biological growth due to the presence of moisture shall not be installed on the *building project*.

10.3.1.6 Construction Activity Pollution Prevention: No-Idling of Construction Vehicles. Vehicle staging areas shall be established for waiting to load or unload materials. These staging areas shall be located 100 ft (30 m) from any *outdoor air* intakes, operable openings, and hospitals, schools, residences, hotels, daycare facilities, elderly housing, and convalescent facilities.

10.3.2 Plans for Operation. This section specifies the items to be included in plans for operation of a *building project* that falls under the requirements of this standard.

10.3.2.1 High Performance Building Operation Plan. A Master Building Plan for Operation shall be developed that meets the requirements specified in Sections 10.3.2.1.1 through 10.3.2.1.4.

10.3.2.1.1 Site Sustainability. A *site* sustainability portion of the Plan for Operation shall be developed and contain the following provisions. When trees and vegetation are used to comply with the shade requirements of Section 5.3.2.1, 5.4 or 5.5, the Plan for Operation shall include the maintenance procedures needed to maintain healthy vegetation growth. The Plan shall also outline the procedures for replacing any vegetation used to comply with the provisions in Section 5.

10.3.2.1.2 Water Use Efficiency. The Plan for Operation shall specify water use *verification* activities for *building projects* to track and assess building water consumption. The Plan shall describe the procedures needed to comply with the requirements outlined below.

10.3.2.1.2.1 Initial Measurement and Verification. Use the water measurement devices and collection/storage infrastructure specified in Section 6.3.3 to collect and store water use data for each device, starting no later than after building acceptance testing has been completed and certificate of occupancy has been issued.

10.3.2.1.2.2 Track and Assess Water Use. The Plan shall specify the procedures for tracking and assessing the *building project* water use, and the frequency for benchmark comparisons. The initial assessment shall be completed after 12 months but no later than 18 months after a certificate of occupancy has been issued. Ongoing assessments shall be completed at least every three years. The Plan shall include the following:

a. Usage Reports: Develop a Plan for collecting *building project* water use data for water sources and subsystems measured in Section 6.3.3.
b. Benchmark Water Performance: Develop a Plan to enter building operating characteristics and water use data into the ENERGY STAR Portfolio Manager. For building parameter inputs into Portfolio Manager (e.g., number of occupants, hours of operation, etc.), use actual average values.
c. Assess Water Use Performance: Develop a Plan to assess *building project* water use efficiency.

10.3.2.1.2.3 Documentation of Water Use. All documents associated with the measurement and *verification* of the building's water use shall be retained by owner for a minimum of three years.

10.3.2.1.3 Energy Efficiency. The Plan for Operation shall specify energy performance *verification* activities for *building projects* to track and assess building energy performance. The Plan shall describe the procedures needed to comply with the requirements outlined below.

10.3.2.1.3.1 Initial Measurement and Verification. Use the energy measurement devices and collection/storage infrastructure specified in Section 7.3.3 to collect and store energy data for each device, starting no later than after acceptance testing has been completed and certificate of occupancy has been issued.

10.3.2.1.3.2 Track and Assess Energy Consumption. The Plan for Operation shall specify the procedures for tracking and assessing the *building project* energy performance, and the frequency for benchmark comparisons. The initial assessment shall be completed after 12 months but no later than 18 months after a certificate of occupancy has been issued. Ongoing assessments shall be completed at least every three years. The Plan shall include the following:

a. Energy Usage Reports: Develop a Plan for collecting *building project* energy data for energy sources and system energy loads measured in Section 7.3.3. The reports shall include the following, as minimum:
   1. Hourly load profile for each day
   2. Monthly average daily load profile
   3. Monthly and annual energy use
   4. Monthly and annual peak demand
b. Track Energy Performance: Develop a Plan to enter building operating characteristics and energy consumption data into the ENERGY STAR Portfolio Manager for those building types addressed by this program to track building performance. For building parameter inputs into Portfolio Manager (e.g., number of occupants, hours of operation, number of PCs, etc.), use actual average values.
c. Assess Energy Performance: Develop a Plan to assess *building project* energy performance.

10.3.2.1.3.3 Documentation of Energy Efficiency. All documents associated with the measurement and *verification* of the building's energy efficiency shall be retained by owner.

10.3.2.1.4 Indoor Environmental Quality. The Plan for Operation shall include the requirements of Section 8 of ANSI/ASHRAE Standard 62.1 and shall describe the procedures for implementing a regular indoor environmental quality measurement and *verification* program after building occupancy, as outlined below.

10.3.2.1.4.1 Outdoor Airflow Measurement. The Plan for Operation shall document procedures for implementing a regular outdoor airflow monitoring program after building occupancy. The Plan shall include minimum verification frequencies of airflows supplied by mechanical ventilation systems at the system level. Verification shall be performed using hand-held airflow measuring instruments appropriate for such measurements or permanently installed airflow measuring stations. Hand-held airflow measuring instruments or airflow measuring stations used for airflow verifications must be calibrated no more than 6 months prior to such verifications. Naturally ventilated systems shall be exempted from this requirement provided that the design parameters, including but not limited to permanent openings or window opening frequency are not modified.

10.3.2.1.4.2 Outdoor Airflow Verification Procedures. The plan procedures shall contain the following requirements:

a. For each mechanical ventilation system where direct outdoor airflow measurement is required according to Section 8.3.1.2, a procedure shall be in place to react when the outdoor airflow is 15% or more lower than *minimum outdoor airflow rate*. It shall be verified that the device that measures *outdoor air* flow rate is actually measuring the flow rate within ±15% of the sensor output reading at the *minimum outdoor airflow rate*. If the sensor is not within ±15%, it shall be recalibrated. *Verification* of outdoor airflow shall be done on a quarterly basis and records maintained onsite. Direct outdoor airflow measurement devices shall be calibrated at the manufacturer's recommended interval or at least annually.
b. For each mechanical ventilation system where direct outdoor airflow measurement is not required according to Section 8.3.1, a procedure shall be in place to verify outdoor airflow and records maintained onsite and shall be made available upon request.

10.3.2.1.4.3 Outdoor Airflow Scheduling. Ventilation systems shall be operated such that spaces are ventilated when these spaces are expected to be occupied.

10.3.2.1.4.4 Outdoor Airflow Documentation. The following documentation shall be maintained concerning outdoor airflow measurement and *verification*.

a. A list of each air system requiring direct *outdoor air* flow measurement.
b. Monitoring procedures and monitoring frequencies for each monitored sensing device, including a description of the specific response measures to be taken if needed.
c. Ventilation systems shall be operated such that spaces are ventilated when these spaces are expected to be occupied.
d. Operation and calibration check procedures, and the records associated with operation checks and recalibration.

10.3.2.1.4.5 Indoor Air Quality. The Plan for Operation shall document procedures for maintaining and monitoring indoor air quality after building occupancy, and shall contain the following:

a. For buildings located in non-attainments areas for $PM_{2.5}$ as defined by the USEPA, air filtration and/or air cleaning equipment as defined in Section 8.3.1.3(a) shall be operated continuously during occupied hours or when the USEPA Air Quality Index exceeds 100 or equivalent designations by the local authorities for $PM_{2.5}$.

   Exception to 10.3.2.1.4.5(a): Spaces without mechanical ventilation.

b. For buildings located in non-attainments areas for ozone as defined by the USEPA, air-cleaning equipment as defined in Section 8.3.1.3(b) shall be operated continuously during occupied hours during the local summer and fall seasons, or when the USEPA Air Quality Index exceeds 100 or equivalent designations by the local authorities for ozone.

   Exception to 10.3.2.1.4.5(b): Spaces without mechanical ventilation.

c. Biennial monitoring of Indoor Air Quality by one of the following methods:
   1. Perform IAQ testing as described in Section 10.3.1.4.
   2. Monitoring occupant perceptions of indoor air quality by any method, including but not limited to occupant questionnaires.
   3. Each building shall have an occupant complaint/response program for IEQ.

10.3.2.1.4.6 Building Green Cleaning Plan. A Green Cleaning Plan shall be developed for the *building project* in compliance with Green Seal Standard, GS-42.

Exception: *Dwelling units* of a *building project*.

10.3.2.1.4.7 Document all measurement and *verification* data.

10.3.2.2 Maintenance Plan. A *Maintenance Plan* shall be developed for mechanical, electrical, plumbing, and fire protection systems, which includes the following:

a. The Plan shall be in accordance with ANSI/ASHRAE/ACCA Standard 180 for HVAC systems in buildings that meet the definition of commercial buildings in ANSI/ASHRAE/ACCA Standard 180.
b. The Plan shall address all elements of Section 4 of ANSI/ASHRAE/ACCA Standard 180 and shall develop required inspection and maintenance tasks similar to Section 5 of ANSI/ASHRAE/ACCA Standard 180 for electrical and plumbing systems in buildings that meet the definition of commercial buildings in ANSI/ASHRAE/ACCA Standard 180.

TABLE 10.3.2.3 Minimum Design Service Life for Buildings

Category	Minimum Service Life	Building Types
Temporary	Up to 10 years	Non-permanent construction buildings (sales offices, bunkhouses) Temporary exhibition buildings
Medium life	25 years	Industrial buildings Stand-alone parking structures
Long life	50 years	All buildings not temporary or medium life, including the parking structures below buildings designed for long life category

c. Documentation of the Plan and of completed maintenance procedures shall be maintained on the building site at all times in:
   1. Electronic format for storage on the building Energy Management System (EMS), Building Management System (BMS), computerized maintenance management system (CMMS) or other computer storage means, or
   2. Maintenance manuals specifically developed and maintained for documenting completed maintenance activities.

10.3.2.3 Service Life Plan. A Service Life Plan that is consistent with the *OPR* shall be developed to estimate to what extent structural, *building envelope* (not mechanical and electrical), and *hardscape* materials will need to be repaired or replaced during the service life of the building. The design service life of the building shall be no less than that determined using Table 10.3.2.3. The estimated service life shall be documented for building assemblies, products, and materials that will need to be inspected, repaired, and/or replaced during the service life of the building. *Site* improvements and *hardscape* shall also be included. Documentation in the Service Life Plan shall include the *building project* design service life and basis for determination, and the following for each assembly or component:

a. Building assembly description
b. Materials or products
c. Design or estimated service life, years
d. Maintenance frequency
e. Maintenance access for components with an estimated service life less than the service life of the building

Provide a Service Life Plan at the completion of design development. The *owner* shall retain a copy of the Service Life Plan for use during the life of building.

10.3.2.4 Transportation Management Plan (TMP). A transportation management plan shall be developed compliant with the following requirements. *Owner* shall retain a copy of the transportation management plan.

10.3.2.4.1 All Building Projects. The Plan shall include the following:

a. Preferred parking for carpools and vanpools with parking facilities.
b. A plan for bicycle transportation.

10.3.2.4.2 Owner-Occupied Building Projects or Portions of Building Projects. For *owner*-occupied buildings, or for the employees in the *owner*-occupied portions of a building, the building *owner* shall offer at least one of the following primary benefits to the *owner's* employees:

a. Incentivize employees to commute using mass transit, vanpool, carpool, or non-motorized forms of transportation.
b. Initiate a telework or flexible work schedule program that reduces by at least 5% the number of commuting trips by the *owner's* employees.
c. Initiate a ridesharing or carpool matching program, either in-house or through an outside organization.

Exception: Multifamily residential *building project.*

In addition, the *owner* shall provide all of the following to the *owner's* employees:

a. Access to an *emergency ride home* for employees, either provided in-house or by an outside organization.
b. A central point of contact in charge of commuter benefits.
c. Maintenance of commuter benefits in a centralized location.
d. Active promotion of commuter benefits to employees.

10.3.2.4.3 Building Tenant. The building *owner*:

a. shall provide a copy of the Plan to tenants within the building.
b. shall not include parking fees in lease rates or shall identify the value of parking in the lease.

10.4 Prescriptive Option. There are no prescriptive options.

10.5 Performance Option. There are no performance options.

11. NORMATIVE REFERENCES

Section numbers indicate where the reference occurs in this document.

Reference	Title	Section
American Society of Heating, Refrigerating and Air-Conditioning Engineers (ASHRAE) **1791 Tullie Circle NE** **Atlanta, GA 30329, United States** **1-404-636-8400; www.ashrae.org**		
ANSI/ASHRAE Standard 55-2010	Thermal Comfort Conditions for Human Occupancy	8.3.2, 10.3.1.2.1
ANSI/ASHRAE Standard 62.1-2010	Ventilation for Acceptable Indoor Air Quality	3.2, 7.4.3.2, 8.3, 10.3.2.1.4, Appendix D
ANSI/ASHRAE/IES Standard 90.1-2010	Energy Standard for Buildings Except Low-Rise Residential Buildings	3.1, 3.2, 5.3.3.1, 5.3.3.3, 7.3.1, 7.4.1, 7.4.2, 7.4.3, 7.4.4, 7.4.5, 7.4.6, 7.4.7, 7.4.8, Appendix A, Appendix C, Appendix D
ANSI/ASHRAE Standard 140-2004	Standard Method of Test for the Evaluation of Building Energy Analysis Computer Programs	Appendix D
ASHRAE Standard 146-2006	Method of Testing and Rating Pool Heaters	Appendix C
ANSI/ASHRAE Standard 169-2006	Weather Data for Building Design Standards	Appendix A
ANSI/ASHRAE/ACCA Standard 180-2008	Standard Practice for Inspection and Maintenance of Commercial Building HVAC Systems	3.2, 10.3.2.2
American National Standards Institute (ANSI) **25 West 43rd Street** **New York, NY 20036, United States** **1-212-642-4900; www.ansi.org**		
ANSI Z21.10.3-1998	Gas Water Heater, Volume 3, Storage, with Input Ratings above 75,000 BTU/h, Circulating with Instantaneous Water Heaters	Appendix C
ANSI Z21.47-2001	Gas-Fired Central Furnaces (Except Direct Vent and Separated Combustion System Furnaces)	Appendix C
ANSI Z83.8-2002	Gas Unit Heaters and Duct Furnaces	Appendix C
American Society of Mechanical Engineers (ASME) **Three Park Avenue** **New York, NY 10016-5990, United States** **1-800-843-2763 and 1-973-882-1170; www.asme.org**		
ASME A112.18.1-2005/CSA B125.1-05	Plumbing Supply Fittings	6.3.2.1
ASME A112.19.2-2008/CSA B45.1-08	Vitreous China Plumbing Fixtures and Hydraulic Requirements for Water Closets and Urinals	6.3.2.1
ASME A112.19.14-2006	Six-Liter Water Closets Equipped With a Dual Flushing Device	6.3.2.1
ASME A112.19.19-2006	Vitreous China Nonwater Urinals	6.3.2.1
ASTM International **100 Barr Harbor Dr.** **West Conshohocken, PA 19428-2959, United States** **1-610-832-9585; www.astm.org**		
ASTM C518-04	Standard Test Method for Steady-State Thermal Transmission Properties by Means of the Heat Flow Meter Apparatus	Appendix C
ASTM C1371-04a	Standard Test Method for Determination of Emittance of Materials Near Room Temperature Using Portable Emissometers	5.3.2.4, Appendix D
ASTM C1549-04	Standard Test Method for Determination of Solar Reflectance Near Ambient Temperature Using a Portable Solar Reflectometer	5.3.2.4, Appendix D

Reference	Title	Section
ASTM D1003-07e1	Standard Test Method for Haze and Luminous Transmittance of Transparent Plastics	8.3.4.2, 8.4.1.2
ASTM D5197	Standard Test Method for Determination of Formaldehyde and Other Carbonyl Compounds in Air (Active Sampler Methodology)	10.3.1.4
ASTM E90-04	Standard Test Method for Laboratory Measurement of Airborne Sound Transmission Loss of Building Partitions and Elements	8.3.3.3
ASTM E408-71(2008)	Standard Test Methods for Total Normal Emittance of Surfaces Using Inspection-Meter Techniques	5.3.2.4, Appendix D
ASTM E413-04	Classification for Rating Sound Insulation	8.3.3.3
ASTM E779-03	Standard Test Method for Determining Air Leakage Rate by Fan Pressurization	Appendix B
ASTM E1332-90 (2003)	Standard Classification for the Determination of Outdoor-Indoor Transmission Class	8.3.3.3
ASTM E1677-05	Standard Specification for an Air Retarder (AR) Material or System for Low-Rise Framed Building Walls	Appendix B
ASTM E1903-97 (2002)	Standard Guide for Environmental Site Assessments: Phase II Environmental Site Assessment Process	3.2
ASTM E1918-06	Standard Test Method for Measuring Solar Reflectance of Horizontal and Low-Sloped Surfaces in the Field	5.3.2.4, Appendix D
ASTM E1980-01	Standard Practice for Calculating Solar Reflectance Index of Horizontal and Low-Sloped Opaque Surfaces	5.3.2.4, Appendix D
ASTM E2178-03	Standard Test Method for Air Permeance of Building Materials	Appendix B
ASTM E2357-05	Standard Test Method for Determining Air Leakage of Air Barrier Assemblies	Appendix B
ASTM F2324-03	Standard Test Method for Prerinse Spray Valves	6.4.2.2

Air-Conditioning, Heating, and Refrigeration Institute (AHRI)
2111 Wilson Blvd, Suite 500
Arlington, VA 22201, United States
1-703-524-8800; www.ahrinet.org

Reference	Title	Section
AHRI 210/240-2008	Performance Rating of Unitary Air-Conditioning & Air-Source Heat Pump Equipment	Appendix C
AHRI 310/380-2004	Standard for Packaged Terminal Air-Conditioners and Heat Pumps	Appendix C
AHRI 340/360-2003	Performance Rating of Commercial and Industrial Unitary Air-Conditioning and Heat Pump Equipment	Appendix C
AHRI 390-2003	Performance Rating of Single Packaged Terminal Air-Conditioners and Heat Pumps	Appendix C
AHRI 550/590-2003	Performance Rating of Water Chilling Packages Using the Vapor Compression Cycle	Appendix C
AHRI 560-2000	Absorption Water Chilling an Water Heating Packages	Appendix C

Association of Home Appliance Manufacturers (AHAM)
1111 19th Street NW, Suite 402
Washington, DC, 20036, United States
1-202-872-5955; www.aham.org

Reference	Title	Section
ANSI/AHAM RAC-1-R2008	Room Air Conditioners	Appendix C

Reference	Title	Section
The Business and Institutional Furniture Manufacturer's Association (BIFMA) **678 Front Avenue NW, Suite 150** **Grand Rapids, MI 49504-5368, United States** **1-616-285-3963; www.bifma.org; email@bifma.org**		
ANSI/BIFMA M7.1-2007	Standard Test Method For Determining VOC Emissions From Office Furniture Systems, Components And Seating	8.5.2, Appendix E
ANSI/BIFMA X7.1-2007	Standard for Formaldehyde and TVOC Emissions of Low-Emitting Office Furniture Systems and Seating	Appendix E
BIFMA e3-2008	Furniture Sustainability Standard	Appendix E
California Air Resources Board (CARB) **1001 "I" Street** **P.O. Box 2815** **Sacramento, CA 95812, United States** **1-916-322-2990; www.arb.ca.gov/homepage.htm**		
No-Added Formaldehyde Based Resins	Airborne Toxic Control Measure to Reduce Formaldehyde Emissions from Composite Wood Products. California Code of Regulations, Title 17, Sections 93120-93120.12	8.5.2
California Department of Public Health (CDPH) **Indoor Air Quality Section** **850 Marina Bay Parkway** **Richmond, CA 94804, United States** **1-510-620-2802; www.cdph.ca.gov/programs/IAQ and www.cal-iaq.org**		
CDPH/EHLB/Standard Method V1.1	Standard Method for the Testing and Evaluation of Volatile Organic Chemical Emissions from Indoor Sources Using Environmental Chambers—Version 1.1	8.4.2.1.1, 8.4.2.2.1, 8.4.2.3, 8.4.2.4, 8.4.2.6, 8.5.2, Table 10.3.1.4, Appendix F
Convention on International Trade in Endangered Species of Wild Fauna and Flora (CITES) **International Environment House** **11 Chemin des Anémones** **CH-1219 Châtelaine, Geneva, Switzerland** **+41-(0)22-917-81-39/40**		
CITES- 1973, amended 1979 and 1983	Convention on International Trade in Endangered Species of Wild Fauna and Flora	9.3.2
Cooling Tower Technology Institute (CTI) **P.O Box 73383** **Houston, TX 77273, United States** **1-281-583-4087; www.cti.org**		
CTI ATC-105 (2/2000)	Acceptance Test Code	Appendix C
CTI STD 201 (1/2009)	Standard for the Certification of Water Cooling Tower Thermal Performance	Appendix C
Green-e **c/o Center for Resource Solutions** **1012 Torney Ave., Second Floor** **San Francisco, CA 94129, United States** **1- 415-561-2100; www.green-e.org**		
Version 1.6, Dec 5, 2008	Green-e Energy National Standard for Renewable Electricity Products	7.4.1.2
Green Seal **1001 Connecticut Avenue, NW, Suite 827** **Washington, DC 20036-5525, United States** **1-202-872-6400; www.greenseal.org**		
GS-11, May 12, 2008	Environmental Standard for Paints and Coatings	8.4.2.2.2

Reference	Title	Section
GS-36, October 19, 2000	Standard for Commercial Adhesives	8.4.2.1.2
GS-42, September 1, 2006	Environmental Standard for Cleaning Services	10.3.2.1.4.6
Illuminating Engineering Society of North America **120 Wall Street, Floor 17** **New York, NY 10005-4001, United States** **1-212-248-5017, www.ies.org**		
TM-15-2007 including addendum "a"	Backlight, Uplight, and Glare (BUG) Ratings	5.3.3.2A
International Association of Plumbing and Mechanical Officials (IAPMO) **5001 East Philadelphia Street** **Ontario, CA 91761, United States** **1-909-472-4100; www.iapmo.org**		
Z124.9-2004	Plastic Urinal Fixtures	6.3.2.1
International Organization for Standardization (ISO) **ISO Central Secretariat, 1 rue de Varembee, Case postale 56** **CH-1211 Geneva 20, Switzerland** **+41-22-749-01-11; www.iso.org**		
ISO-13256-1-1998	Water-source heat pumps -- Testing and rating for performance — Part 1: Water-to-air and brine-to-air heat pumps	Appendix C
ISO 14044 – 2006	Environmental management — Life cycle assessment — Requirements and guidelines	9.5.1, 9.5.1.2
ISO/IEC Guide 59-1994	Code of Good Practice for Standardization	9.4.1.3.1
Irrigation Association (IA) **6540 Arlington Boulevard** **Falls Church, VA 22042-6638, United States** **1-703-536-7080; www.irrigation.org**		
Smart Water Application Technology (SWAT) Climatological Based Controllers 8th Draft Testing Protocol – November 2006	Smart Water Application Technology (SWAT), Turf and Landscape Irrigation Equipment Climatologically Based Controllers	6.3.1.3
National Electric Manufacturers Association (NEMA) **1300 North 17th Street, Suite 1752** **Rosslyn, VA 22209, United States** **1-703-841-3200; www.nema.org**		
ANSI/NEMA MG-1-2006	Motors and Generators	Appendix C
National Fire Protection Association **1 Battery March Park** **Quincy, MA 02169-7471** **United States** **1-617-770-0700;** **www.nfpa.org**		
NFPA 70 -2011	National Electrical Code	5.3.3
South Coast Air Quality Management District (SCAQMD) **California Air Resources Board** **1001 "I" Street** **P.O. Box 2815** **Sacramento, CA 95812, United States** **1-916-322-2990; www.arb.ca.gov**		
SCAQMD Rule 1113, Amended July 13. 2007	Architectural Coatings	8.4.2.2
SCAQMD Rule 1168, Amended January 7, 2005	Adhesive and Sealant Applications	8.4.2.1

Reference	Title	Section
Underwriters Laboratories (UL) **2600 N.W. Lake Rd.** **Camas, WA 98607-8542, United States** **1-877-854-3577; www.ul.com**		
UL 727-2006	Standard for Oil Fired Central Furnaces	Appendix C
UL 731-1995	Standard for Oil-Fired Unit Heaters	Appendix C
United States Congress **Washington, DC 20515, United States** **1-202-224-3121; http://frwebgate.access.gpo.gov/cgi-bin/getdoc.cgi?dbname=109_cong_bills&docid=f:h6enr.txt.pdf and www.govtrack.us/data/us/bills.text/110/h/h6.pdf**		
EPAct 2005 HR6 Public Law 109-58	The Energy Policy Act (EPAct) of 2005	7.4.3.1, 7.4.7.3
EISA 2007 HR6 Public Law 110-140	The Energy Independence and Security Act of 2007	7.4.3, 7.4.7
United States Department of Agriculture (USDA) **BioPreferred Program** **1400 Independence Avenue, SW** **Washington, DC 20250, United States** **1-877-251-6522 and 1-202-720-2791; www.usda.gov/biopreferred**		
7 CFR Part 2902, March 16, 2006 (Round 1), May 14, 2008 (Rounds 2, 3 and 4), October 27, 2009 (Round 5), October 18, 2010 (Round 6) and July 22, 2011 (Round 7)	Designation of Biobased Items for Federal Procurement	9.4.1.3
United States Department of Energy (USDOE) **Energy Information Administration** **Washington, DC 20585, United States** **1-202-586-5000; www.eia.doe.gov/emeu/cbecs/contents.html and http://tonto.eia.doe.gov/state**		
EIA Average Energy Prices	State and U.S. Historical Data	Appendix D
Title 10 – Energy Chapter II – Department of Energy – Part 430	Energy Conservation Program for Consumer Products	Appendix C
Title 10 – Energy Chapter II – Department of Energy – Part 431	Energy Efficiency Program for Certain Commercial and Industrial Equipment	Appendix C
United States Environmental Protection Agency (EPA) **Ariel Rios Building** **1200 Pennsylvania Avenue, NW** **Washington, DC 20460, United States** **1-919-541-0800; www.epa.gov** **ENERGY STAR ® 1-888-782-7937** **WaterSense 1-866-987-7367 and 1-202-564-2660**		
Clean Air Act of 1970 and as amended in 1990	Clean Air Act	9.5.1.2
Code of Federal Regulations, Title 40 Part 50 (40 CFR 50), as amended July 1, 2004	National Primary and Secondary Ambient Air Quality Standards	8.3.1.3, 9.5.1.2
January 21, 2005	NPDES General Permit for Stormwater Discharges From Construction Activities	10.3.1.3
Version 5.0, July 1, 2009	ENERGY STAR Program Requirements for Computers	7.4.7
Version 1.1, July 1, 2009	ENERGY STAR Program Requirements for Imaging Equipment	7.4.7
Version 1.2, July 1, 2009	ENERGY STAR Program Requirements and Criteria for Room Air Conditioners	7.4.7
Version 4.0, January 1, 2009	ENERGY STAR Program Requirements for ASHPs and Central Air Conditioners	7.4.7

Reference	Title	Section
Version 2.1, April 1, 2002	ENERGY STAR Program Requirements for Boilers	7.4.7
Version 1.2, January 22, 2010	ENERGY STAR Program Requirements for Water Coolers	7.4.7
Version 4.0, December 2, 2008	ENERGY STAR Program Requirements for CFLs	7.4.7
Version 5.1, July 1, 2009	ENERGY STAR Program Requirements for Clothes Washers	6.3.2.2, 7.4.7
Version 1.1, October 11, 2007	ENERGY STAR Program Requirements for Commercial Dishwashers	6.4.2.2, 7.4.7
Version 1.0, August 15, 2003	ENERGY STAR Program Requirements for Commercial Fryers	7.4.7
Version 1.0, January 1, 2008	ENERGY STAR Program Requirements for Commercial Ice Machines	6.4.2.2, 7.4.7
Version 1.0, August 1, 2003	ENERGY STAR Program Requirements for Commercial Steam Cookers	7.4.7
Version 5.0, October 30, 2009	ENERGY STAR Program Requirements for Displays	7.4.7
Version 2.1, July 30, 2010	ENERGY STAR Program Requirements for Audio and Video	7.4.7
Version 2.0, June 1, 2008	ENERGY STAR Program Requirements for Dehumidifiers	7.4.7
Version 4.1, August 11, 2010	ENERGY STAR Program Requirements Product Specification for Residential Dishwashers	6.3.2.2, 7.4.7
Version 2.0, October 1, 2008	ENERGY STAR Program Requirements for Furnaces	7.4.7
Version 4.1, August 11, 2009	ENERGY STAR Program Requirements for Geothermal Heat Pumps	7.4.7
Version 1.0, August 15, 2003	ENERGY STAR Program Requirements for Hot Food Holding Cabinets	7.4.7
January 1 2006	ENERGY STAR Program Requirements for Products with Battery Charger Systems (BCSs)	7.4.7
Version 2.0, July 1, 2007	ENERGY STAR Program Requirements for Refrigerated Beverage Vending Machines	7.4.7
Version 1.0, April 28, 2008	ENERGY STAR Program Requirements for Refrigerators and Freezers	7.4.7
Version 2.3, January 1, 2009	ENERGY STAR Program Requirements for Residential Ceiling Fans	7.4.7
Version 1.0, January 1, 2009	ENERGY STAR Program Requirements for Residential Water Heaters	7.4.7
Version 2.0, December 31, 2007	ENERGY STAR Program Requirements for Roof Products	5.3.2.3
Version 1.0, July 1 2004	ENERGY STAR Program Requirements for Room Air Cleaners	7.4.7
Version 2.2 January 15, 2009	ENERGY STAR Program Requirements for Residential Ventilating Fans	7.4.7
Version 2.0, November 1, 2008	ENERGY STAR Program Requirements for Single-Voltage AC-DC and AC-AC Power Supplies	7.4.7
Version 2.0, April 1, 2009	ENERGY STAR Program Requirements for Commercial Refrigerators and Freezers	7.4.7
Version 2. 1, November 1, 2008	ENERGY STAR Program Requirements for Telephony	7.4.7
Version 4.2, April 30, 2010	ENERGY STAR Program Requirements for Televisions	7.4.7
Version 1.0, October 4, 2007	WaterSense Tank-Type High-Efficiency Lavatory Specification	6.3.2.1
Version 1.0, January 24, 2007	WaterSense Tank-Type High-Efficiency Toilet Specification	6.3.2.1

Reference	Title	Section
EPA 402-R-93-071, September 1993	U.S. EPA Map of Radon Zones	8.3.5
EPA 430-R-07-002, April 2007	Inventory of U.S. Greenhouse Gas Emissions and Sinks: 1990–2005	
Version 2.0, January 1, 2009	ENERGY STAR Program Requirements for Set-Top Boxes	7.4.7
Version 4.1, August 1, 2008	ENERGY STAR Program Requirements for Residential Light Fixtures	7.4.7
Version 1.2, August 31, 2010	ENERGY STAR Program Requirements for Integral LED Lamps	7.4.7
Version 1.0, May 8, 2009	ENERGY STAR Program Requirements for Commercial Griddles	7.4.7
Version 1.0, May 16, 2009	ENERGY STAR Program Requirements for Commercial Ovens	7.4.7
Version 1.0, October 15, 2010	ENERGY STAR Program Requirements for Computer Servers	7.4.7
United States Environmental Protection Agency (EPA) **Atmospheric Research and Exposure Assessment Laboratory** **Research Triangle Park, NC 27711, United States** **1-919-541-2258; www.epa.gov**		
EPA 625/R-96/0106, January 1999	Compendium of Methods for the Determination of Toxic Organic Pollutants in Ambient Air, Sections TO-1, TO-11, TO-17	10.3.1.4
World Trade Organization (WTO) **Centre William Rappard** **Rue de Lausanne 154,** **CH-1211 Geneva 21, Switzerland** **41-22-739-51-11; www.wto.org**		
WTO TBT-1994	WTO Technical Barriers to Trade (TBT) Agreement Annex 3 Code of Good Practice for the Preparation, Adoption and Application of Standards	9.4.1.3.1

(This is a normative appendix and is part of this standard.)

NORMATIVE APPENDIX A— PRESCRIPTIVE BUILDING ENVELOPE TABLES

Tables A-1 through A-9 appear twice in this appendix. The nine tables are shown first with I-P units, followed by nine tables with SI units.

For *climate zones*, see Section 5.1.4 of ANSI/ASHRAE/IES Standard 90.1 and Normative Appendix B of ANSI/ASHRAE Standard 169.

a. For the United States, the ANSI/ASHRAE Standard 169 *climate zone* map is reproduced above. A list of counties and their respective *climate zones* can be found in Table B1 in ANSI/ASHRAE Standard 169.
b. For Canada, see Table B2 in ANSI/ASHRAE Standard 169.
c. For available international locations (outside the US and Canada), see Table B3 in ANSI/ASHRAE Standard 169.
d. For locations not provided in Table B2 or B3, see Table B4 (reproduced below) in ANSI/ASHRAE Standard 169 for the international *climate zone* definitions.

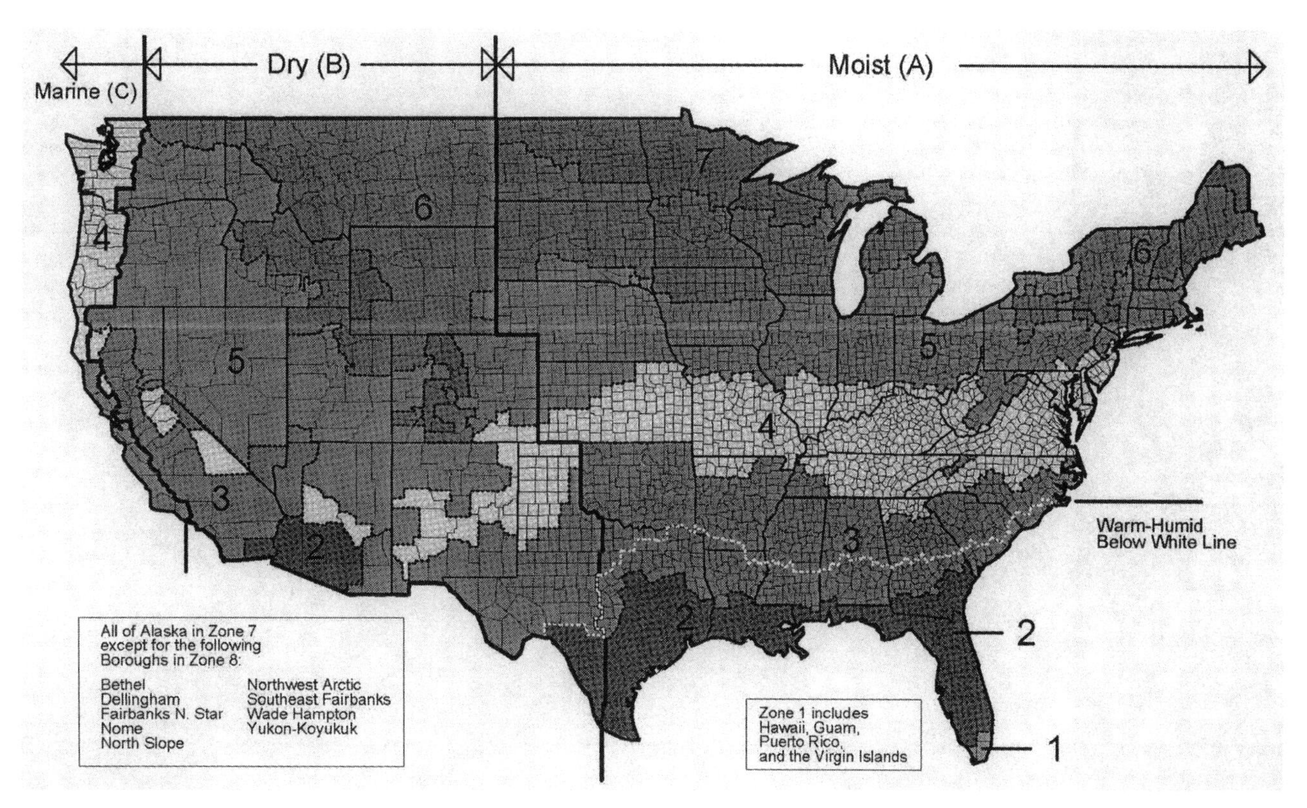

International Climate Zone Definitions

Climate Zone Number	Name	Thermal Criteria (I-P)	Thermal Criteria (SI)
1	Very Hot – Humid (1A), Dry (1B)	9000 < CDD50°F	5000 < CDD10°C
2	Hot – Humid (2A), Dry (2B)	6300 < CDD50°F ≤ 9000	3500 < CDD10°C ≤5000
3A, 3B	Warm – Humid (3A), Dry (3B)	4500 < CDD50°F ≤ 6300	2500 < CDD10°C ≤ 3500
3C	Warm – Marine (3C)	CDD50°F ≤ 4500 and HDD65°F ≤ 3600	CDD10°C ≤ 2500 and HDD18°C ≤ 2000
4A, 4B	Mixed – Humid (4A), Dry (4B)	CDD50°F ≤ 4500 and 3600 < HDD65°F ≤ 5400	2500 ≤ CDD10°C and 2000 < HDD18°C ≤ 3000
4C	Mixed – Marine (4C)	3600 < HDD65°F ≤ 5400	2000 < HDD18°C ≤ 3000
5A, 5B, 5C	Cool– Humid (5A), Dry (5B), Marine (5C)	5400 < HDD65°F ≤ 7200	3000 < HDD18°C ≤ 4000
6A, 6B	Cold – Humid (6A), Dry (6B)	7200 < HDD65°F ≤ 9000	4000 < HDD18°C ≤ 5000
7	Very Cold	9000 < HDD65°F ≤ 12600	5000 < HDD18°C ≤ 7000
8	Subarctic	12600 < HDD65°F	7000 < HDD18°C

TABLE A-1 (Supersedes Table 5.5-1 in ANSI/ASHRAE/IES Standard 90.1) Building Envelope Requirements for Climate Zone 1 (A, B) (I-P)

	Nonresidential		Residential		Semiheated	
	Assembly	Insulation	Assembly	Insulation	Assembly	Insulation
Opaque Elements	Max.	Min. R-Value	Max.	Min. R-Value	Max.	Min.R-Value
Roofs						
Insulation Entirely above Deck	U-0.048	R-20.0 ci	U-0.039	R-25.0 ci	U-0.173	R-5.0 ci
Metal Building	U-0.044	R-19.0 + R-11.0 Ls[d]	U-0.035	R-19.0 + R-11.0 Ls	U-0.082	R-19.0
Attic and Other	U-0.027	R-38.0	U-0.021	R-49.0	U-0.053	R-19.0
Walls, Above Grade						
Mass	U-0.151[a]	R-5.7 ci[a]	U-0.123	R-7.6 ci	U-0.151[a]	R-5.7 ci[a]
Metal Building	U-0.079	R-13.0 + R-6.5 ci	U-0.079	R-13.0 + R-6.5 ci	U-0.147	R-19.0
Steel Framed	U-0.077	R-13.0 + R-5.0 ci	U-0.077	R-13.0 + R-5.0 ci	U-0.124	R-13.0
Wood Framed and Other	U-0.064	R-13.0 + R-3.8 ci	U-0.064	R-13.0 + R-3.8 ci	U-0.089	R-13.0
Wall, Below Grade						
Below Grade Wall	C-1.140	NR	C-1.140	NR	C-1.140	NR
Floors						
Mass	U-0.137	R-4.2 ci	U-0.137	R-4.2 ci	U-0.322	NR
Steel Joist	U-0.052	R-19.0	U-0.052	R-19.0	U-0.350	NR
Wood Framed and Other	U-0.051	R-19.0	U-0.051	R-19.0	U-0.282	NR
Slab-On-Grade Floors						
Unheated	F-0.730	NR	F-0.730	NR	F-0.730	NR
Heated	F-0.640	R-7.5 for 12 in. + R-5 ci below	F-0.640	R-7.5 for 12 in. + R-5 ci below	F-1.020	R-7.5 for 12 in.
Opaque Doors						
Swinging	U-0.600		U-0.600		U-0.600	
Non-Swinging	U-0.500		U-0.500		U-0.500	

	Assembly	Assembly	Assembly	Assembly	Assembly	Assembly
Fenestration	Max. U	Max. SHGC	Max. U	Max. SHGC	Max. U	Max. SHGC
Vertical Fenestration, 0%–40% of Wall						
Nonmetal framing: all[b] Metal fr: curtainwall/storefront[c] Metal framing: entrance door[c] Metal framing: all other[c]	U-1.20 U-1.20 U-1.20 U-1.20	SHGC-0.25 all	U-1.20 U-1.20 U-1.20 U-1.20	SHGC-0.25 all	U-1.20 U-1.20 U-1.20 U-1.20	SHGC-NR all
Skylight with Curb, Glass, % of Roof						
0%–2.0%	U_{all}-0.71	$SHGC_{all}$-0.19	U_{all}-0.71	$SHGC_{all}$-0.16	U_{all}-1.98	$SHGC_{all}$-NR
2.1%–5.0%	U_{all}-0.71	$SHGC_{all}$-0.19	U_{all}-0.71	$SHGC_{all}$-0.16	U_{all}-1.98	$SHGC_{all}$-NR
Skylight with Curb, Plastic, % of Roof						
0%–2.0%	U_{all}-1.12	$SHGC_{all}$-0.27	U_{all}-1.12	$SHGC_{all}$-0.27	U_{all}-1.90	$SHGC_{all}$-NR
2.1%–5.0%	U_{all}-1.12	$SHGC_{all}$-0.27	U_{all}-1.12	$SHGC_{all}$-0.27	U_{all}-1.90	$SHGC_{all}$-NR
Skylight without Curb, All, % of Roof						
0%–2.0%	U_{all}-0.57	$SHGC_{all}$-0.19	U_{all}-0.57	$SHGC_{all}$-0.19	U_{all}-1.36	$SHGC_{all}$-NR
2.1%–5.0%	U_{all}-0.57	$SHGC_{all}$-0.19	U_{all}-0.57	$SHGC_{all}$-0.19	U_{all}-1.36	$SHGC_{all}$-NR

The following definitions apply: ci = continuous insulation, Ls = *liner system*, NR = no (insulation) requirement.

[a] Mass *walls* with a heat capacity greater than 12 Btu/ft^2·°F which are unfinished or finished only on the interior do not need to be insulated.

[b] Nonmetal framing includes framing materials other than metal with or without metal reinforcing or cladding.

[c] Metal framing includes metal framing with or without thermal break. The all other subcategory includes operable windows, fixed windows, and non-entrance doors.

[d] Liner system without thermal spacer blocks for this case only.

TABLE A-2 (Supersedes Table 5.5-2 in ANSI/ASHRAE/IES Standard 90.1) Building Envelope Requirements for Climate Zone 2 (A, B) (I-P)

	Nonresidential		Residential		Semiheated	
	Assembly	Insulation	Assembly	Insulation	Assembly	Insulation
Opaque Elements	Max.	Min. R-Value	Max.	Min. R-Value	Max.	Min.R-Value
Roofs						
Insulation Entirely above Deck	U-0.039	R-25.0 ci	U-0.039	R-25.0 ci	U-0.173	R-5.0 ci
Metal Building	U-0.035	R-19.0 + R-11.0 Ls	U-0.035	R-19.0 + R-11.0 Ls	U-0.068	R-13.0 +R- 19.0
Attic and Other	U-0.021	R-49.0	U-0.021	R-49.0	U-0.053	R-19.0
Walls, Above Grade						
Mass	U-0.123	R-7.6 ci	U-0.104	R-9.5 ci	U-0.151[a]	R-5.7 ci[a]
Metal Building	U-0.079	R-13.0 + R-6.5 ci	U-0.052	R-13.0 + R-13.0 ci	U-0.147	R-19.0
Steel Framed	U-0.077	R-13.0 + R-5.0 ci	U-0.055	R-13.0 + R-10.0 ci	U-0.084	R-13.0 + R-3.8 ci
Wood Framed and Other	U-0.064	R-13.0 + R-3.8 ci	U-0.064	R-13.0 + R-3.8 ci	U-0.064	R-13.0 + R-3.8 ci
Wall, Below Grade						
Below Grade Wall	C-1.140	NR	C-1.140	NR	C-1.140	NR
Floors						
Mass	U-0.107	R-6.3 ci	U-0.087	R-8.3 ci	U-0.322	NR
Steel Joist	U-0.038	R-30.0	U-0.038	R-30.0	U-0.052	R-19.0
Wood Framed and Other	U-0.033	R-30.0	U-0.026	R-30.0 + R-7.5 ci	U-0.051	R-19.0
Slab-On-Grade Floors						
Unheated	F-0.730	NR	F-0.730	NR	F-0.730	NR
Heated	F-0.640	R-7.5 for 12 in. + R-5 ci below	F-0.640	R-7.5 for 12 in. + R-5 ci below	F-1.020	R-7.5 for 12 in.
Opaque Doors						
Swinging	U-0.600		U-0.600		U-0.600	
Non-Swinging	U-0.500		U-0.400		U-0.500	

	Assembly	Assembly	Assembly	Assembly	Assembly	Assembly
Fenestration	Max. U	Max. SHGC	Max. U	Max. SHGC	Max. U	Max. SHGC
Vertical Fenestration, 0%–40% of Wall						
Nonmetal framing: all[b]	U-0.75	SHGC-0.25 all	U-0.75	SHGC-0.25 all	U-1.20	SHGC-NR all
Metal fr: curtainwall/storefront[c]	U-0.70		U-0.70		U-1.20	
Metal framing: entrance door[c]	U-1.10		U-1.10		U-1.20	
Metal framing: all other[c]	U-0.75		U-0.75		U-1.20	
Skylight with Curb, Glass, % of Roof						
0%–2.0%	U_{all}-0.71	$SHGC_{all}$-0.19	U_{all}-0.71	$SHGC_{all}$-0.16	U_{all}-1.98	$SHGC_{all}$-NR
2.1%–5.0%	U_{all}-0.71	$SHGC_{all}$-0.19	U_{all}-0.71	$SHGC_{all}$-0.16	U_{all}-1.98	$SHGC_{all}$-NR
Skylight with Curb, Plastic, % of Roof						
0%–2.0%	U_{all}-1.12	$SHGC_{all}$-0.27	U_{all}-1.12	$SHGC_{all}$-0.27	U_{all}-1.90	$SHGC_{all}$-NR
2.1%–5.0%	U_{all}-1.12	$SHGC_{all}$-0.27	U_{all}-1.12	$SHGC_{all}$-0.27	U_{all}-1.90	$SHGC_{all}$-NR
Skylight without Curb, All, % of Roof						
0%–2.0%	U_{all}-0.57	$SHGC_{all}$-0.19	U_{all}-0.57	$SHGC_{all}$-0.19	U_{all}-1.36	$SHGC_{all}$-NR
2.1%–5.0%	U_{all}-0.57	$SHGC_{all}$-0.19	U_{all}-0.57	$SHGC_{all}$-0.19	U_{all}-1.36	$SHGC_{all}$-NR

The following definitions apply: ci = continuous insulation, Ls = *liner system*, NR = no (insulation) requirement.
[a] Mass *walls* with a heat capacity greater than 12 Btu/ft^2·°F which are unfinished or finished only on the interior do not need to be insulated.
[b] Nonmetal framing includes framing materials other than metal with or without metal reinforcing or cladding.
[c] Metal framing includes metal framing with or without thermal break. The all other subcategory includes operable windows, fixed windows, and non-entrance doors.

TABLE A-3 (Supersedes Table 5.5-3 in ANSI/ASHRAE/IES Standard 90.1) Building Envelope Requirements for Climate Zone 3 (A, B, C) (I-P)

	Nonresidential		Residential		Semiheated	
Opaque Elements	Assembly Max.	Insulation Min. R-Value	Assembly Max.	Insulation Min. R-Value	Assembly Max.	Insulation Min.R-Value
Roofs						
Insulation Entirely above Deck	U-0.039	R-25.0 ci	U-0.039	R-25.0 ci	U-0.119	R-7.6 ci
Metal Building	U-0.035	R-19.0 + R-11.0 Ls	U-0.035	R-19.0 + R-11.0 Ls	U-0.068	R-13.0 + R- 19.0
Attic and Other	U-0.021	R-49.0	U-0.021	R-49.0	U-0.034	R-30.0
Walls, Above Grade						
Mass	U-0.104	R-9.5 ci	U-0.090	R-11.4 ci	U-0.151[a]	R-5.7 ci[a]
Metal Building	U-0.079	R-13.0 + R-6.5 ci	U-0.052	R-13.0 + R-13.0 ci	U-0.079	R-13.0 + R-6.5 ci
Steel Framed	U-0.077	R-13.0 + R-5.0 ci	U-0.055	R-13.0 + R-10.0 ci	U-0.084	R-13.0 + R-3.8 ci
Wood Framed and Other	U-0.064	R-13.0 + R-3.8 ci	U-0.064	R-13.0 + R-3.8 ci	U-0.064	R-13.0 + R-3.8 ci
Wall, Below Grade						
Below Grade Wall	C-1.140	NR	C-1.140	NR	C-1.140	NR
Floors						
Mass	U-0.107	R-6.3 ci	U-0.087	R-8.3 ci	U-0.332	NR
Steel Joist	U-0.038	R-30.0	U-0.038	R-30.0	U-0.052	R-19.0
Wood Framed and Other	U-0.033	R-30.0	U-0.026	R-30.0 + R-7.5 ci	U-0.051	R-19.0
Slab-On-Grade Floors						
Unheated	F-0.730	NR	F-0.730	NR	F-0.730	NR
Heated	F-0.640	R-7.5 for 12 in. + R-5 ci below	F-0.640	R-7.5 for 12 in. + R-5 ci below	F-1.020	R-7.5 for 12 in.
Opaque Doors						
Swinging	U-0.600		U-0.600		U-0.600	
Non-Swinging	U-0.500		U-0.400		U-0.500	
Fenestration	**Assembly Max. U**	**Assembly Max. SHGC**	**Assembly Max. U**	**Assembly Max. SHGC**	**Assembly Max. U**	**Assembly Max. SHGC**
Vertical Fenestration, 0%–40% of Wall						
Nonmetal framing: all[b]	U-0.45	SHGC-0.25 all	U-0.45	SHGC-0.25 all	U-0.55	SHGC-NR all
Metal fr: curtainwall/storefront[c]	U-0.50		U-0.50		U-0.60	
Metal framing: entrance door[c]	U-0.80		U-0.80		U-0.80	
Metal framing: all other[c]	U-0.55		U-0.55		U-0.65	
Skylight with Curb, Glass, % of Roof						
0%–2.0%	U_{all}-0.69	$SHGC_{all}$-0.19	U_{all}-0.69	$SHGC_{all}$-0.16	U_{all}-1.98	$SHGC_{all}$-NR
2.1%–5.0%	U_{all}-0.69	$SHGC_{all}$-0.19	U_{all}-0.69	$SHGC_{all}$-0.16	U_{all}-1.98	$SHGC_{all}$-NR
Skylight with Curb, Plastic, % of Roof						
0%–2.0%	U_{all}-0.69	$SHGC_{all}$-0.27	U_{all}-0.69	$SHGC_{all}$-0.27	U_{all}-1.90	$SHGC_{all}$-NR
2.1%–5.0%	U_{all}-0.69	$SHGC_{all}$-0.27	U_{all}-0.69	$SHGC_{all}$-0.27	U_{all}-1.90	$SHGC_{all}$-NR
Skylight without Curb, All, % of Roof						
0%–2.0%	U_{all}-0.45	$SHGC_{all}$-0.19	U_{all}-0.45	$SHGC_{all}$-0.19	U_{all}-1.36	$SHGC_{all}$-NR
2.1%–5.0%	U_{all}-0.45	$SHGC_{all}$-0.19	U_{all}-0.45	$SHGC_{all}$-0.19	U_{all}-1.36	$SHGC_{all}$-NR

The following definitions apply: ci = continuous insulation, Ls = *liner system*, NR = no (insulation) requirement.

[a] Mass *walls* with a heat capacity greater than 12 Btu/ft^2·°F which are unfinished or finished only on the interior do not need to be insulated.

[b] Nonmetal framing includes framing materials other than metal with or without metal reinforcing or cladding.

[c] Metal framing includes metal framing with or without thermal break. The all other subcategory includes operable windows, fixed windows, and non-entrance doors.

TABLE A-4 (Supersedes Table 5.5-4 in ANSI/ASHRAE/IES Standard 90.1) Building Envelope Requirements for Climate Zone 4 (A, B, C) (I-P)

	Nonresidential		Residential		Semiheated	
	Assembly	Insulation	Assembly	Insulation	Assembly	Insulation
Opaque Elements	Max.	Min. R-Value	Max.	Min. R-Value	Max.	Min.R-Value
Roofs						
Insulation Entirely above Deck	U-0.039	R-25.0 ci	U-0.039	R-25.0 ci	U-0.119	R-7.6 ci
Metal Building	U-0.035	R-19.0 + R-11.0 Ls	U-0.035	R-19.0 + R-11.0 Ls	U-0.068	R-13.0 +R- 19.0
Attic and Other	U-0.021	R-49.0	U-0.021	R-49.0	U-0.034	R-30.0
Walls, Above Grade						
Mass	U-0.090	R-11.4 ci	U-0.080	R-13.3 ci	U-0.151[a]	R-5.7 ci[a]
Metal Building	U-0.052	R-13.0 + R-13.0 ci	U-0.052	R-13.0 + R-13.0 ci	U-0.079	R-13.0 + R-6.5 ci
Steel Framed	U-0.055	R-13.0 + R-10.0 ci	U-0.055	R-13.0 + R-10.0 ci	U-0.084	R-13.0 + R-3.8 ci
Wood Framed and Other	U-0.064	R-13.0 + R-3.8 ci	U-0.051	R-13.0 + R-7.5 ci	U-0.064	R-13.0 + R-3.8 ci
Wall, Below Grade						
Below Grade Wall	C-0.119	R-7.5 ci	C-0.092	R-10.0 ci	C-0.119	R-7.5 ci
Floors						
Mass	U-0.074	R-10.4 ci	U-0.064	R-12.5 ci	U-0.107	R-6.3 ci
Steel Joist	U-0.032	R-38.0	U-0.032	R-38.0	U-0.052	R-19.0
Wood Framed and Other	U-0.026	R-30.0 + R-7.5 ci	U-0.026	R-30.0 + R-7.5 ci	U-0.051	R-19.0
Slab-On-Grade Floors						
Unheated	F-0.540	R-10 for 24 in.	F-0.520	R-15 for 24 in.	F-0.540	R-10 for 24 in.
Heated	F-0.550	R-10.0 for 24 in. + R-5 ci below	F-0.550	R-10.0 for 24 in. + R-5 ci below	F-0.950	R-7.5 for 24 in.
Opaque Doors						
Swinging	U-0.600		U-0.600		U-0.600	
Non-Swinging	U-0.400		U-0.400		U-0.500	
	Assembly	Assembly	Assembly	Assembly	Assembly	Assembly
Fenestration	Max. U	Max. SHGC	Max. U	Max. SHGC	Max. U	Max. SHGC
Vertical Fenestration, 0-40% of Wall						
Nonmetal framing: all[b]	U-0.30	SHGC-0.35 all	U-0.30	SHGC-0.40 all	U-0.55	SHGC-NR all
Metal fr: curtainwall/storefront[c]	U-0.40		U-0.40		U-0.60	
Metal framing: entrance door[c]	U-0.75		U-0.75		U-0.80	
Metal framing: all other[c]	U-0.45		U-0.45		U-0.65	
Skylight with Curb, Glass, % of Roof						
0%–2.0%	U_{all}-0.69	$SHGC_{all}$-0.32	U_{all}-0.69	$SHGC_{all}$-0.19	U_{all}-1.98	$SHGC_{all}$-NR
2.1%–5.0%	U_{all}-0.69	$SHGC_{all}$-0.32	U_{all}-0.69	$SHGC_{all}$-0.19	U_{all}-1.98	$SHGC_{all}$-NR
Skylight with Curb, Plastic, % of Roof						
0%–2.0%	U_{all}-0.69	$SHGC_{all}$-0.34	U_{all}-0.69	$SHGC_{all}$-0.27	U_{all}-1.90	$SHGC_{all}$-NR
2.1%–5.0%	U_{all}-0.69	$SHGC_{all}$-0.34	U_{all}-0.69	$SHGC_{all}$-0.27	U_{all}-1.90	$SHGC_{all}$-NR
Skylight without Curb, All, % of Roof						
0%–2.0%	U_{all}-0.45	$SHGC_{all}$-0.32	U_{all}-0.45	$SHGC_{all}$-0.19	U_{all}-1.36	$SHGC_{all}$-NR
2.1%–5.0%	U_{all}-0.45	$SHGC_{all}$-0.32	U_{all}-0.45	$SHGC_{all}$-0.19	U_{all}-1.36	$SHGC_{all}$-NR

The following definitions apply: ci = continuous insulation, Ls = *liner system*, NR = no (insulation) requirement.
[a] Mass *walls* with a heat capacity greater than 12 Btu/ft^2·°F which are unfinished or finished only on the interior do not need to be insulated.
[b] Nonmetal framing includes framing materials other than metal with or without metal reinforcing or cladding.
[c] Metal framing includes metal framing with or without thermal break. The all other subcategory includes operable windows, fixed windows, and non-entrance doors.

TABLE A-5 (Supersedes Table 5.5-5 in ANSI/ASHRAE/IES Standard 90.1) Building Envelope Requirements for Climate Zone 5 (A, B, C) (I-P)

	Nonresidential		Residential		Semiheated	
	Assembly	**Insulation**	**Assembly**	**Insulation**	**Assembly**	**Insulation**
Opaque Elements	**Max.**	**Min. R-Value**	**Max.**	**Min. R-Value**	**Max.**	**Min.R-Value**
Roofs						
Insulation Entirely above Deck	U-0.039	R-25.0 ci	U-0.039	R-25.0 ci	U-0.093	R-10.0 ci
Metal Building	U-0.035	R-19.0 + R-11.0 Ls	U-0.035	R-19.0 + R-11.0 Ls	U-0.068	R-13.0 +R- 19.0
Attic and Other	U-0.021	R-49.0	U-0.021	R-49.0	U-0.034	R-30.0
Walls, Above Grade						
Mass	U-0.080	R-13.3 ci	U-0.071	R-15.2 ci	U-0.123	R-7.6 ci
Metal Building	U-0.052	R-13.0 + R-13.0 ci	U-0.052	R-13.0 + R-13.0 ci	U-0.079	R-13.0 + R-6.5 ci
Steel Framed	U-0.055	R-13.0 + R-10.0 ci	U-0.055	R-13.0 + R-10.0 ci	U-0.084	R-13.0 +R-3.8 ci
Wood Framed and Other	U-0.051	R-13.0 + R-7.5 ci	U-0.045	R-13.0 + R-10.0 ci	U-0.064	R-13.0 + R-3.8 ci
Wall, Below Grade						
Below Grade Wall	C-0.092	R-10.0 ci	C-0.092	R-10.0 ci	C-0.119	R-7.5 ci
Floors						
Mass	U-0.064	R-12.5 ci	U-0.057	R-14.6 ci	U-0.107	R-6.3 ci
Steel Joist	U-0.032	R-38.0	U-0.032	R-38.0	U-0.038	R-30.0
Wood Framed and Other	U-0.026	R-30.0 + R-7.5 ci	U-0.026	R-30.0 + R-7.5 ci	U-0.033	R-30.0
Slab-On-Grade Floors						
Unheated	F-0.540	R-10 for 24 in.	F-0.520	R-15 for 24 in.	F-0.540	R-10 for 24 in.
Heated	F-0.440	R-15.0 for 36 in. + R-5 ci below	F-0.440	R-15.0 for 36 in. + R-5 ci below	F-0.900	R-10 for 24 in.
Opaque Doors						
Swinging	U-0.400		U-0.400		U-0.600	
Non-Swinging	U-0.400		U-0.400		U-0.500	

	Assembly	**Assembly**	**Assembly**	**Assembly**	**Assembly**	**Assembly**
Fenestration	**Max. U**	**Max. SHGC**	**Max. U**	**Max. SHGC**	**Max. U**	**Max. SHGC**
Vertical Fenestration, 0%–40% of Wall						
Nonmetal framing: all[b]	U-0.25	SHGC-0.35 all	U-0.25	SHGC-0.40 all	U-0.55	SHGC-NR all
Metal fr: curtainwall/storefront[c]	U-0.35		U-0.35		U-0.60	
Metal framing: entrance door[c]	U-0.70		U-0.70		U-0.80	
Metal framing: all other[c]	U-0.45		U-0.45		U-0.65	
Skylight with Curb, Glass, % of Roof						
0%–2.0%	U_{all}-0.67	$SHGC_{all}$-0.36	U_{all}-0.67	$SHGC_{all}$-0.36	U_{all}-1.98	$SHGC_{all}$-NR
2.1%–5.0%	U_{all}-0.67	$SHGC_{all}$-0.36	U_{all}-0.67	$SHGC_{all}$-0.36	U_{all}-1.98	$SHGC_{all}$-NR
Skylight with Curb, Plastic, % of Roof						
0-2%–0%	U_{all}-0.69	$SHGC_{all}$-0.34	U_{all}-0.69	$SHGC_{all}$-0.34	U_{all}-1.90	$SHGC_{all}$-NR
2.1%–5.0%	U_{all}-0.69	$SHGC_{all}$-0.34	U_{all}-0.69	$SHGC_{all}$-0.34	U_{all}-1.90	$SHGC_{all}$-NR
Skylight without Curb, All, % of Roof						
0%–2.0%	U_{all}-0.45	$SHGC_{all}$-0.36	U_{all}-0.45	$SHGC_{all}$-0.36	U_{all}-1.36	$SHGC_{all}$-NR
2.1%–5.0%	U_{all}-0.45	$SHGC_{all}$-0.36	U_{all}-0.45	$SHGC_{all}$-0.36	U_{all}-1.36	$SHGC_{all}$-NR

The following definitions apply: ci = continuous insulation, Ls = *liner system*, NR = no (insulation) requirement.

[a] Mass *walls* with a heat capacity greater than 12 $Btu/ft^2 \cdot {}^\circ F$ which are unfinished or finished only on the interior do not need to be insulated.

[b] Nonmetal framing includes framing materials other than metal with or without metal reinforcing or cladding.

[c] Metal framing includes metal framing with or without thermal break. The all other subcategory includes operable windows, fixed windows, and non-entrance doors.

TABLE A-6 (Supersedes Table 5.5-6 in ANSI/ASHRAE/IES Standard 90.1) Building Envelope Requirements for Climate Zone 6 (A, B) (I-P)

	Nonresidential		Residential		Semiheated	
	Assembly	Insulation	Assembly	Insulation	Assembly	Insulation
Opaque Elements	Max.	Min. R-Value	Max.	Min. R-Value	Max.	Min.R-Value
Roofs						
Insulation Entirely above Deck	U-0.032	R-30.0 ci	U-0.032	R-30.0 ci	U-0.063	R-15.0 ci
Metal Building	U-0.031	R-25.0 + R-11.0 Ls	U-0.031	R-25.0 + R-11.0 Ls	U-0.068	R-13.0 +R- 19.0
Attic and Other	U-0.021	R-49.0	U-0.021	R-49.0	U-0.027	R-38.0
Walls, Above Grade						
Mass	U-0.071	R-15.2 ci	U-0.060	R-20.0 ci	U-0.104	R-9.5 ci
Metal Building	U-0.052	R-13.0 + R-13.0 ci	U-0.052	R-13.0 + R-13.0 ci	U-0.079	R-13.0 + R-6.5 ci
Steel Framed	U-0.055	R-13.0 + R-10.0 ci	U-0.055	R-13.0 + R-10.0 ci	U-0.084	R-13.0 + R-3.8 ci
Wood Framed and Other	U-0.045	R-13.0 + R-10.0 ci	U-0.045	R-13.0 + R-10.0 ci	U-0.064	R-13.0 + R-3.8 ci
Wall, Below Grade						
Below Grade Wall	C-0.092	R-10.0 ci	C-0.092	R-10.0 ci	C-0.119	R-7.5 ci
Floors						
Mass	U-0.057	R-14.6 ci	U-0.051	R-16.7 ci	U-0.107	R-6.3 ci
Steel Joist	U-0.032	R-38.0	U-0.023	R-38.0 + R-12.5 ci	U-0.038	R-30.0
Wood Framed and Other	U-0.026	R-30.0 + R-7.5 ci	U-0.026	R-30.0 + R-7.5 ci	U-0.033	R-30.0
Slab-On-Grade Floors						
Unheated	F-0.520	R-15 for 24 in.	F-0.510	R-20 for 24 in.	F-0.540	R-10 for 24 in.
Heated	F-0.440	R-15.0 for 36 in. + R-5 ci below	F-0.440	R-15.0 for 36 in. + R-5 ci below	F-0.900	R-10 for 24 in.
Opaque Doors						
Swinging	U-0.400		U-0.400		U-0.600	
Non-Swinging	U-0.400		U-0.400		U-0.500	

	Assembly	Assembly	Assembly	Assembly	Assembly	Assembly
Fenestration	Max. U	Max. SHGC	Max. U	Max. SHGC	Max. U	Max. SHGC
Vertical Fenestration,0%–40% of Wall						
Nonmetal framing: all[b]	U-0.25	SHGC-0.40 all	U-0.25	SHGC-0.40 all	U-0.45	SHGC-NR all
Metal fr: curtainwall/storefront[c]	U-0.35		U-0.35		U-0.50	
Metal framing: entrance door[c]	U-0.70		U-0.70		U-0.80	
Metal framing: all other[c]	U-0.45		U-0.45		U-0.55	
Skylight with Curb, Glass,% of Roof						
0%–2.0%	U_{all}-0.67	$SHGC_{all}$-0.46	U_{all}-0.67	$SHGC_{all}$-0.46	U_{all}-1.98	$SHGC_{all}$-NR
2.1%–5.0%	U_{all}-0.67	$SHGC_{all}$-0.46	U_{all}-0.67	$SHGC_{all}$-0.46	U_{all}-1.98	$SHGC_{all}$-NR
Skylight with Curb, Plastic,% of Roof						
0%–2.0%	U_{all}-0.69	$SHGC_{all}$-0.49	U_{all}-0.69	$SHGC_{all}$-0.49	U_{all}-1.90	$SHGC_{all}$-NR
2.1%–5.0%	U_{all}-0.69	$SHGC_{all}$-0.49	U_{all}-0.69	$SHGC_{all}$-0.49	U_{all}-1.90	$SHGC_{all}$-NR
Skylight without Curb, All,% of Roof						
0%–2.0%	U_{all}-0.45	$SHGC_{all}$-0.46	U_{all}-0.45	$SHGC_{all}$-0.39	U_{all}-1.36	$SHGC_{all}$-NR
2.1%–5.0%	U_{all}-0.45	$SHGC_{all}$-0.46	U_{all}-0.45	$SHGC_{all}$-0.39	U_{all}-1.36	$SHGC_{all}$-NR

The following definitions apply: ci = continuous insulation, Ls = *liner system*, NR = no (insulation) requirement.

[a] Mass *walls* with a heat capacity greater than 12 Btu/ft^2·°F which are unfinished or finished only on the interior do not need to be insulated.

[b] Nonmetal framing includes framing materials other than metal with or without metal reinforcing or cladding.

[c] Metal framing includes metal framing with or without thermal break. The all other subcategory includes operable windows, fixed windows, and non-entrance doors.

TABLE A-7 (Supersedes Table 5.5-7 in ANSI/ASHRAE/IES Standard 90.1)
Building Envelope Requirements for Climate Zone 7 (I-P)

	Nonresidential		Residential		Semiheated	
Opaque Elements	**Assembly Max. U**	**Assembly Max. SHGC**	**Assembly Max. U**	**Assembly Max. SHGC**	**Assembly Max. U**	**Assembly Max. SHGC**
Roofs						
Insulation Entirely above Deck	U-0.028	R-35.0 ci	U-0.028	R-35.0 ci	U-0.063	R-15.0 ci
Metal Building	U-0.029	R-30.0 + R-11.0 Ls	U-0.029	R-30.0 + R-11.0 Ls	U-0.068	R-13.0 + R-19.0
Attic and Other	U-0.017	R-60.0	U-0.017	R-60.0	U-0.027	R-38.0
Walls, Above Grade						
Mass	U-0.060	R-20.0 ci	U-0.060	R-20.0 ci	U-0.090	R-11.4 ci
Metal Building	U-0.052	R-13.0 + R-13.0 ci	U-0.039	R-13.0 + R-19.5 ci	U-0.079	R-13.0 + R-6.5 ci
Steel Framed	U-0.055	R-13.0 + R-10.0 ci	U-0.037	R-13.0 + R-18.8 ci	U-0.084	R-13.0 + R-3.8 ci
Wood Framed and Other	U-0.045	R-13.0 + R-10.0 ci	U-0.045	R-13.0 + R-10.0 ci	U-0.064	R-13.0 + R-3.8 ci
Wall, Below Grade						
Below Grade Wall	C-0.092	R-10.0 ci	C-0.075	R-12.5 ci	C-0.119	R-7.5 ci
Floors						
Mass	U-0.043	R-20.0 ci	U-0.043	R-20.0 ci	U-0.087	R-8.3 ci
Steel Joist	U-0.032	R-38.0	U-0.023	R-38.0 + R-12.5 ci	U-0.038	R-30.0
Wood Framed and Other	U-0.026	R-30.0 + R-7.5 ci	U-0.026	R-30.0 + R-7.5 ci	U-0.033	R-30.0
Slab-On-Grade Floors						
Unheated	F-0.300	R-15 for 24 in. + R-5 ci below	F-0.300	R-15 for 24 in. + R-5 ci below	F-0.540	R-10 for 24 in.
Heated	F-0.373	R-20.0 for 36 in. + R-5 ci below	F-0.373	R-20.0 for 36 in. + R-5 ci below	F-0.688	R-20 for 48 in.
Opaque Doors						
Swinging	U-0.400		U-0.400		U-0.600	
Non-Swinging	U-0.400		U-0.400		U-0.500	
Fenestration	**Assembly Max. U**	**Assembly Max. SHGC**	**Assembly Max. U**	**Assembly Max. SHGC**	**Assembly Max. U**	**Assembly Max. SHGC**
Vertical Fenestration, 0%–40% of Wall						
Nonmetal framing: all[b]	U-0.25	SHGC-0.45 all	U-0.25	SHGC-NR all	U-0.45	SHGC-NR all
Metal fr: curtainwall/storefront[c]	U-0.30		U-0.30		U-0.50	
Metal framing: entrance door[c]	U-0.70		U-0.70		U-0.80	
Metal framing: all other[c]	U-0.35		U-0.35		U-0.55	
Skylight with Curb, Glass, % of Roof						
0%–2.0%	U_{all}-0.67	$SHGC_{all}$-0.46	U_{all}-0.67	$SHGC_{all}$-0.46	U_{all}-1.98	$SHGC_{all}$-NR
2.1%–5.0%	U_{all}-0.67	$SHGC_{all}$-0.46	U_{all}-0.67	$SHGC_{all}$-0.46	U_{all}-1.98	$SHGC_{all}$-NR
Skylight with Curb, Plastic, % of Roof						
0%–2.0%	U_{all}-0.69	$SHGC_{all}$-0.50	U_{all}-0.69	$SHGC_{all}$-0.50	U_{all}-1.90	$SHGC_{all}$-NR
2.1%–5.0%	U_{all}-0.69	$SHGC_{all}$-0.50	U_{all}-0.69	$SHGC_{all}$-0.50	U_{all}-1.90	$SHGC_{all}$-NR
Skylight without Curb, All, % of Roof						
0%–2.0%	U_{all}-0.45	$SHGC_{all}$-0.46	U_{all}-0.45	$SHGC_{all}$-0.46	U_{all}-1.36	$SHGC_{all}$-NR
2.1%–5.0%	U_{all}-0.45	$SHGC_{all}$-0.46	U_{all}-0.45	$SHGC_{all}$-0.46	U_{all}-1.36	$SHGC_{all}$-NR

The following definitions apply: ci = continuous insulation, Ls = *liner system*, NR = no (insulation) requirement.
[a] Mass *walls* with a heat capacity greater than 12 Btu/ft^2·°F which are unfinished or finished only on the interior do not need to be insulated.
[b] Nonmetal framing includes framing materials other than metal with or without metal reinforcing or cladding.
[c] Metal framing includes metal framing with or without thermal break. The all other subcategory includes operable windows, fixed windows, and non-entrance doors.

TABLE A-8 (Supersedes Table 5.5-8 in ANSI/ASHRAE/IES Standard 90.1) Building Envelope Requirements for Climate Zone 8 (I-P)

	Nonresidential		Residential		Semiheated	
	Assembly	Insulation	Assembly	Insulation	Assembly	Insulation
Opaque Elements	Max.	Min. R-Value	Max.	Min. R-Value	Max.	Min.R-Value
Roofs						
Insulation Entirely above Deck	U-0.028	R-35.0 ci	U-0.028	R-35.0 ci	U-0.048	R-20.0 ci
Metal Building	U-0.029	R-30.0 + R11.0 Ls	U-0.029	R-30.0 + R11.0 Ls	U-0.044	R-19.0 + R11.0 Ls[d]
Attic and Other	U-0.017	R-60.0	U-0.017	R-60.0	U-0.027	R-38.0
Walls, Above Grade						
Mass	U-0.060	R-20.0 ci	U-0.043	R-31.3 ci	U-0.080	R-13.3 ci
Metal Building	U-0.052	R-13.0 + R-13.0 ci	U-0.031	R-13.0 + R-26 ci	U-0.052	R-13.0 + R-13 ci
Steel Framed	U-0.055	R-13.0 + R-10.0 ci	U-0.033	R-13.0 + R-21.9 ci	U-0.064	R-13.0 + R-7.5 ci
Wood Framed and Other	U-0.045	R-13.0 + R-10.0 ci	U-0.032	R-13.0 + R-18.8 ci	U-0.064	R-13.0 + R-3.8 ci
Wall, Below Grade						
Below Grade Wall	C-0.092	R-10.0 ci	C-0.063	R-15.0 ci	C-0.119	R-7.5 ci
Floors						
Mass	U-0.043	R-20.0 ci	U-0.043	R-20.0 ci	U-0.064	R-12.5 ci
Steel Joist	U-0.023	R-38.0 + R-12.5 ci	U-0.023	R-38.0 + R-12.5 ci	U-0.038	R-30.0
Wood Framed and Other	U-0.026	R-30.0 + R-7.5 ci	U-0.026	R-30.0 + R-7.5 ci	U-0.026	R-30.0 + R-7.5 ci
Slab-On-Grade Floors						
Unheated	F-0.300	R-15 for 24 in. + R-5 ci below	F-0.300	R-15 for 24 in. + R-5 ci below	F-0.540	R-10 for 24 in.
Heated	F-0.373	R-20.0 for 36 in. + R-5 ci below	F-0.373	R-20.0 for 36 in. + R-5 ci below	F-0.688	R-20 for 48 in.
Opaque Doors						
Swinging	U-0.400		U-0.400		U-0.400	
Non-Swinging	U-0.400		U-0.400		U-0.400	

	Assembly	Assembly	Assembly	Assembly	Assembly	Assembly
Fenestration	Max. U	Max. SHGC	Max. U	Max. SHGC	Max. U	Max. SHGC
Vertical Fenestration, 0%–40% of Wall						
Nonmetal framing: all[b]	U-0.25	SHGC-0.45 all	U-0.25	SHGC-NR all	U-0.45	SHGC-NR all
Metal fr: curtainwall/storefront[c]	U-0.30		U-0.30		U-0.50	
Metal framing: entrance door[c]	U-0.70		U-0.70		U-0.80	
Metal framing: all other[c]	U-0.35		U-0.35		U-0.55	
Skylight with Curb, Glass, % of Roof						
0%–2.0%	U_{all}-0.58	$SHGC_{all}$-NR	U_{all}-0.58	$SHGC_{all}$-NR	U_{all}-1.30	$SHGC_{all}$-NR
2.1%–5.0%	U_{all}-0.58	$SHGC_{all}$-NR	U_{all}-0.58	$SHGC_{all}$-NR	U_{all}-1.30	$SHGC_{all}$-NR
Skylight with Curb, Plastic, % of Roof						
0%–2.0%	U_{all}-0.58	$SHGC_{all}$-NR	U_{all}-0.58	$SHGC_{all}$-NR	U_{all}-1.10	$SHGC_{all}$-NR
2.1%–5.0%	U_{all}-0.58	$SHGC_{all}$-NR	U_{all}-0.58	$SHGC_{all}$-NR	U_{all}-1.10	$SHGC_{all}$-NR
Skylight without Curb, All, % of Roof						
0%–2.0%	U_{all}-0.45	$SHGC_{all}$-NR	U_{all}-0.45	$SHGC_{all}$-NR	U_{all}-0.81	$SHGC_{all}$-NR
2.1%–5.0%	U_{all}-0.45	$SHGC_{all}$-NR	U_{all}-0.45	$SHGC_{all}$-NR	U_{all}-0.81	$SHGC_{all}$-NR

The following definitions apply: ci = continuous insulation, Ls = *liner system*, NR = no (insulation) requirement.

[a] Mass *walls* with a heat capacity greater than 12 Btu/ft^2·°F which are unfinished or finished only on the interior do not need to be insulated.

[b] Nonmetal framing includes framing materials other than metal with or without metal reinforcing or cladding.

[c] Metal framing includes metal framing with or without thermal break. The all other subcategory includes operable windows, fixed windows, and non-entrance doors.

[d] Liner system without thermal spacer blocks for this case only.

TABLE A-9 (Supersedes Table A2.4.2 in ANSI/ASHRAE/IES Standard 90.1)
Single-Rafter Roof Requirements (I-P)

Climate Zone	Minimum Insulation R-Value or Maximum Assembly U-Factor		
	Nonresidential	Residential	Semiheated
1	R-38 U-0.029	R-38 + R10 ci U-0.022	R-19 U-0.055
2	R-38 + R10 ci U-0.022	R-38 + R10 ci U-0.022	R-19 U-0.055
3, 4, 5	R-38 + R10 ci U-0.022	R-38 + R10 ci U-0.022	R-30 U-0.036
6	R-38 + R10 ci U-0.022	R-38 + R10 ci U-0.022	R-38 U-0.029
7, 8	R-38 + R15 ci U-0.020	R-38 + R15 ci U-0.020	R-38 U-0.029

TABLE A-1 (Supersedes Table 5.5-1 in ANSI/ASHRAE/IES Standard 90.1)
Building Envelope Requirements for Climate Zone 1 (A, B) (SI)

	Nonresidential		Residential		Semiheated	
	Assembly	Insulation	Assembly	Insulation	Assembly	Insulation
Opaque Elements	Max.	Min. R-Value	Max.	Min. R-Value	Max.	Min.R-Value
Roofs						
Insulation Entirely above Deck	U-0.27	R-3.5 ci	U-0.22	R-4.4 ci	U-0.98	R-0.9 ci
Metal Building	U-0.25	R-3.3 +R-1.9 Ls[d]	U-0.20	R-3.3 + R-1.9 Ls	U-0.47	R-3.3
Attic and Other	U-0.15	R-6.7	U-0.12	R-8.6	U-0.30	R-3.3
Walls, Above Grade						
Mass	U-0.86 [a]	R-1.0 ci [a]	U-0.70	R-1.3 ci	U-0.86 [a]	R-1.0 ci [a]
Metal Building	U-0.45	R-2.3 + R-1.1 ci	U-0.45	R-2.3 + R-1.1 ci	U-0.84	R-3.3
Steel Framed	U-0.43	R-2.3 + R-0.9 ci	U-0.43	R-2.3 + R-0.9 ci	U-0.71	R-2.3
Wood Framed and Other	U-0.36	R-2.3 + R-0.7 ci	U-0.36	R-2.3 + R-0.7 ci	U-0.50	R-2.3
Wall, Below Grade						
Below Grade Wall	C-6.47	NR	C-6.47	NR	C-6.47	NR
Floors						
Mass	U-0.78	R-0.7 ci	U-0.78	R-0.7 ci	U-1.83	NR
Steel Joist	U-0.30	R-3.3	U-0.30	R-3.3	U-1.99	NR
Wood Framed and Other	U-0.29	R-3.3	U-0.29	R-3.3	U-1.60	NR
Slab-On-Grade Floors						
Unheated	F-1.26	NR	F-1.26	NR	F-1.26	NR
Heated	F-1.11	R-1.3 for 300 mm + R-0.9 ci below	F-1.11	R-1.3 for 300 mm + R-0.9 ci below	F-1.77	R-1.3 for 300mm
Opaque Doors						
Swinging	U-3.41		U-3.41		U-3.41	
Non-Swinging	U-2.84		U-2.84		U-2.84	
	Assembly	Assembly	Assembly	Assembly	Assembly	Assembly
Fenestration	Max. U	Max. SHGC	Max. U	Max. SHGC	Max. U	Max. SHGC
Vertical Fenestration,0-40% of Wall						
Nonmetal framing: all[b]	U-6.81	SHGC-0.25 all	U-6.81	SHGC-0.25 all	U-6.81	SHGC-NR all
Metal fr: curtainwall/storefront[c]	U-6.81		U-6.81		U-6.81	
Metal framing: entrance door[c]	U-6.81		U-6.81		U-6.81	
Metal framing: all other[c]	U-6.81		U-6.81		U-6.81	
Skylight with Curb, Glass,% of Roof						
0%–2.0%	U_{all}-4.03	$SHGC_{all}$-0.19	U_{all}-4.03	$SHGC_{all}$-0.16	U_{all}-11.24	$SHGC_{all}$-NR
2.1%–5.0%	U_{all}-4.03	$SHGC_{all}$-0.19	U_{all}-4.03	$SHGC_{all}$-0.16	U_{all}-11.24	$SHGC_{all}$-NR
Skylight with Curb, Plastic,% of Roof						
0%–2.0%	U_{all}-6.36	$SHGC_{all}$-0.27	U_{all}-6.36	$SHGC_{all}$-0.27	U_{all}-10.79	$SHGC_{all}$-NR
2.1%–5.0%	U_{all}-6.36	$SHGC_{all}$-0.27	U_{all}-6.36	$SHGC_{all}$-0.27	U_{all}-10.79	$SHGC_{all}$-NR
Skylight without Curb, All,% of Roof						
0%–2.0%	U_{all}-3.24	$SHGC_{all}$-0.19	U_{all}-3.24	$SHGC_{all}$-0.19	U_{all}-7.72	$SHGC_{all}$-NR
2.1%–5.0%	U_{all}-3.24	$SHGC_{all}$-0.19	U_{all}-3.24	$SHGC_{all}$-0.19	U_{all}-7.72	$SHGC_{all}$-NR

The following definitions apply: ci = continuous insulation, Ls = *liner system*, NR = no (insulation) requirement
[a] Mass *walls* with a heat capacity greater than 245 kJ/m^2·K which are unfinished or finished only on the interior do not need to be insulated.
[b] Nonmetal framing includes framing materials other than metal with or without metal reinforcing or cladding.
[c] Metal framing includes metal framing with or without thermal break. The all other subcategory includes operable windows, fixed windows, and non-entrance doors.
[d] Liner system without thermal spacer blocks for this case only.

TABLE A-2 (Supersedes Table 5.5-2 in ANSI/ASHRAE/IES Standard 90.1) Building Envelope Requirements for Climate Zone 2 (A, B) (SI)

	Nonresidential		Residential		Semiheated	
Opaque Elements	**Assembly Max.**	**Insulation Min. R-Value**	**Assembly Max.**	**Insulation Min. R-Value**	**Assembly Max.**	**Insulation Min.R-Value**
Roofs						
Insulation Entirely above Deck	U-0.22	R-4.4 ci	U-0.22	R-4.4 ci	U-0.98	R-0.9 ci
Metal Building	U-0.20	R-3.3 + R-1.9 Ls	U-0.20	R-3.3 + R-1.9 Ls	U-0.39	R-2.3 + R-3.3
Attic and Other	U-0.12	R-8.6	U-0.12	R-8.6	U-0.30	R-3.3
Walls, Above Grade						
Mass	U-0.70	R-1.3 ci	U-0.59	R-1.7 ci	U-0.86 [a]	R-1.0 ci [a]
Metal Building	U-0.45	R-2.3 + R-1.1 ci	U-0.30	R-2.3 + R-2.3 ci	U-0.84	R-3.3
Steel Framed	U-0.43	R-2.3 + R-0.9 ci	U-0.31	R-2.3 + R-1.8 ci	U-0.48	R-2.3 + R-0.7 ci
Wood Framed and Other	U-0.36	R-2.3 + R-0.7 ci	U-0.36	R-2.3 + R-0.7 ci	U-0.36	R-2.3 + R-0.7 ci
Wall, Below Grade						
Below Grade Wall	C-6.47	NR	C-6.47	NR	C-6.47	NR
Floors						
Mass	U-0.61	R-1.1 ci	U-0.50	R-1.5	U-1.83	NR
Steel Joist	U-0.21	R-5.3	U-0.21	R-5.3	U-0.30	R-3.3
Wood Framed and Other	U-0.19	R-5.3	U-0.15	R-5.3 + R-1.3 ci	U-0.29	R-3.3
Slab-On-Grade Floors						
Unheated	F-1.26	NR	F-1.26	NR	F-1.26	NR
Heated	F-1.11	R-1.3 for 300 mm + R-0.9 ci below	F-1.11	R-1.3 for 300 mm + R-0.9 ci below	F-1.77	R-1.3 for 300mm
Opaque Doors						
Swinging	U-3.41		U-3.41		U-3.41	
Non-Swinging	U-2.84		U-2.27		U-2.84	
Fenestration	**Assembly Max. U**	**Assembly Max. SHGC**	**Assembly Max. U**	**Assembly Max. SHGC**	**Assembly Max. U**	**Assembly Max. SHGC**
Vertical Fenestration, 0%–40% of Wall						
Nonmetal framing: all[b]	U-4.26	SHGC-0.25 all	U-4.26	SHGC-0.25 all	U-6.81	SHGC-NR all
Metal fr: curtainwall/storefront[c]	U-3.97		U-3.97		U-6.81	
Metal framing: entrance door[c]	U-6.25		U-6.25		U-6.81	
Metal framing: all other[c]	U-4.26		U-4.26		U-6.81	
Skylight with Curb, Glass,% of Roof						
0%–2.0%	U_{all}-4.03	$SHGC_{all}$-0.19	U_{all}-4.03	$SHGC_{all}$-0.16	U_{all}-11.24	$SHGC_{all}$-NR
2.1%–5.0%	U_{all}-4.03	$SHGC_{all}$-0.19	U_{all}-4.03	$SHGC_{all}$-0.16	U_{all}-11.24	$SHGC_{all}$-NR
Skylight with Curb, Plastic,% of Roof						
0%–2.0%	U_{all}-6.36	$SHGC_{all}$-0.27	U_{all}-6.36	$SHGC_{all}$-0.27	U_{all}-10.79	$SHGC_{all}$-NR
2.1%–5.0%	U_{all}-6.36	$SHGC_{all}$-0.27	U_{all}-6.36	$SHGC_{all}$-0.27	U_{all}-10.79	$SHGC_{all}$-NR
Skylight without Curb, All,% of Roof						
0%–2.0%	U_{all}-3.24	$SHGC_{all}$-0.19	U_{all}-3.24	$SHGC_{all}$-0.19	U_{all}-7.72	$SHGC_{all}$-NR
2.1%–5.0%	U_{all}-3.24	$SHGC_{all}$-0.19	U_{all}-3.24	$SHGC_{all}$-0.19	U_{all}-7.72	$SHGC_{all}$-NR

The following definitions apply: ci = continuous insulation, Ls = *liner system*, NR = no (insulation) requirement

a Mass *walls* with a heat capacity greater than 245 $kJ/m^2{\cdot}K$ which are unfinished or finished only on the interior do not need to be insulated.

[b] Nonmetal framing includes framing materials other than metal with or without metal reinforcing or cladding.

[c] Metal framing includes metal framing with or without thermal break. The all other subcategory includes operable windows, fixed windows, and non-entrance doors.

TABLE A-3 (Supersedes Table 5.5-3 in ANSI/ASHRAE/IES Standard 90.1) Building Envelope Requirements for Climate Zone 3 (A, B, C) (SI)

	Nonresidential		Residential		Semiheated	
	Assembly	Insulation	Assembly	Insulation	Assembly	Insulation
Opaque Elements	Max.	Min. R-Value	Max.	Min. R-Value	Max.	Min.R-Value
Roofs						
Insulation Entirely above Deck	U-0.22	R-4.4 ci	U-0.22	R-4.4 ci	U-0.68	R-1.3 ci
Metal Building	U-0.20	R-3.3 + R-1.9 Ls	U-0.20	R-3.3 + R-1.9 Ls	U-0.39	R-2.3 + R-3.3
Attic and Other	U-0.12	R-8.6	U-0.12	R-8.6	U-0.19	R-5.3
Walls, Above Grade						
Mass	U-0.59	R-1.7 ci	U-0.51	R-2.0 ci	U-0.86 [a]	R-1.0 ci [a]
Metal Building	U-0.45	R-2.3 + R-1.1 ci	U-0.30	R-2.3 + R-2.3 ci	U-0.45	R-2.3 + R-1.1 ci
Steel Framed	U-0.43	R-2.3 + R-0.9 ci	U-0.31	R-2.3 + R-1.8 ci	U-0.48	R-2.3 + R-0.7 ci
Wood Framed and Other	U-0.36	R-2.3 + R-0.7 ci	U-0.36	R-2.3 + R-0.7 ci	U-0.36	R-2.3 + R-0.7 ci
Wall, Below Grade						
Below Grade Wall	C-6.47	NR	C-6.47	NR	C-6.47	NR
Floors						
Mass	U-0.61	R-1.1 ci	U-0.50	R-1.5 ci	U-1.83	NR
Steel Joist	U-0.21	R-5.3	U-0.21	R-5.3	U-0.30	R-3.3
Wood Framed and Other	U-0.19	R-5.3	U-0.15	R-5.3 + R-1.3 ci	U-0.29	R-3.3
Slab-On-Grade Floors						
Unheated	F-1.26	NR	F-1.26	NR	F-1.26	NR
Heated	F-1.11	R-1.3 for 300 mm + R-0.9 ci below	F-1.11	R-1.3 for 300 mm + R-0.9 ci below	F-1.77	R-1.3 for 300mm
Opaque Doors						
Swinging	U-3.41		U-3.41		U-3.41	
Non-Swinging	U-2.84		U-2.27		U-2.84	

	Assembly	Assembly	Assembly	Assembly	Assembly	Assembly
Fenestration	Max. U	Max. SHGC	Max. U	Max. SHGC	Max. U	Max. SHGC
Vertical Fenestration, 0%–40% of Wall						
Nonmetal framing: all[b]	U-2.56	SHGC-0.25 all	U-2.56	SHGC-0.25 all	U-3.12	SHGC-NR all
Metal fr: curtainwall/storefront[c]	U-2.84		U-2.84		U-3.41	
Metal framing: entrance door[c]	U-4.54		U-4.54		U-4.54	
Metal framing: all other[c]	U-3.12		U-3.12		U-3.69	
Skylight with Curb, Glass, % of Roof						
0%–2.0%	U_{all}-3.92	$SHGC_{all}$-0.19	U_{all}-3.92	$SHGC_{all}$-0.16	U_{all}-11.24	$SHGC_{all}$-NR
2.1%–5.0%	U_{all}-3.92	$SHGC_{all}$-0.19	U_{all}-3.92	$SHGC_{all}$-0.16	U_{all}-11.24	$SHGC_{all}$-NR
Skylight with Curb, Plastic, % of Roof						
0%–2.0%	U_{all}-3.92	$SHGC_{all}$-0.27	U_{all}-3.92	$SHGC_{all}$-0.27	U_{all}-10.79	$SHGC_{all}$-NR
2.1%–5.0%	U_{all}-3.92	$SHGC_{all}$-0.27	U_{all}-3.92	$SHGC_{all}$-0.27	U_{all}-10.79	$SHGC_{all}$-NR
Skylight without Curb, All, % of Roof						
0%–2.0%	U_{all}-2.56	$SHGC_{all}$-0.19	U_{all}-2.56	$SHGC_{all}$-0.19	U_{all}-7.72	$SHGC_{all}$-NR
2.1%–5.0%	U_{all}-2.56	$SHGC_{all}$-0.19	U_{all}-2.56	$SHGC_{all}$-0.19	U_{all}-7.72	$SHGC_{all}$-NR

The following definitions apply: ci = continuous insulation, Ls = *liner system*, NR = no (insulation) requirement

[a] Mass *walls* with a heat capacity greater than 245 $kJ/m^2 \cdot K$ which are unfinished or finished only on the interior do not need to be insulated.

[b] Nonmetal framing includes framing materials other than metal with or without metal reinforcing or cladding.

[c] Metal framing includes metal framing with or without thermal break. The all other subcategory includes operable windows, fixed windows, and non-entrance doors.

TABLE A-4 (Supersedes Table 5.5-4 in ANSI/ASHRAE/IES Standard 90.1) Building Envelope Requirements for Climate Zone 4 (A, B, C) (SI)

	Nonresidential		Residential		Semiheated	
Opaque Elements	Assembly Max.	Insulation Min. R-Value	Assembly Max.	Insulation Min. R-Value	Assembly Max.	Insulation Min.R-Value
Roofs						
Insulation Entirely above Deck	U-0.22	R-4.4 ci	U-0.22	R-4.4 ci	U-0.68	R-1.3 ci
Metal Building	U-0.20	R-3.3 + R-1.9 Ls	U-0.20	R-3.3 + R-1.9 Ls	U-0.39	R-2.3 + R-3.3
Attic and Other	U-0.12	R-8.6	U-0.12	R-8.6	U-0.19	R-5.3
Walls, Above Grade						
Mass	U-0.51	R-2.0 ci	U-0.45	R-2.3 ci	U-0.86 [a]	R-1.0 ci [a]
Metal Building	U-0.30	R-2.3 + R-2.3 ci	U-0.30	R-2.3 + R-2.3 ci	U-0.45	R-2.3 + R-1.1 ci
Steel Framed	U-0.31	R-2.3 + R-1.8 ci	U-0.31	R-2.3 + R-1.8 ci	U-0.48	R-2.3 + R-0.7 ci
Wood Framed and Other	U-0.36	R-2.3 + R-0.7 ci	U-0.29	R-2.3 + R-1.3 ci	U-0.36	R-2.3 + R-0.7 ci
Wall, Below Grade						
Below Grade Wall	C-0.68	R-1.3 ci	C-0.52	R-1.8 ci	C-0.68	R-1.3 ci
Floors						
Mass	U-0.42	R-1.8 ci	U-0.36	R-2.2 ci	U-0.61	R-1.1 ci
Steel Joist	U-0.18	R-6.7	U-0.18	R-6.7	U-0.30	R-3.3
Wood Framed and Other	U-0.15	R-5.3 + R-1.3 ci	U-0.15	R-5.3 + R-1.3 ci	U-0.29	R-3.3
Slab-On-Grade Floors						
Unheated	F-0.93	R-1.8 for 600 mm	F-0.90	R-2.6 for 600 mm	F-0.93	R-1.8 for 600 mm
Heated	F-0.95	R-1.8 for 600 mm + R-0.9 ci below	F-0.95	R-1.8 for 600 mm + R-0.9 ci below	F-1.64	R-1.3 for 600 mm
Opaque Doors						
Swinging	U-3.41		U-3.41		U-3.41	
Non-Swinging	U-2.27		U-2.27		U-2.84	

Fenestration	Assembly Max. U	Assembly Max. SHGC	Assembly Max. U	Assembly Max. SHGC	Assembly Max. U	Assembly Max. SHGC
Vertical Fenestration, 0%–40% of Wall						
Nonmetal framing: all[b]	U-1.70	SHGC-0.35 all	U-1.70	SHGC-0.40 all	U-3.12	SHGC-NR all
Metal fr: curtainwall/storefront[c]	U-2.27		U-2.27		U-3.41	
Metal framing: entrance door[c]	U-4.26		U-4.26		U-4.54	
Metal framing: all other[c]	U-2.56		U-2.56		U-3.69	
Skylight with Curb, Glass, % of Roof						
0%–2.0%	U_{all}-3.92	$SHGC_{all}$-0.32	U_{all}-3.92	$SHGC_{all}$-0.19	U_{all}-11.24	$SHGC_{all}$-NR
2.1%–5.0%	U_{all}-3.92	$SHGC_{all}$-0.32	U_{all}-3.92	$SHGC_{all}$-0.19	U_{all}-11.24	$SHGC_{all}$-NR
Skylight with Curb, Plastic, % of Roof						
0%–2.0%	U_{all}-3.92	$SHGC_{all}$-0.34	U_{all}-3.92	$SHGC_{all}$-0.27	U_{all}-10.79	$SHGC_{all}$-NR
2.1%–5.0%	U_{all}-3.92	$SHGC_{all}$-0.34	U_{all}-3.92	$SHGC_{all}$-0.27	U_{all}-10.79	$SHGC_{all}$-NR
Skylight without Curb, All, % of Roof						
0%–2.0%	U_{all}-2.56	$SHGC_{all}$-0.32	U_{all}-2.56	$SHGC_{all}$-0.19	U_{all}-7.72	$SHGC_{all}$-NR
2.1%–5.0%	U_{all}-2.56	$SHGC_{all}$-0.32	U_{all}-2.56	$SHGC_{all}$-0.19	U_{all}-7.72	$SHGC_{all}$-NR

The following definitions apply: ci = continuous insulation, Ls = *liner system*, NR = no (insulation) requirement

[a] Mass *walls* with a heat capacity greater than 245 $kJ/m^2 \cdot K$ which are unfinished or finished only on the interior do not need to be insulated.

[b] Nonmetal framing includes framing materials other than metal with or without metal reinforcing or cladding.

[c] Metal framing includes metal framing with or without thermal break. The all other subcategory includes operable windows, fixed windows, and non-entrance doors.

TABLE A-5 (Supersedes Table 5.5-5 in ANSI/ASHRAE/IES Standard 90.1) Building Envelope Requirements for Climate Zone 5 (A, B, C) (SI)

	Nonresidential		Residential		Semiheated	
	Assembly	Insulation	Assembly	Insulation	Assembly	Insulation
Opaque Elements	Max.	Min. R-Value	Max.	Min. R-Value	Max.	Min.R-Value
Roofs						
Insulation Entirely above Deck	U-0.22	R-4.4 ci	U-0.22	R-4.4 ci	U-0.53	R-1.8 ci
Metal Building	U-0.20	R-3.3 + R-1.9 Ls	U-0.20	R-3.3 + R-1.9 Ls	U-0.39	R-2.3 + R-3.3
Attic and Other	U-0.12	R-8.6	U-0.12	R-8.6	U-0.19	R-5.3
Walls, Above Grade						
Mass	U-0.45	R-2.3 ci	U-0.40	R-2.7 ci	U-0.70	R-1.3 ci
Metal Building	U-0.30	R-2.3 + R-2.3 ci	U-0.30	R-2.3 + R-2.3 ci	U-0.45	R-2.3 + R-1.1 ci
Steel Framed	U-0.31	R-2.3 + R-1.8 ci	U-0.31	R-2.3 + R-1.8 ci	U-0.48	R-2.3 + R-0.7 ci
Wood Framed and Other	U-0.29	R-2.3 + R-1.3 ci	U-0.26	R-2.3 + R-1.8 ci	U-0.36	R-2.3 + R-0.7 ci
Wall, Below Grade						
Below Grade Wall	C-0.52	R-1.8 ci	C-0.52	R-1.8 ci	C-0.68	R-1.3 ci
Floors						
Mass	U-0.36	R-2.2 ci	U-0.32	R-2.6 ci	U-0.61	R-1.1 ci
Steel Joist	U-0.18	R-6.7	U-0.18	R-6.7	U-0.21	R-5.3
Wood Framed and Other	U-0.15	R-5.3 + R-1.3 ci	U-0.15	R-5.3 + R-1.3 ci	U-0.19	R-5.3
Slab-On-Grade Floors						
Unheated	F-0.93	R-1.8 for 600 mm	F-0.90	R-2.6 for 600 mm	F-0.93	R-1.8 for 600 mm
Heated	F-0.76	R-2.6 for 900 mm + R-0.9 ci below	F-0.76	R-2.6 for 900 mm + R-0.9 ci below	F-1.56	R-1.8 for 600 mm
Opaque Doors						
Swinging	U-2.27		U-2.27		U-3.41	
Non-Swinging	U-2.27		U-2.27		U-2.84	

	Assembly	Assembly	Assembly	Assembly	Assembly	Assembly
Fenestration	Max. U	Max. SHGC	Max. U	Max. SHGC	Max. U	Max. SHGC
Vertical Fenestration, 0%–40% of Wall						
Nonmetal framing: all[b] Metal fr: curtainwall/storefront[c] Metal framing: entrance door[c] Metal framing: all other[c]	U-1.42 U-1.99 U-3.97 U-2.56	SHGC-0.35 all	U-1.42 U-1.99 U-3.97 U-2.56	SHGC-0.40 all	U-3.12 U-3.41 U-4.54 U-3.69	SHGC-NR all
Skylight with Curb, Glass, % of Roof						
0%–2.0%	U_{all}-3.80	$SHGC_{all}$-0.36	U_{all}-3.80	$SHGC_{all}$-0.36	U_{all}-11.24	$SHGC_{all}$-NR
2.1%–5.0%	U_{all}-3.80	$SHGC_{all}$-0.36	U_{all}-3.80	$SHGC_{all}$-0.36	U_{all}-11.24	$SHGC_{all}$-NR
Skylight with Curb, Plastic, % of Roof						
0%–2.0%	U_{all}-3.92	$SHGC_{all}$-0.34	U_{all}-3.92	$SHGC_{all}$-0.34	U_{all}-10.79	$SHGC_{all}$-NR
2.1%–5.0%	U_{all}-3.92	$SHGC_{all}$-0.34	U_{all}-3.92	$SHGC_{all}$-0.34	U_{all}-10.79	$SHGC_{all}$-NR
Skylight without Curb, All, % of Roof						
0%–2.0%	U_{all}-2.56	$SHGC_{all}$-0.36	U_{all}-2.56	$SHGC_{all}$-0.36	U_{all}-7.72	$SHGC_{all}$-NR
2.1%–5.0%	U_{all}-2.56	$SHGC_{all}$-0.36	U_{all}-2.56	$SHGC_{all}$-0.36	U_{all}-7.72	$SHGC_{all}$-NR

The following definitions apply: ci = continuous insulation, Ls = *liner system*, NR = no (insulation) requirement

[a] Mass *walls* with a heat capacity greater than 245 kJ/m^2·K which are unfinished or finished only on the interior do not need to be insulated.

[b] Nonmetal framing includes framing materials other than metal with or without metal reinforcing or cladding.

[c] Metal framing includes metal framing with or without thermal break. The all other subcategory includes operable windows, fixed windows, and non-entrance doors.

TABLE A-6 (Supersedes Table 5.5-6 in ANSI/ASHRAE/IES Standard 90.1) Building Envelope Requirements for Climate Zone 6 (A, B) (SI)

	Nonresidential		Residential		Semiheated	
	Assembly	Insulation	Assembly	Insulation	Assembly	Insulation
Opaque Elements	Max.	Min. R-Value	Max.	Min. R-Value	Max.	Min.R-Value
Roofs						
Insulation Entirely above Deck	U-0.18	R-5.3 ci	U-0.18	R-5.3 ci	U-0.36	R-2.6 ci
Metal Building	U-0.18	R-4.4 + R 1.9 Ls	U-0.18	R-4.4 + R 1.9 Ls	U-0.39	R-2.3 + R-3.3
Attic and Other	U-0.12	R-8.6	U-0.12	R-8.6	U-0.15	R-6.7
Walls, Above Grade						
Mass	U-0.40	R-2.7 ci	U-0.34	R-3.5 ci	U-0.59	R-1.7 ci
Metal Building	U-0.30	R-2.3 + R-2.3 ci	U-0.30	R-2.3 + R-2.3 ci	U-0.45	R-2.3 + R-1.1 ci
Steel Framed	U-0.31	R-2.3 + R-1.8 ci	U-0.31	R-2.3 + R-1.8 ci	U-0.48	R-2.3 + R-0.7 ci
Wood Framed and Other	U-0.26	R-2.3 + R-1.8 ci	U-0.26	R-2.3 + R-1.8 ci	U-0.36	R-2.3 + R-0.7 ci
Wall, Below Grade						
Below Grade Wall	C-0.52	R-1.8 ci	C-0.52	R-1.8 ci	C-0.68	R-1.3 ci
Floors						
Mass	U-0.32	R-2.6 ci	U-0.29	R-2.9 ci	U-0.61	R-1.1 ci
Steel Joist	U-0.18	R-6.7	U-0.13	R-6.7 + R-2.2 ci	U-0.21	R-5.3
Wood Framed and Other	U-0.15	R-5.3 + R-1.3 ci	U-0.15	R-5.3 + R-1.3 ci	U-0.19	R-5.3
Slab-On-Grade Floors						
Unheated	F-0.90	R-2.6 for 600 mm	F-0.88	R-3.5 for 600 mm	F-0.93	R-1.8 for 600 mm
Heated	F-0.76	R-2.6 for 900 mm + R-0.9 ci below	F-0.76	R-2.6 for 900 mm + R-0.9 ci below	F-1.56	R-1.8 for 600 mm
Opaque Doors						
Swinging	U-2.27		U-2.27		U-3.41	
Non-Swinging	U-2.27		U-2.27		U-2.84	
	Assembly	Assembly	Assembly	Assembly	Assembly	Assembly
Fenestration	Max. U	Max. SHGC	Max. U	Max. SHGC	Max. U	Max. SHGC
Vertical Fenestration, 0%–40% of Wall						
Nonmetal framing: all[b]	U-1.42	SHGC-0.40 all	U-1.42	SHGC-0.40 all	U-2.56	SHGC-NR all
Metal fr: curtainwall/storefront[c]	U-1.99		U-1.99		U-2.84	
Metal framing: entrance door[c]	U-3.97		U-3.97		U-4.54	
Metal framing: all other[c]	U-2.56		U-2.56		U-3.12	
Skylight with Curb, Glass, % of Roof						
0%–2.0%	U_{all}-3.80	$SHGC_{all}$-0.46	U_{all}-3.80	$SHGC_{all}$-0.46	U_{all}-11.24	$SHGC_{all}$-NR
2.1%–5.0%	U_{all}-3.80	$SHGC_{all}$-0.46	U_{all}-3.80	$SHGC_{all}$-0.46	U_{all}-11.24	$SHGC_{all}$-NR
Skylight with Curb, Plastic, % of Roof						
0%–2.0%	U_{all}-3.92	$SHGC_{all}$-0.49	U_{all}-3.92	$SHGC_{all}$-0.49	U_{all}-10.79	$SHGC_{all}$-NR
2.1%–5.0%	U_{all}-3.92	$SHGC_{all}$-0.49	U_{all}-3.92	$SHGC_{all}$-0.49	U_{all}-10.79	$SHGC_{all}$-NR
Skylight without Curb, All, % of Roof						
0%–2.0%	U_{all}-2.56	$SHGC_{all}$-0.46	U_{all}-2.56	$SHGC_{all}$-0.39	U_{all}-7.72	$SHGC_{all}$-NR
2.1%–5.0%	U_{all}-2.56	$SHGC_{all}$-0.46	U_{all}-2.56	$SHGC_{all}$-0.39	U_{all}-7.72	$SHGC_{all}$-NR

The following definitions apply: ci = continuous insulation, Ls = *liner system*, NR = no (insulation) requirement

[a] Mass *walls* with a heat capacity greater than 245 $kJ/m^2 \cdot K$ which are unfinished or finished only on the interior do not need to be insulated.

[b] Nonmetal framing includes framing materials other than metal with or without metal reinforcing or cladding.

[c] Metal framing includes metal framing with or without thermal break. The all other subcategory includes operable windows, fixed windows, and non-entrance doors.

TABLE A-7 (Supersedes Table 5.5-7 in ANSI/ASHRAE/IES Standard 90.1) Building Envelope Requirements for Climate Zone 7 (SI)

	Nonresidential		Residential		Semiheated	
Opaque Elements	**Assembly Max. U**	**Assembly Max. SHGC**	**Assembly Max. U**	**Assembly Max. SHGC**	**Assembly Max. U**	**Assembly Max. SHGC**
Roofs						
Insulation Entirely above Deck	U-0.16	R-6.2 ci	U-0.16	R-6.2 ci	U-0.36	R-2.6 ci
Metal Building	U-0.16	R-5.3 + R-1.9 Ls	U-0.16	R-5.3 + R-1.9 Ls	U-0.39	R-2.3 + R-3.3
Attic and Other	U-0.10	R-10.6	U-0.10	R-10.6	U-0.15	R-6.7
Walls, Above Grade						
Mass	U-0.34	R-3.5 ci	U-0.34	R-3.5 ci	U-0.51	R-2.0 ci
Metal Building	U-0.30	R-2.3 + R-2.3 ci	U-0.22	R-2.3 + R-3.4 ci	U-0.45	R-2.3 + R-1.1 ci
Steel Framed	U-0.31	R-2.3 + R-1.8 ci	U-0.21	R-2.3 + R-3.3 ci	U-0.48	R-2.3 + R-0.7 ci
Wood Framed and Other	U-0.26	R-2.3 + R-1.8 ci	U-0.26	R-2.3 + R-1.8 ci	U-0.36	R-2.3 + R-0.7 ci
Wall, Below Grade						
Below Grade Wall	C-0.52	R-1.8 ci	C-0.42	R-2.2 ci	C-0.68	R-1.3 ci
Floors						
Mass	U-0.25	R-3.5 ci	U-0.25	R-3.5 ci	U-0.50	R-1.5 ci
Steel Joist	U-0.18	R-6.7	U-0.13	R-6.7 + R-2.2 ci	U-0.21	R-5.3
Wood Framed and Other	U-0.15	R-5.3 + R-1.3 ci	U-0.15	R-5.3 + R-1.3 ci	U-0.19	R-5.3
Slab-On-Grade Floors						
Unheated	F-0.52	R-2.6 for 600 mm + R-0.9 ci below	F-0.52	R-2.6 for 600 mm + R-0.9 ci below	F-0.93	R-1.8 for 600 mm
Heated	F-0.65	R-3.5 for 900 mm + R-0.9 ci below	F-0.65	R-3.5 for 900 mm + R-0.9 ci below	F-1.19	R-3.5 for1200 mm
Opaque Doors						
Swinging	U-2.27		U-2.27		U-3.41	
Non-Swinging	U-2.27		U-2.27		U-2.84	
Fenestration	**Assembly Max. U**	**Assembly Max. SHGC**	**Assembly Max. U**	**Assembly Max. SHGC**	**Assembly Max. U**	**Assembly Max. SHGC**
Vertical Fenestration, 0%–40% of Wall						
Nonmetal framing: all[b]	U-1.42	SHGC-0.45 all	U-1.42	SHGC-NR all	U-2.56	SHGC-NR all
Metal fr: curtainwall/storefront[c]	U-1.70		U-1.70		U-2.84	
Metal framing: entrance door[c]	U-3.97		U-3.97		U-4.54	
Metal framing: all other[c]	U-1.99		U-1.99		U-3.12	
Skylight with Curb, Glass, % of Roof						
0%–2.0%	U_{all}-3.80	$SHGC_{all}$-0.46	U_{all}-3.80	$SHGC_{all}$-0.46	U_{all}-11.24	$SHGC_{all}$-NR
2.1%–5.0%	U_{all}-3.80	$SHGC_{all}$-0.46	U_{all}-3.80	$SHGC_{all}$-0.46	U_{all}-11.24	$SHGC_{all}$-NR
Skylight with Curb, Plastic, % of Roof						
0%–2.0%	U_{all}-3.92	$SHGC_{all}$-0.50	U_{all}-3.92	$SHGC_{all}$-0.50	U_{all}-10.79	$SHGC_{all}$-NR
2.1%–5.0%	U_{all}-3.92	$SHGC_{all}$-0.50	U_{all}-3.92	$SHGC_{all}$-0.50	U_{all}-10.79	$SHGC_{all}$-NR
Skylight without Curb, All, % of Roof						
0%–2.0%	U_{all}-2.56	$SHGC_{all}$-0.46	U_{all}-2.56	$SHGC_{all}$-0.46	U_{all}-7.72	$SHGC_{all}$-NR
2.1%–5.0%	U_{all}-2.56	$SHGC_{all}$-0.46	U_{all}-2.56	$SHGC_{all}$-0.46	U_{all}-7.72	$SHGC_{all}$-NR

The following definitions apply: ci = continuous insulation, Ls = *liner system*, NR = no (insulation) requirement

[a] Mass *walls* with a heat capacity greater than 245 $kJ/m^2 \cdot K$ which are unfinished or finished only on the interior do not need to be insulated.

[b] Nonmetal framing includes framing materials other than metal with or without metal reinforcing or cladding.

[c] Metal framing includes metal framing with or without thermal break. The all other subcategory includes operable windows, fixed windows, and non-entrance doors.

TABLE A-8 (Supersedes Table 5.5-8 in ANSI/ASHRAE/IES Standard 90.1) Building Envelope Requirements for Climate Zone 8 (SI)

	Nonresidential		Residential		Semiheated	
Opaque Elements	**Assembly Max.**	**Insulation Min. R-Value**	**Assembly Max.**	**Insulation Min. R-Value**	**Assembly Max.**	**Insulation Min.R-Value**
Roofs						
Insulation Entirely above Deck	U-0.16	R-6.2 ci	U-0.16	R-6.2 ci	U-0.27	R-3.5 ci
Metal Building	U-0.16	R-5.3 + R-1.9 Ls	U-0.16	R-5.3 + R-1.9 Ls	U-0.25	R-3.3 + R-1.9 Ls[d]
Attic and Other	U-0.10	R-10.6	U-0.10	R-10.6	U-0.15	R-6.7
Walls, Above Grade						
Mass	U-0.34	R-3.5 ci	U-0.24	R-5.5 ci	U-0.45	R-2.3 ci
Metal Building	U-0.30	R-2.3 + R-2.3 ci	U-0.18	R-2.3 + R-4.6 ci	U-0.30	R-2.3 + R-2.3 ci
Steel Framed	U-0.31	R-2.3 + R-1.8 ci	U-0.19	R-2.3 + R-3.9 ci	U-0.37	R-2.3 + R-1.3 ci
Wood Framed and Other	U-0.26	R-2.3 + R-1.8 ci	U-0.18	R-2.3 + R-3.3 ci	U-0.36	R-2.3 + R-0.7 ci
Wall, Below Grade						
Below Grade Wall	C-0.52	R-1.8 ci	C-0.36	R-2.6 ci	C-0.68	R-1.3 ci
Floors						
Mass	U-0.25	R-3.5 ci	U-0.25	R-3.5 ci	U-0.36	R-2.2 ci
Steel Joist	U-0.13	R-6.7 + R-2.2 ci	U-0.13	R-6.7 + R-2.2 ci	U-0.21	R-5.3
Wood Framed and Other	U-0.15	R-5.3 + R-1.3 ci	U-0.15	R-5.3 + R-1.3 ci	U-0.15	R-5.3 + R-1.3 ci
Slab-On-Grade Floors						
Unheated	F-0.52	R-2.6 for 600 mm + R-0.9 ci below	F-0.52	R-2.6 for 600 mm + R-0.9 ci below	F-0.93	R-1.8 for 600 mm
Heated	F-0.65	R-3.5 for 900 mm + R-0.9 ci below	F-0.65	R-3.5 for 900 mm + R-0.9 ci below	F-1.19	R-3.5 for1200 mm
Opaque Doors						
Swinging	U-2.27		U-2.27		U-2.27	
Non-Swinging	U-2.27		U-2.27		U-2.27	

Fenestration	**Assembly Max. U**	**Assembly Max. SHGC**	**Assembly Max. U**	**Assembly Max. SHGC**	**Assembly Max. U**	**Assembly Max. SHGC**
Vertical Fenestration, 0%–40% of Wall						
Nonmetal framing: all[b] Metal fr: curtainwall/storefront[c] Metal framing: entrance door[c] Metal framing: all other[c]	U-1.42 U-1.70 U-3.97 U-1.99	SHGC-0.45 all	U-1.42 U-1.70 U-3.97 U-1.99	SHGC-NR all	U-2.56 U-2.84 U-4.54 U-3.12	SHGC-NR all
Skylight with Curb, Glass, % of Roof						
0%–2.0%	U_{all}-3.29	$SHGC_{all}$-NR	U_{all}-3.29	$SHGC_{all}$-NR	U_{all}-7.38	$SHGC_{all}$-NR
2.1%–5.0%	U_{all}-3.29	$SHGC_{all}$-NR	U_{all}-3.29	$SHGC_{all}$-NR	U_{all}-7.38	$SHGC_{all}$-NR
Skylight with Curb, Plastic, % of Roof						
0%–2.0%	U_{all}-3.29	$SHGC_{all}$-NR	U_{all}-3.29	$SHGC_{all}$-NR	U_{all}-6.25	$SHGC_{all}$-NR
2.1%–5.0%	U_{all}-3.29	$SHGC_{all}$-NR	U_{all}-3.29	$SHGC_{all}$-NR	U_{all}-6.25	$SHGC_{all}$-NR
Skylight without Curb, All, % of Roof						
0%–2.0%	U_{all}-2.56	$SHGC_{all}$-NR	U_{all}-2.56	$SHGC_{all}$-NR	U_{all}-4.60	$SHGC_{all}$-NR
2.1%–5.0%	U_{all}-2.56	$SHGC_{all}$-NR	U_{all}-2.56	$SHGC_{all}$-NR	U_{all}-4.60	$SHGC_{all}$-NR

The following definitions apply: ci = continuous insulation, Ls = *liner system*, NR = no (insulation) requirement.
[a] Mass *walls* with a heat capacity greater than 245 $kJ/m^2 \cdot K$ which are unfinished or finished only on the interior do not need to be insulated.
[b] Nonmetal framing includes framing materials other than metal with or without metal reinforcing or cladding.
[c] Metal framing includes metal framing with or without thermal break. The all other subcategory includes operable windows, fixed windows, and non-entrance doors.
[d] Liner system without thermal spacer blocks for this case only.-

TABLE A-9 (Supersedes Table A2.4.2 in ANSI/ASHRAE/IES Standard 90.1)
Single-Rafter Roof Requirements (SI)

Climate Zone	Minimum Insulation R-Value or Maximum Assembly U-Factor		
	Nonresidential	Residential	Semiheated
1	R-6.7 U-0.165	R-6.7 + R-1.8 ci U-0.112	R-3.3 U-0.312
2	R-6.7 + R-1.8 ci U-0.112	R-6.7 + R-1.8 ci U-0.112	R-3.3 U-0.312
3, 4, 5	R-6.7 + R-1.8 ci U-0.112	R-6.7 + R-1.8 ci U-0.112	R-5.3 U-0.204
6	R-6.7 + R-1.8 ci U-0.112	R-6.7 + R-1.8 ci U-0.112	R-6.7 U-0.165
7, 8	R-6.7 + R-2.6 ci U-0.111	R-6.7 + R-2.6 ci U-0.111	R-6.7 U-0.165

(This is a normative appendix and is part of this standard.)

NORMATIVE APPENDIX B— PRESCRIPTIVE CONTINUOUS AIR BARRIER

B1. CHARACTERISTICS

The *continuous air barrier* shall have the following characteristics:

a. It shall be continuous throughout the envelope (at the lowest *floor*, exterior *walls*, and ceiling or *roof*), with all joints and seams sealed and with sealed connections between all transitions in planes and changes in materials and at all penetrations.
b. The air barrier component of each assembly shall be joined and sealed in a flexible manner to the air barrier component of adjacent assemblies, allowing for the relative movement of these assemblies and components.
c. It shall be capable of withstanding positive and negative combined design wind, fan, and stack pressures on the air barrier without damage or displacement, and shall transfer the load to the structure. It shall not displace adjacent materials under full load.
d. It shall be installed in accordance with the *manufacturer's* instructions and in such a manner as to achieve the performance requirements.
e. Where lighting *fixtures* with ventilation holes or other similar objects are to be installed in such a way as to penetrate the *continuous air barrier*, provisions shall be made to maintain the integrity of the *continuous air barrier*.

Exception: Buildings that comply with (c) below are not required to comply with either (a) or (e) above.

B2. COMPLIANCE

Compliance of the *continuous air barrier* for the *opaque building envelope* shall be demonstrated by one of the following:

a. **Materials.** Using individual materials that have an air permeability not to exceed 0.004 cfm/ft^2 under a pressure differential of 0.3 in. water (1.57 lb/ft^2) (0.02 L/s·m^2 under a pressure differential of 75 Pa) when tested in accordance with ASTM E2178. These materials comply with this requirement when all joints are sealed and the above section on characteristics are met:
   1. Plywood—minimum 3/8 in. (10 mm)
   2. Oriented strand board—minimum 3/8 in. (10 mm)
   3. Extruded polystyrene insulation board—minimum 3/4 in. (19 mm)
   4. Foil-back urethane insulation board—minimum 3/4 in. (19 mm)
   5. Exterior or interior gypsum board—minimum 1/2 in. (12 mm)
   6. Cement board—minimum 1/2 in. (12 mm)
   7. Built up roofing membrane
   8. Modified bituminous *roof* membrane
   9. Fully adhered single-ply *roof* membrane
   10. A Portland cement/sand parge, or gypsum plaster minimum 5/8 in. (16 mm) thick
   11. Cast-in-place and precast concrete
   12. Fully grouted concrete block masonry
   13. Sheet steel

b. **Assemblies.** Using assemblies of materials and components that have an average air leakage not to exceed 0.04 cfm/ft^2 under a pressure differential of 0.3 in. water (1.57 lb/ft^2) (0.2 L/s·m^2 under a pressure differential of 75 Pa) when tested in accordance with ASTM E2357 or ASTM E1677. These assemblies comply with this requirement when all joints are sealed and the above section on characteristics are met:
   1. Concrete masonry *walls* coated with:
      a. one application of block filler and two applications of a paint or sealer coating, or
      b. a Portland cement/sand parge, stucco or plaster minimum 1/2 in. (12 mm) thick.

c. **Building.** Testing the completed building and demonstrating that the air leakage rate of the *building envelope* does not exceed 0.4 cfm/ft^2 under a pressure differential of 0.3 in. water (1.57 lb/ft^2) (2.0 L/s·m^2 under a pressure differential of 75 Pa) in accordance with ASTM E779 or an equivalent approved method.

(This is a normative appendix and is part of this standard.)

NORMATIVE APPENDIX C— PRESCRIPTIVE EQUIPMENT EFFICIENCY TABLES

Informative Note: The first 15 tables are in I-P units, followed by15 tables in SI units.

TABLE C-1 (Supersedes Table 6.8.1A in ANSI/ASHRAE/IES Standard 90.1) Electrical-Operated Unitary Air Conditioners and Condensing Units (I-P)

Equipment Type	Size Category	Heating Section Type	Sub-Category or Rating Conditions	Minimum Efficiency	Test Procedure[a]
Air conditioners, air-cooled	<65,000 Btu/h	All	Split systems	14.0 SEER 12.0 EER	ARI 210/240
			Single packaged	14.0 SEER 11.6 EER	
Through-the-wall, air-cooled	<30,000 Btu/h	All	Split systems	12.0 SEER	
			Single packaged	12.0 SEER	
Small-duct high velocity, air-cooled	<65,000 Btu/h	All	Split systems	10 SEER	
Air conditioners, air-cooled	≥65,000 Btu/h and <135,000 Btu/h	Electric resistance (or none)	Split systems and single package	11.5 EER 12.0 IEER	ARI 340/360
		All other	Split systems and single package	11.3 EER 11.8 IEER	
	≥135,000 Btu/h and <240,000 Btu/h	Electric resistance (or none)	Split systems and single package	11.5 EER 12.0 IEER	
		All other	Split systems and single package	11.3 EER 11.8 IEER	
	≥240,000 Btu/h and <760,000 Btu/h	Electric resistance (or none)	Split systems and single package	10.0 EER 10.5 IEER	
		All other	Split systems and single package	9.8 EER 10.3 IEER	
	≥760,000 Btu/h	Electric resistance (or none)	Split systems and single package	9.7 EER 10.2 IEER	
		All other	Split systems and single package	9.5 EER 10.0 IEER	
Air conditioners, water and evaporatively cooled	<65,000 Btu/h	All	Split systems and single package	14.0 EER 14.3 IEER	ARI 210/240
	≥65,000 Btu/h and <135,000 Btu/h	Electric resistance (or none)	Split systems and single package	14.0 EER 14.3 IEER	ARI 340/360
		All other	Split systems and single package	13.8 EER 14.1 IEER	
	≥135,000 Btu/h and <240,000 Btu/h	Electric resistance (or none)	Split systems and single package	14.0 EER 14.3 IEER	
		All other	Split systems and single package	13.8 EER 14.1 IEER	
	≥240,000 Btu/h	Electric resistance (or none)	Split systems and single package	14.0 EER 14.0 IEER	
		All other	Split systems and single package	13.8 EER 13.8 IEER	
Condensing units, air-cooled	≥135,000 Btu/h			Not applicable match with indoor coil	ARI 365
Condensing, water or evaporatively cooled	≥135,000 Btu/h			Not applicable match with indoor coil	

a. Section 11 contains a complete specification of the referenced test procedures, including year version of the test procedure.

**TABLE C-2 (Supersedes Table 6.8.1B in ANSI/ASHRAE/IES Standard 90.1)
Electrically-Operated Unitary and Applied Heat Pumps Minimum Efficiency Requirements (I-P)**

Equipment Type	Size Category	Heating Section Type	Sub-Category or Rating Conditions	Minimum Efficiency	Test Procedure[a]
Air conditioners, air-cooled (cooling mode)	<65,000 Btu/h	All	Split systems	14.0 SEER 12.0 EER	ARI 210/240
			Single packaged	14.0 SEER 11.6 EER	
Through-the-wall, air-cooled (cooling mode)	<30,000 Btu/h	All	Split systems	12.0 SEER	
			Single packaged	12.0 SEER	
Small-duct high velocity, air-cooled (cooling mode)	<65,000 Btu/h	All	Split systems	10.0 SEER	
Air conditioners, air-cooled (cooling mode)	≥65,000 Btu/h and <135,000 Btu/h	Electric resistance (or none)	Split systems and single package	11.3 EER 11.8 IEER	ARI 340/360
		All other	Split systems and single package	11.1EER 11.6 IEER	
	≥135,000 Btu/h and <240,000 Btu/h	Electric resistance (or none)	Split systems and single package	11.3 EER 11.8 IEER	
		All other	Split systems and single package	11.1EER 11.6 IEER	
	≥240,000 Btu/h	Electric resistance (or none)	Split systems and single package	9.8 EER 9.8 IEER	
		All other	Split systems and single package	9.6 EER 9.6 IEER	
Water-source (cooling mode)	<17,000 Btu/h	All	86°F entering water	14.0 EER	ISO-13256-1
	≥17,000 Btu/h and <65,000 Btu/h	All	86°F entering water	14.0 EER	
	>65,000 Btu/h and <135,000 Btu/h	All	86°F entering water	14.0 EER	
Groundwater-source (cooling mode)	<135,000 Btu/h	All	59°F entering water	16.2 EER	
			77°F entering water	13.4 EER	
Air conditioners, air-cooled (heating mode)	<65,000 Btu/h	All	Split systems	8.5 HSPF	ARI210/240
			Single packaged	8.0 HSPF	
Through-the-wall, air-cooled (heating mode)	<30,000 Btu/h	All	Split systems	7.4 HSPF	
			Single packaged	7.4 HSPF	
Small-duct high velocity, air-cooled (heating mode)	<65,000 Btu/h	All	Split systems	6.8 HSPF	
Air-cooled (heating mode)	≥65,000 Btu/h and <135,000 Btu/h (cooling capacity)		47°F DB/43°F WB *Outdoor air*	3.3 COP	ARI 340/360
			17°F DB/15°F WB *Outdoor air*	2.2 COP	
	≥135,000 Btu/h (cooling capacity)		47°F DB/43°F WB *Outdoor air*	3.2 COP	
			17°F DB/15°F WB *Outdoor air*	2.0 COP	
Water-source (heating mode)	<135,000 Btu/h (cooling capacity)		68°F entering water	4.2 COP	ISO-13256-1
Groundwater-source (heating mode)	<135,000 Btu/h (cooling capacity)		50°F entering water	3.6 COP	
			32°F entering fluid	3.1 COP	

a. Section 11 contains a complete specification of the referenced test procedures, including year version of the test procedure.

TABLE C-3 (Supersedes Table 6.8.1C in ANSI/ASHRAE/IES Standard 90.1) Water Chilling Packages—Minimum Efficiency Requirements[a] (I-P)

Equipment Type	Size Category	Units	Minimum Efficiency[c] (I-P)				Test Procedure[b]
			Path A		Path B[d]		
			Full Load	IPLV	Full Load	IPLV	
Air-cooled chillers with condenser, electrically operated	<150 tons	EER	10.000	12.500	NA	NA	ARI 550/590
	≥150 tons	EER	10.000	12.750	NA	NA	
Air-cooled without condenser, electrical operated	All capacities	EER	Condenserless units shall be rated with matched condensers				ARI 550/590
Water-cooled, electrically operated, positive displacement (reciprocating)	All capacities	kw/ton	Reciprocating units required to comply with water cooled positive displacement requirements				ARI 550/590
Water-cooled electrically operated, positive displacement	<75 tons	kw/ton	0.780	0.630	0.800	0.600	ARI 550/590
	≥75 tons and <150 tons	kw/ton	0.775	0.615	0.790	0.586	
	≥150 tons and <300 tons	kw/ton	0.680	0.580	0.718	0.540	
	≥300 tons	kw/ton	0.620	0.540	0.639	0.490	
Water-cooled electrically operated, centrifugal[a]	<150 tons	kw/ton	0.634	0.596	0.639	0.450	ARI 550/590
	≥150 tons and <300 tons	kw/ton	0.634	0.596	0.639	0.450	
	≥300 tons and <600 tons	kw/ton	0.576	0.549	0.600	0.400	
	≥600 tons	kw/ton	0.570	0.539	0.590	0.400	
Air-cooled absorption single effect[g]	All capacities	COP	0.600	NR[f]	NA[e]	NA[e]	ARI 560
Water-cooled absorption single effect[g]	All capacities	COP	0.700	NR[f]	NA[e]	NA[e]	
Absorption double effect indirect-fired	All capacities	COP	1.000	1.050	NA[e]	NA[e]	
Absorption double effect direct fired	All capacities	COP	1.000	1.000	NA[e]	NA[e]	

a. The chiller equipment requirements do not apply for chillers used in low-temperature applications where the design leaving fluid temperature is <40°F
b Section 11 contains a complete specification of the referenced test procedure, including the referenced year version of the test procedure
c. Compliance with this standard can be obtained by meeting the minimum requirements of Path A or Path B. However both the full load and IPLV must be met to fulfill the requirements of Path A and Path B
d. Path B is intended for applications with significant operating time at part load. All Path B machines shall be equipped with demand limiting capable controls
e. NA means that this requirement is not applicable and can not be used for compliance
f. NR means that for this category there are no minimum requirements
g. Only allowed to be used in heat recovery applications
h. Packages that are not designed for operation at ARI Standard 550/590 test conditions (and, thus, cannot be tested to meet the requirements of Table C-3) of 44°F leaving chilled-water temperature and 85°F entering condenser-water temperature with 3 gpm/ton condenser-water flow shall have maximum full-load kW/ton and *NPLV* ratings adjusted using the following equation:

Adjusted maximum full load kW/ton rating = (full load kW/ton from Table C-3) / K_{adj}

Adjusted maximum NPLV rating = (IPLV from Table C-3) / K_{adj}

where

K_{adj} = $6.174722 - 0.303668(X) + 0.00629466(X)^2 - 0.000045780(X)^3$
X = DT_{std} + LIFT (°F)
DT_{std} = (24 + (full load kW/ton from Table C-3) · 6.83) / flow (°F)
Flow = condenser-water flow (gpm) / cooling full load capacity (tons)
LIFT = CEWT – CLWT (°F)
CEWT = full load entering condenser-water temperature (°F)
CLWT = full load leaving chilled-water temperature (°F)

The adjusted full load and *NPLV* values are only applicable over the following full-load design ranges:

- minimum leaving chilled-water temperature: 38°F
- maximum condenser entering water temperature: 102°F
- condenser-water flow: 1 to 6 gpm/ton
- $X \geq 39$°F and ≤ 60°F

TABLE C-4 (Supersedes Table 6.8.1D in ANSI/ASHRAE/IES Standard 90.1) Electrically Operated Packaged Terminal Air Conditioners, Packaged Terminal Heat Pumps, Single Packaged Vertical Air Conditioners, Single Packaged Vertical Heat Pumps, Room Air Conditioners, and Room Air Conditioner Heat Pumps—Minimum Efficiency Requirements (I-P)

Equipment Type	Size Category (Input)	Subcategory or Rating Condition	Minimum Efficiency	Test Procedure[a]
PTAC (cooling mode) new construction	<7,000 Btu/h	95°F DB *Outdoor air*	11.9 EER	ARI 310/380
	≥7,000 Btu/h and <10,000 Btu/h	95°F DB *Outdoor air*	11.3 EER	
	≥10,000 Btu/h and <13,000 Btu/h	95°F DB *Outdoor air*	10.7 EER	
	≥13,000 Btu/h	95°F DB *Outdoor air*	9.5 EER	
PTAC (cooling mode) replacement[b]	<7,000 Btu/h	95°F DB *Outdoor air*	11.9 EER	ARI 310/380
	≥7,000 Btu/h and <10,000 Btu/h	95°F DB *Outdoor air*	11.3 EER	
	≥10,000 Btu/h and <13,000 Btu/h	95°F DB *Outdoor air*	10.7 EER	
	≥13,000 Btu/h	95°F DB *Outdoor air*	9.5 EER	
PTHP (cooling mode) new construction	<7,000 Btu/h	95°F DB *Outdoor air*	11.7 EER	ARI 310/380
	≥7,000 Btu/h and <10,000 Btu/h	95°F DB *Outdoor air*	11.1 EER	
	≥10,000 Btu/h and <13,000 Btu/h	95°F DB *Outdoor air*	10.5 EER	
	≥13,000 Btu/h	95°F DB *Outdoor air*	9.3 EER	
PTHP (heating mode) new construction	All capacities	95°F DB *Outdoor air*	2.8 COP	ARI 310/380
PTHP (cooling mode) replacement[b]	<7,000 Btu/h	95°F DB *Outdoor air*	11.7 EER	ARI 310/380
	≥7,000 Btu/h and <10,000 Btu/h	95°F DB *Outdoor air*	11.1 EER	
	≥10,000 Btu/h and <13,000 Btu/h	95°F DB *Outdoor air*	10.5 EER	
	≥13,000 Btu/h	95°F DB *Outdoor air*	9.3 EER	
PTHP (heating mode) replacement[b]	All capacities	95°F DB *Outdoor air*	2.8 COP	ARI 310/380

a. Section 11 contains a complete specification of the referenced test procedures, including year version of the test procedure.
b. Replacement units shall be factory labeled as follows: "MANUFACTURED FOR REPLACEMENT APPLICATIONS ONLY; NOT TO BE INSTALLED IN NEW CONSTRUCTION PROJECTS." Replacement efficiencies apply only to units with existing sleeves less than 16 in. high and less than 42 in. wide.

TABLE C-5 (Supersedes Table 6.8.1D in ANSI/ASHRAE/IES Standard 90.1)
Electrically Operated Packaged Terminal Air Conditioners, Packaged Terminal Heat Pumps, Single Packaged Vertical Air Conditioners, Single Packaged Vertical Heat Pumps, Room Air Conditioners, and Room Air Conditioner Heat Pumps—Minimum Efficiency Requirements (I-P)

Equipment Type	Size Category (Input)	Subcategory or Rating Condition	Minimum Efficiency	Test Procedure[a]
SPVAC (cooling mode)	<65,000 Btu/h	95°F DB/75°F WB *Outdoor air*	10.0 EER 13.5 IPLV	ARI 390
	≥65,000 Btu/h and <135,000 Btu/h	95°F DB/75°F WB *Outdoor air*	11.5 EER	
	≥135,000 Btu/h and <240,000 Btu/h	95°F DB/75°F WB *Outdoor air*	11.5 EER	
SPVHP (cooling mode)	<65,000 Btu/h	95°F DB/75°F WB *Outdoor air*	10.0 EER 13.5 IPLV	
	≥65,000 Btu/h and <135,000 Btu/h	95°F DB/75°F WB *Outdoor air*	11.5 EER	
	≥135,000 Btu/h and <240,000 Btu/h	95°F DB/75°F WB *Outdoor air*	11.5 EER	
SPVHP (heating mode)	<65,000 Btu/h	47°F DB/43°F WB *Outdoor air*	3.0 COP	
	≥65,000 Btu/h and <135,000 Btu/h	47°F DB/43°F WB *Outdoor air*	3.0 COP	
	≥135,000 Btu/h and <240,000 Btu/h	47°F DB/43°F WB *Outdoor air*	2.9 COP	
Room air conditioners, with louvered sides	<6000 Btu/h		10.7 SEER	ANSI/ AHAM RAC-1
	≥6000 Btu/h and <8000 Btu/h		10.7 EER	
	≥8000 Btu/h and <14,000 Btu/h		10.8 EER	
	≥14000 Btu/h and <20,000 Btu/h		10.7 EER	
	≥20,000 Btu/h		9.3 EER	
Room air conditioners, without louvered sides	<8000 Btu/h		9.9 EER	
	≥8000 Btu/h and <20,000 Btu/h		9.3 EER	
	≥20,000 Btu/h		9.3 EER	
Room air conditioner heat pump with louvered sides	<20,000 Btu/h		9.9 EER	
	≥20,000 Btu/h		9.3 EER	
Room air conditioner heat pump without louvered sides	<14,000 Btu/h		9.3 EER	
	≥14,000 Btu/h		8.8 EER	
Room air conditioner, casement only	All capacities		9.6 EER	
Room air conditioner, casement-slider	All capacities		10.4 EER	

a. Section 11 contains a complete specification of the referenced test procedure, including the referenced year version of the test procedure.

TABLE C-6 (Supersedes Table 6.8.1E in ANSI/ASHRAE/IES Standard 90.1) Warm Air Furnace and Combustion Warm Air Furnaces/Air-Conditioning Units, Warm Air Duct Furnaces, and Unit Heaters (I-P)

Equipment Type	Size Category (Input)	Subcategory or Rating Condition	Test Procedure[b]	Minimum Efficiency[a]
Warm air furnace, gas-fired (weatherized)	<225,000 Btu/h	Maximum capacity[d]	DOE 10 CFR Part 430 or ANSI Z21.47	78% AFUE or 80% E_t[c,e]
	>225,000 Btu/h	Maximum capacity[d]	ANSI Z21.47	80% E_c[c,e]
Warm air furnace, gas-fired (non-weatherized)	<225,000 Btu/h	Maximum capacity[d]	DOE 10 CFR Part 430 or ANSI Z21.47	90% AFUE or 92% E_t[c,e]
	>225,000 Btu/h	Maximum capacity[d]	ANSI Z21.47	92% E_c[c,e]
Warm air furnace, oil-fired (weatherized)	<225,000 Btu/h	Maximum capacity[d]	DOE 10 CFR Part 430 or UL 727	78% AFUE or 80% E_t[c,e]
	>225,000 Btu/h	Maximum capacity[d]	UL 727	81% E_t[e]
Warm air furnace, oil-fired (non-weatherized)	<225,000 Btu/h	Maximum capacity[d]	DOE 10 CFR Part 430 or UL 727	85% AFUE or 87%% E_t[c,e]
	>225,000 Btu/h	Maximum capacity[d]	UL 727	87% E_t[e]
Warm air duct furnaces, gas-fired (weatherized)	All capacities	Maximum capacity[d]	ANSI Z83.9	80% E_c[f]
Warm air duct furnaces, gas-fired (non-weatherized)	All capacities	Maximum capacity[d]	ANSI Z83.9	90% E_c[f]
Warm air unit heaters, gas fired (non-weatherized)	All capacities	Maximum capacity[d]	ANSI Z83.8	90% E_c[f,g]
Warm air unit heaters, oil-fired (non-weatherized)	All capacities	Maximum capacity[d]	UL 731	90% E_c[f,g]

a E_t = thermal efficiency. See test procedure for detailed discussions
b Section 11 contains a complete specification of the referenced test procedure, including the referenced year version of the test procedure.
c Combustion units not covered by NAECA (3-phase power or cooling capacity greater than or equal to 65,000 Btu/h) is allowed to comply with either rating
d Minimum and maximum ratings as provided for and allowed by the unit's controls
e Units shall also include an interrupted or intermittent ignition device (IID), have jacket losses not exceeding 0.75% of the input rating, and have either power venting or flue damper. A vent damper is an acceptable alternative to the flue damper for those furnaces where combustion air is drawn from the *conditioned space*.
f E_c = combustion efficiency (100% less flue losses) See test procedures for detailed discussion
g As of August 8, 2008, according to the Energy Policy Act of 2005, units shall also include an interrupted or intermittent ignition devices (IID) and have either power venting or automatic flue dampers. A vent damper is an acceptable alternative to a flue damper for those unit heaters where combustion air is drawn from the *conditioned space*.

TABLE C-7 (Supersedes Table 6.8.1F in ANSI/ASHRAE/IES Standard 90.1)
Gas- and Oil-Fired Boilers—Minimum Efficiency Requirements (I-P)

Equipment Type[a]	Subcategory or Rating Condition	Size Category (Input)	Efficiency[b,c]	Test Procedure
Boilers, hot water	Gas-fired	<300,000 Btu/h	89% AFUE[f]	10 CFR Part 430
		≥300,000 Btu/h and ≤2,500,000 Btu/h[d]	89% E_t[f]	10 CFR Part 431
		>2,500,000 Btu/h[a]	91% E_c[f]	
	Oil-fired[e]	<300,000 Btu/h	89% AFUE[f]	10 CFR Part 430
		≥300,000 Btu/h and ≤2,500,000 Btu/h[d]	89% E_t[f]	10 CFR Part 431
		>2,500,000 Btu/h[a]	91% E_c[f]	
Boilers, steam	Gas-fired	<300,000 Btu/h	75% AFUE	10 CFR Part 430
	Gas-fired all, except natural draft	≥300,000 Btu/h and ≤2,500,000 Btu/h[d]	79% E_t	10 CFR Part 431
		>2,500,000 Btu/h[a]	79% E_t	
	Gas-fired natural draft	≥300,000 Btu/h and ≤2,500,000 Btu/h[d]	77% E_t	
		>2,500,000 Btu/h[a]	77% E_t	
	Oil-fired[e]	<300,000 Btu/h	80% E_t	10 CFR Part 430
		≥300,000 Btu/h and ≤2,5000,000 Btu/h[d]	81% E_t	10 CFR Part 431
		>2,500,000 Btu/h[a]	81% E_t	

a These requirements apply to boilers with rated input of 8,000,000 Btu/h or less that are not packaged boilers, and to all packaged boilers. Minimum efficiency requirements for boilers cover all capacities of packaged boilers.
b E_c = thermal efficiency (100% less flue losses). See reference document for detailed information.
c E_t = thermal efficiency. See reference document for detailed information.
d Maximum capacity - minimum and maximum ratings as provided for and allowed by the unit's controls
e Includes oil fired (residual)
f Systems shall be designed with lower operating return hot water temperatures (<130°F) and use hot water reset to take advantage of the much higher efficiencies of condensing boilers

TABLE C-8 (Supersedes Table 6.8.1G in ANSI/ASHRAE/IES Standard 90.1)
Performance Requirements for Heat Rejection Equipment (I-P)

Equipment Type	Total System Heat Rejection Capacity at Rated Conditions	Rating Standard	Rating Conditions	Performance Required[a,b]
Open-loop propeller or axial fan cooling towers[a]	All	CTI ATC-105 and CTI STD-201	95°F entering water 85°F leaving water 75°F wb entering air	>40 gpm/hp
Closed-loop propeller or axial fan cooling towers[b]	All	CTI ATC-105 and CTI STD-201	102°F entering water 90°F leaving water 75°F wb entering air	>15 gpm/hp
Open-loop centrifugal fan cooling towers [a]	All	CTI ATC-105 and CTI STD-201	95°F entering water 85°F leaving water 75°F wb entering air	>22 gpm/hp
Closed-loop centrifugal fan cooling towers[b]	All	CTI ATC-105 and CTI STD-201	102°F entering water 90°F leaving water 75°F wb entering air	>8 gpm/hp
Air-cooled condensers	All	ARI 460		Not applicable, air-cooled condenser shall be matched to the HVAC system and rated per Table C-3

a For purposes of this table, open circuit cooling tower performance is defined as the water flow rating of the tower at the thermal rating condition listed in this table divided by the fan nameplate rated motor nameplate power.

b For purposes of this table, closed circuit cooling tower performance is defined as the process water flow rating of the tower at the thermal rating condition listed in this table divided by the sum of the fan motor nameplate power and the integral spray pump motor nameplate power.

TABLE C-9 (Supersedes Table 6.8.2A in ANSI/ASHRAE/IES Standard 90.1)
Minimum Duct Insulation R-Value[a] Cooling and Heating Only Supply Ducts and Return Ducts (I-P)

Climate Zone	Duct Location						
	Exterior	Ventilated Attic	Unvented Attic Above Insulated Ceiling	Unvented Attic with *Roof* Insulation[a]	Unconditioned Space[b]	Indirectly Conditioned Space[c]	Buried
Heating Ducts Only							
1, 2	None	None	None	None	None	None	None
3	R-6	None	None	None	R-6	None	None
4	R.6	None	None	None	R-6	None	None
5	R-8	R-6	None	None	R-6	None	R-6
6	R-8	R-8	R-6	None	R-6	None	R-6
7	R-10	R-8	R-8	None	R-6	None	R-6
8	R-10	R-10	R-8	None	R-8	None	R-8
Cooling Only Ducts							
1	R-6	R-8	R-10	R-6	R-6	None	R-6
2	R-6	R-8	R-10	R-6	R-6	None	R-6
3	R-6	R-8	R-8	R-6	R-3.5	None	None
4	R-3.5	R-6	R-8	R-3.5	R-3.5	None	None
5, 6	R-3.5	R-3.5	R-6	R-3.5	R-3.5	None	None
7, 8	R-1.9	R-3.5	R-3.5	R-3.5	R-3.5	None	None
Return Ducts							
1 to 8	R-6	R-6	R-6	None	None	None	None

a Insulation R-values, measured in (h·ft^2·°F)/Btu, are for the insulation as installed and do not include film resistance. The required minimum thicknesses do not consider water vapor transmission and possible surface condensation. Where exterior *walls* are used as plenum *walls*, *wall* insulation shall be as required by the most restrictive condition of this table or Section 7.4.2. Insulation resistance measured on a horizontal plane in accordance with ASTM C518 at a mean temperature of 75°F at the installed thickness.
b Includes crawl spaces, both ventilated and nonventilated.
c Includes return air plenums with or without exposed *roofs* above.

TABLE C-10 (Supersedes Table 6.8.2B in ANSI/ASHRAE/IES Standard 90.1)
Minimum Duct Insulation R-Value[a], Combined Heating and Cooling Supply Ducts and Return Ducts (I-P)

Climate Zone	Duct Location						
	Exterior	Ventilated Attic	Unvented Attic Above Insulated Ceiling	Unvented Attic w/ *Roof* Insulation[a]	Unconditioned Space[b]	Indirectly Conditioned Space[c]	Buried
Supply Ducts							
1	R-8	R-8	R-10	R-6	R-6	None	R-6
2	R-8	R-8	R-8	R-6	R-8	None	R-6
3	R-8	R-8	R-8	R-6	R-8	None	R-6
4	R-8	R-8	R-8	R-6	R-8	None	R-6
5	R-8	R-8	R-8	R-3.5	R-8	None	R-6
6	R-10	R-8	R-8	R-3.5	R-8	None	R-6
7	R-10	R-8	R-8	R-3.5	R-8	None	R-6
8	R-10	R11	R11	R-3.5	R-8	None	R-8
Return Ducts							
1 to 8	R-6	R-6	R-6	None	None	None	None

a Insulation R-values, measured in $(h \cdot ft^2 \cdot {}^\circ F)/Btu$, are for the insulation as installed and do not include film resistance. The required minimum thicknesses do not consider water vapor transmission and possible surface condensation. Where exterior *walls* are used as plenum *walls*, *wall* insulation shall be as required by the most restrictive condition of this table or Section 7.4.2. Insulation resistance measured on a horizontal plane in accordance with ASTM C518 at a mean temperature of 75°F at the installed thickness.
b Includes crawl spaces, both ventilated and non-ventilated.
c Includes return air plenums with or without exposed *roofs* above.

TABLE C-11 (Supersedes Table 7.8 in ANSI/ASHRAE/IES Standard 90.1) Performance Requirements for Water Heating Equipment (I-P)

Equipment Type	Size Category (Input)	Subcategory or Rating Condition	Performance Required [a]	Test Procedure [b]
Electric water heaters	12 kW	Resistance ≥20 gal	$EF \geq 0.97 - 0.00132V$	DOE 10 CFR Part 430
	>12 kW	Resistance ≥20 gal	$SL \leq 20 + 35\sqrt{V}$, Btu/h	ANSI Z21.10.3
	All sizes	Heat Pump	$EF \geq 2.0$	DOE 10 CFR Part 430
Gas storage water heaters	≤75,000 Btu/h	≥20 gal	$EF \geq 0.67$	DOE 10 CFR Part 430
	>75,000 Btu/h	<4000 (Btu/h)/gal	$E_t \geq 80\%$ and $SL \leq (Q/800 + 110\sqrt{V})$, Btu/h	ANSI Z21.10.3
Gas instantaneous water heaters	>50,000 Btu/h and <200,000 Btu/h	≥4000 (Btu/h)/gal and <2 gal	$EF \geq 0.82$	DOE 10 CFR Part 430
	≥200,000 Btu/h[c]	≥4000 (Btu/h)/gal and <10 gal	$E_t \geq 80\%$	ANSI Z21.10.3
	≥200,000 Btu/h	4000 (Btu/h)/gal and ≥10 gal	$E_t \geq 80\%$ and $SL \leq (Q/800 + 110\sqrt{V})$, Btu/h	
Oil storage water heaters	≤105,000 Btu/h	≥20 gal	$EF \geq 0.59\text{-}0.0019V$	DOE 10 CFR Part 430
	>105,000 Btu/h	<4000 (Btu/h)/gal	$E_t \geq 78\%$ and $SL \leq (Q/800 + 110\sqrt{V})$, Btu/h	ANSI Z21.10.3
Oil instantaneous water heaters	≤210,000 Btu/h	≥4000 (Btu/h)/gal and <2 gal	$EF \geq 0.59\text{-}0.0019V$	DOE 10 CFR Part 430
	>210,000 Btu/h	≥4000 (Btu/h)/gal and <10 gal	$E_t \geq 80\%$	ANSI Z21.10.3
	>210,000 Btu/h	≥4000 (Btu/h)/gal and ≥10 gal	$E_t \geq 78\%$ and $SL \leq (Q/800 + 110\sqrt{V})$, Btu/h	
Hot-water supply boilers, gas and oil	300,000 Btu/h and <12,500,000 Btu/h	≥4000 (Btu/h)/gal and <10 gal	$E_t \geq 80\%$	ANSI Z21.10.3
Hot-water supply boilers, gas		≥4000 (Btu/h)/gal and ≥10 gal	$E_t \geq 80\%$ and $SL \leq (Q/800 + 110\sqrt{V})$, Btu/h	
Hot-water supply boilers, oil		≥4000 (Btu/h)/gal and ≥10 gal	$E_t \geq 78\%$ and $SL \leq (Q/800 + 110\sqrt{V})$, Btu/h	
Pool heaters, oil and gas	All sizes		$E_t \geq 78\%$	ASHRAE 146
Heat pump pool heaters	All sizes		≥4.0 COP	ASHRAE 146
Unfired storage tanks	All sizes		≥R-12.5	(none)

a Energy factor (EF) and thermal efficiency (Et) are minimum requirements, while standby loss (SL) is maximum Btu/h based on a 70°F temperature difference between stored water and ambient requirements. In the EF equation, V is the rated volume in gallons. In the SL equation, V is the rated volume in gallons and Q is the nameplate input rate in Btu/h.
b Section 11 contains a complete specification, including the year version, of the referenced test procedure.
c Instantaneous water heaters with input rates below 200,000 Btu/h shall comply with these requirements if the water heater is designed to heat water to temperatures 180°F or higher.

TABLE C-12 Minimum Nominal Efficiency for General Purpose Design A and Design B Motors[a] (I-P)

	Minimum Nominal Full-Load Efficiency (%)					
	Open Motors			Enclosed Motors		
Number of Poles ==>	2	4	6	2	4	6
Synchronous Speed (RPM) ==>	3600	1800	1200	3600	1800	1200
Motor Horsepower						
1	77.0	85.5	82.5	77.0	85.5	82.5
1.5	84.0	86.5	86.5	84.0	86.5	87.5
2	85.5	86.5	87.5	85.5	86.5	88.5
3	85.5	89.5	88.5	86.5	89.5	89.5
5	86.5	89.5	89.5	88.5	89.5	89.5
7.5	88.5	91.0	90.2	89.5	91.7	91.0
10	89.5	91.7	91.7	90.2	91.7	91.0
15	90.2	93.0	91.7	91.0	92.4	91.7
20	91.0	93.0	92.4	91.0	93.0	91.7
25	91.7	93.6	93.0	91.7	93.6	93.0
30	91.7	94.1	93.6	91.7	93.6	93.0
40	92.4	94.1	94.1	92.4	94.1	94.1
50	93.0	94.5	94.1	93.0	94.5	94.1
60	93.6	95.0	94.5	93.6	95.0	94.5
75	93.6	95.0	94.5	93.6	95.4	94.5
100	93.6	95.4	95.0	94.1	95.4	95.0
125	94.1	95.4	95.0	95.0	95.4	95.0
150	94.1	95.8	95.4	95.0	95.8	95.8
200	95.0	95.8	95.4	95.4	96.2	95.8
250	95.0	95.8	95.4	95.8	96.2	95.8
300	95.4	95.8	95.4	95.8	96.2	95.8
350	95.4	95.8	95.4	95.8	96.2	95.8
400	95.8	95.8	95.8	95.8	96.2	95.8
450	95.8	96.2	96.2	95.8	96.2	95.8
500	95.8	96.2	96.2	95.8	96.2	95.8

a. Nominal efficiencies shall be established in accordance with NEMA Standard MG1. Design A and Design B are National Electric Manufacturers Association (NEMA) design class designations for fixed frequency small and medium AC squirrel-cage induction motors.

TABLE C-13 Commercial Refrigerator and Freezers (I-P)

Equipment Type	Application	Energy Use Limit (kW/h per day)
Refrigerators with solid doors	Holding temperature	0.10 V + 2.04
Refrigerators with transparent doors		0.12 V + 3.34
Freezers with solid doors		0.40 V + 1.38
Freezers with transparent doors		0.75 V + 4.10
Refrigerators/freezers with solid doors		The greater of 0.12 V + 3.34 or 0.70
Commercial refrigerators	Pulldown	0.126 V + 3.51

V means the chiller or frozen compartment volume (ft^3) as defined in the Association of Home Appliance Manufacturers Standard HRF1-1979

TABLE C-14 Commercial Clothes Washers (I-P)

Product	MER	WF
All commercial clothes washers	1.72	8

MER = Modified Energy Factor, a combination of Energy Factor and MEF = Modified Energy Factor, a combination of Energy Factor and Remaining Moisture Content. MEF measures energy consumption of the total laundry cycle (washing and drying). It indicates how many cubic feet of laundry can be washed and dried with one kWh of electricity; the higher the number, the greater the efficiency.

TABLE C-1 (Supersedes Table 6.8.1A in ANSI/ASHRAE/IES Standard 90.1)
Electrical Operated Unitary Air Conditioners and Condensing Units (SI)

Equipment Type	Size Category	Heating Section Type	Sub-Category or Rating Conditions	Minimum Efficiency	Test Procedure[a]
Air conditioners, air-cooled	<19 kW	All	Split systems	4.10 SCOP 3.52 COP	ARI 210/240
			Single packaged	4.10 SCOP 3.52 COP	
Through-the-wall, air-cooled	<9 kW	All	Split systems	3.52 SCOP	
			Single packaged	3.52 SCOP	
Small-duct high velocity, air-cooled	<19kW	All	Split systems	2.93 SCOP	
Air conditioners, air-cooled	≥19 kW and <40 kW	Electric resistance (or none)	Split systems and single package	3.37 COP 3.52 ICOP	ARI 340/360
		All other	Split systems and single package	3.31 COP 3.46 ICOP	
	≥40 kW and <70 kW	Electric resistance (or none)	Split systems and single package	3.37 COP 3.52 ICOP	
		All other	Split systems and single package	3.31 COP 3.46 ICOP	
	≥70 kW and <223 kW	Electric resistance (or none)	Split systems and single package	2.93 COP 3.08 ICOP	
		All other	Split systems and single package	2.87 COP 3.02 ICOP	
	≥223 kW	Electric resistance (or none)	Split systems and single package	2.84 COP 2.99 ICOP	
		All other	Split systems and single package	2.78 COP 2.93 ICOP	
Air conditioners, water and evaporatively cooled	<19kW	All	Split systems and single package	4.10 COP 4.19 ICOP	ARI 210/240
	≥19 kW and <40 kW	Electric resistance (or none)	Split systems and single package	4.10 COP 4.19 ICOP	ARI 340/360
		All other	Split systems and single package	4.04 COP 4.13 ICOP	
	≥40 kW and <70 kW	Electric resistance (or none)	Split systems and single package	4.10 COP 4.19 ICOP	
		All other	Split systems and single package	3.81 COP 4.13 ICOP	
	≥70 kW	Electric resistance (or none)	Split systems and single package	4.10 COP 4.10 ICOP	
		All other	Split systems and single package	3.81 COP 3.81 ICOP	
Condensing units, air-cooled	≥40 kW			Not applicable match with indoor coil	ARI 365
Condensing, water or evaporatively cooled	≥40 kW			Not applicable match with indoor coil	

a. Section 11 contains a complete specification of the referenced test procedures, including year version of the test procedure.

TABLE C-2 (Supersedes Table 6.8.1B in ANSI/ASHRAE/IES Standard 90.1)
Electrically Operated Unitary and Applied Heat Pumps Minimum Efficiency Requirements (SI)

Equipment Type	Size Category	Heating Section Type	Sub-Category or Rating Conditions	Minimum Efficiency	Test Procedure[a]
Air conditioners, air-cooled (cooling mode)	<19 kW	All	Split systems	4.10 $SCOP_C$ 3.52 COP_C	ARI 210/240
			Single packaged	4.10 $SCOP_C$ 3.40 COP_C	
Through-the-wall, air-cooled (cooling mode)	<9 kW	All	Split systems	3.52 $SCOP_C$	
			Single packaged	3.52 $SCOP_C$	
Small-duct high velocity, air-cooled (cooling mode)	<19 kW	All	Split systems	2.93 $SCOP_C$	
Air conditioners, air-cooled (cooling mode)	≥19 kW and <40 kW	Electric resistance (or none)	Split systems and single package	3.31 COP_C 3.46 $ICOP_C$	ARI 340/360
		All other	Split systems and single package	3.25 COP_C 3.40 $ICOP_C$	
	≥40 kW and <70 kW	Electric resistance (or none)	Split systems and single package	3.31 COP_C 3.46 $ICOP_C$	
		All other	Split systems and single package	3.25 COP_C 3.40 $ICOP_C$	
	≥70 kW	Electric resistance (or none)	Split systems and single package	2.87 COP_C 2.87 $ICOP_C$	
		All other	Split systems and single package	2.81 COP_C 2.81 $ICOP_C$	
Water-source (cooling mode)	<5 kW	All	30°C entering water	4.10 COP_C	ISO-13256-1
	≥5 kW and <19 kW	All	30°C entering water	4.10 COP_C	
	>19 kW and <40 kW	All	30°C entering water	4.10 COP_C	
Groundwater-source (cooling mode)	<40 kW	All	15°C entering water	4.75 COP_C	
		All	25°C entering water	13.4 COP_C	

a. Section 11 contains a complete specification of the referenced test procedures, including year version of the test procedure.

TABLE C-3 (Supersedes Table 6.8.1C in ANSI/ASHRAE/IES Standard 90.1)
Water Chilling Packages—Minimum Efficiency Requirements[a] (SI)

Equipment Type	Size Category	Units	Minimum Efficiency[c]				Test Procedure[b]
			Path A		Path B[d]		
			Full Load	IPLV	Full Load	IPLV	
Air-cooled chillers with condenser, electrically operated	<528 kW	COP	2.931	3.664	NA	NA	ARI 550/590
	≥528 kW	COP	2.931	3.737	NA	NA	
Air-cooled without condenser, electrical operated	All capacities	COP	Condenserless units shall be rated with matched condensers				ARI 550/590
Water-cooled, electrically operated, positive displacement (reciprocating)	All capacities	COP	Reciprocating units required to comply with water cooled positive displacement requirements				ARI 550/590
Water-cooled electrically operated, positive displacement	<264 kW	COP	4.509	5.583	4.396	5.862	ARI 550/590
	≥264 kW and <528 kW	COP	4.538	5.719	4.452	6.002	
	≥528 kW and <1055 kW	COP	5.172	6.064	4.898	6.513	
	≥1055 kW	COP	5.673	6.513	5.504	7.178	
Water-cooled electrically operated, centrifugal	<528 kW	COP	5.547	5.901	5.504	7.816	ARI 550/590
	≥528 kW and <1055 kW	COP	5.547	5.901	5.504	7.816	
	≥1055 kW and <2110 kW	COP	6.106	6.406	5.862	8.792	
	≥2110 kW	COP	6.170	6.525	5.961	8.792	
Air-cooled absorption single effect[g]	All capacities	COP	0.600	NR[f]	NA[e]	NA[e]	ARI 560
Water-cooled absorption single effect[g]	All capacities	COP	0.700	NR[f]	NA[e]	NA[e]	
Absorption double effect indirect-fired	All capacities	COP	1.000	1.050	NA[e]	NA[e]	
Absorption double effect direct fired	All capacities	COP	1.000	1.000	NA[e]	NA[e]	

a. The chiller equipment requirements do not apply for chillers used in low-temperature applications where the design leaving fluid temperature is <2.4 C
b Section 11 contains a complete specification of the referenced test procedure, including the referenced year version of the test procedure
c. Compliance with this standard can be obtained by meeting the minimum requirements of Path A or Path B. However both the full load and IPLV must be met to fulfill the requirements of Path A or Path B
d. Path B is intended for applications with significant operating time at part load. All path B machines shall be equipped with demand limiting capable controls
e. NA means that this requirement is not applicable and can not be used for compliance
f. NR means that for this category there are no minimum requirements
g. Only allowed to be used in heat recovery applications
h. Packages that are not designed for operation at ARI Standard 550/590 test conditions (and, thus, cannot be tested to meet the requirements of Table C-3) of 6.7°C leaving chilled-water temperature and 29.4°C entering condenser-water temperature with 0.054 L/s·kW condenser-water flow shall have maximum full-load kW/ton and *NPLV* ratings adjusted using the following equation:

Adjusted minimum full load COP rating = (full load COP from Table C-3) · K_{adj}
Adjusted maximum NPLV rating = (IPLV from Table C-3) · K_{adj}

where

K_{adj} = $6.174722 - 0.5466024(X) + 0.020394698(X)^2 - 0.000266989(X)^3$
X = DT_{std} + LIFT (°C)
DT_{std} = (0.267114 + 0.267088/(full load COP from Table C-3)) / Flow (°C)
Flow = condenser-water flow (L/s) / cooling full load capacity (kW)
LIFT = CEWT – CLWT (°C)
CEWT = full load entering condenser-water temperature (°C)
CLWT = full load leaving chilled-water temperature (°C)

The adjusted full load and *NPLV* values are only applicable over the following full-load design ranges:

- minimum leaving chilled-water temperature: 3.3°C
- maximum condenser entering water temperature: 39°C
- condenser-water flow: 0.036 to 0.0721 L/s·kW
- $X \geq 21.7$°C and ≤ 33.3°C

TABLE C-4 (Supersedes Table 6.8.1D in ANSI/ASHRAE/IES Standard 90.1)
Electrically Operated Packaged Terminal Air Conditioners, Packaged Terminal Heat Pumps, Single Packaged Vertical Air Conditioners, Single Packaged Vertical Heat Pumps, Room Air Conditioners and Room Air Conditioners Heat Pumps—Minimum Efficiency Requirements (SI)

Equipment Type	Size Category (Input)	Subcategory or Rating Condition	Minimum Efficiency	Test Procedure[a]
PTAC (cooling mode) new construction	<2.0 kW	35 C DB *Outdoor air*	3.49 COP_C	ARI 310/380
	≥2.0 kW and <2.9 kW	35 C DB *Outdoor air*	3.31 COP_C	
	≥2.9 kW and <3.8 kW	35 C DB *Outdoor air*	3.14 COP_C	
	≥ 3.8 kW	35 C DB *Outdoor air*	3.48 COP_C	
PTAC (cooling mode) replacement[b]	<2.0 kW	35 C DB *Outdoor air*	3.49 COP_C	ARI 310/380
	≥2.0 kW and <2.9 kW	35 C DB *Outdoor air*	3.31 COP_C	
	≥2.9 kW and <3.8 kW	35 C DB *Outdoor air*	3.14 COP_C	
	≥ 3.8 kW	35 C DB *Outdoor air*	3.48 COP_C	
PTHP (cooling mode) new construction	<2.0 kW	35 C DB *Outdoor air*	3.48 COP_C	ARI 310/380
	≥2.0 kW and <2.9 kW	35 C DB *Outdoor air*	3.48 COP_C	
	≥2.9 kW and <3.8 kW	35 C DB *Outdoor air*	3.48 COP_C	
	≥3.8 kW	35 C DB *Outdoor air*	3.48 COP_C	
PTHP (heating mode) new construction	All capacities	35 C DB *Outdoor air*	2.8 COP_H	ARI 310/380
PTHP (cooling mode) replacement[b]	<2.0 kW	35 C DB *Outdoor air*	3.43 COP_C	ARI 310/380
	≥2.0 kW and <2.9 kW	35 C DB *Outdoor air*	3.25 COP_C	
	≥2.9 kW and <3.8 kW	35 C DB *Outdoor air*	3.08 COP_C	
	≥3.8 kW	35 C DB *Outdoor air*	2.73 COP_C	
PTHP (heating mode) replacement[b]	All capacities	35 C DB *Outdoor air*	2.8 COP_H	ARI 310/380

a. Section 11 contains a complete specification of the referenced test procedures, including year version of the test procedure.
b. Replacement units shall be factory labeled as follows: "MANUFACTURED FOR REPLACEMENT APPLICATIONS ONLY; NOT TO BE INSTALLED IN NEW CONSTRUCTION PROJECTS." Replacement efficiencies apply only to units with existing sleeves less than 16 in. high and less than 42 in. wide.

TABLE C-5 (Supersedes Table 6.8.1D in ANSI/ASHRAE/IES Standard 90.1)
Electrically Operated Packaged Terminal Air Conditioners, Packaged Terminal Heat Pumps, Single Packaged Vertical Air Conditioners, Single Packaged Vertical Heat Pumps, Room Air Conditioners, and Room Air Conditioner Heat Pumps—Minimum Efficiency Requirements (SI)

Equipment Type	Size Category (Input)	Subcategory or Rating Condition	Minimum Efficiency	Test Procedure[a]
SPVAC (cooling mode)	<19 kW	35 C DB/23.9 C WB *Outdoor air*	2.93 COP_c	ARI 390
	≥19 kW and <40 kW Btu/h	35 C DB/23.9 C WB *Outdoor air*	3.37 COP_C	
	≥40kW and <70 kW	35 C DB/23.9 C WB *Outdoor air*	3.37 COP_C	
SPVHP (cooling mode)	<19 kW	35 C DB/23.9 C WB *Outdoor air*	2.93 COP_c	
	≥19 kW and <40 kW Btu/h	35 C DB/23.9 C WB *Outdoor air*	3.37 COP_C	
	≥40kW and <70 kW	35 C DB/23.9 C WB *Outdoor air*	3.37 COP_C	
SPVHP (heating mode)	<19 kW	8.3 CF DB/6.1 C WB *Outdoor air*	3.0 COP_H	
	≥19 kW and <40 kW Btu/h	8.3 CF DB/6.1 C WB *Outdoor air*	3.0 COP_H	
	≥40kW and <70 kW	8.3 CF DB/6.1 C WB *Outdoor air*	2.9 COP_H	
Room air conditioners, with louvered sides	<1.8 kW		3.14 $SCOP_C$	ANSI/ AHAM RAC-1
	≥1.8 kW and <2.3 kW		3.14 COP_C	
	≥2.3 kW and <4.1 kW		3.17 COP_C	
	≥4.1 kW and <5.9 kW		3.14 COP_C	
	≥5.9 kW		2.73 COP_C	
Room air conditioners, without louvered sides	<2.3 kW		2.90 COP_C	
	≥2.3 kW and <5.9 kW		2.73 COP_C	
	≥5.9 kW		2.73 COP_C	
Room air conditioner heat pump with louvered sides	<5.9 kW		2.90 COP_C	
	≥5.9 kW		2.73 COP_C	
Room air conditioner heat pump without louvered sides	<4.1 kW		2.73 COP_C	
	≥4.1 kW		2.58 COP_C	
Room air conditioner, casement only	All Capacities		2.81 COP_C	
Room air conditioner, casement-slider	All Capacities		3.05 COP_C	

a. Section 11 contains a complete specification of the referenced test procedure, including the referenced year version of the test procedure.

TABLE C-6 (Supersedes Table 6.8.1E in ANSI/ASHRAE/IES Standard 90.1) Warm Air Furnace and Combustion Warm Air Furnaces/Air Conditioning Units, Warm Air Duct Furnaces and Unit Heaters (SI)

Equipment Type	Size Category (Input)	Subcategory or Rating Condition	Test Procedure[b]	Minimum Efficiency[a]
Warm air furnace, gas-fired (weatherized)	<65.9 kW	Maximum capacity[d]	DOE 10 CFR Part 430 or ANSI Z21.47	78% AFUE or 80% E_t[c,e]
	>65.9 kW	Maximum capacity[d]	ANSI Z21.47	80% E_c[c,e]
Warm air furnace, gas-fired (non-weatherized)	<65.9 kW	Maximum capacity[d]	DOE 10 CFR Part 430 or ANSI Z21.47	90% AFUE or 92% E_t[c,e]
	>65.9 kW	Maximum capacity[d]	ANSI Z21.47	92% E E_c[c,e]
Warm air furnace, oil-fired (weatherized)	<65.9 kW	Maximum capacity[d]	DOE 10 CFR Part 430 or UL 727	78% AFUE or 80% E_t[c,e]
	>65.9 kW	Maximum capacity[d]	UL 727	81% E_t[e]
Warm air furnace, oil-fired (non-weatherized)	<65.9 kW	Maximum capacity[d]	DOE 10 CFR Part 430 or UL 727	85% AFUE or 87%% E_t[c,e]
	>65.9 kW	Maximum capacity[d]	UL 727	87% E_t[e]
Warm air duct furnaces, gas-fired (weatherized)	All capacities	Maximum capacity[d]	ANSI Z83.9	80% E_c[f]
Warm air duct furnaces, gas-fired (non-weatherized)	All capacities	Maximum capacity[d]	ANSI Z83.9	90% E_c[f]
Warm air unit heaters, gas-fired (non-weatherized)	All capacities	Maximum capacity[d]	ANSI Z83.8	90% E_c[f,g]
Warm air unit heaters, oil-fired (non-weatherized)	All capacities	Maximum capacity[d]	UL 731	90% E_c[f,g]

a. E_t = thermal efficiency. See test procedure for detailed discussions.
b. Section 11 contains a complete specification of the referenced test procedure, including the referenced year version of the test procedure.
c. Combustion units not covered by NAECA (3-phase power or cooling capacity greater than or equal to 19.0 kW) may comply with either rating.
d. Minimum and maximum ratings as provided for and allowed by the unit's controls.
e. Units shall also include an interrupted or intermittent ignition device (IID), have jacket losses not exceeding 0.75% of the input rating, and have either power venting or flue damper. A vent damper is an acceptable alternative to the fuel damper for those furnaces where combustion air is drawn from the *conditioned space*.
f. E_c = combustion efficiency (100% less flue losses) See test procedures for detailed discussion.
g. As of August 8, 2008, according to the Energy Policy Act of 2005, units shall also include an interrupted or intermittent ignition devices (IID) and have either power venting or automatic flue dampers. A vent damper is an acceptable alternative to a flue damper for those unit heaters where combustion air is drawn from the *conditioned space*.

TABLE C-7 (Supersedes Table 6.8.1F in ANSI/ASHRAE/IES Standard 90.1)
Gas and Oil Fired Boilers—Minimum Efficiency Requirements (SI)

Equipment Type[a]	Subcategory or Rating Condition	Size Category (Input)	Efficiency[b,c]	Test Procedure[g]
Boilers, hot water	Gas-fired	<87.9 kW	89% AFUE[f]	10 CFR Part 430
		≥87.9 kW and <732.7 kW[d]	89% E_t[f]	10 CFR Part 431
		≥732.7 kW[a]	91% E_c[f]	
	Oil-fired[e]	<87.9 kW	89% AFUE[f]	10 CFR Part 430
		≥87.9 kW and <732.7 kW[d]	89% E_t[f]	10 CFR Part 431
		≥732.7 kW[a]	91% E_c[f]	
Boilers, steam	Gas-fired	<87.9 kW	75% AFUE	10 CFR Part 430
	Gas-fired all, except natural draft	≥87.9 kW and <732.7 kW[d]	79% E_t	10 CFR Part 431
		≥732.7 kW[a]	79% E_t	
	Gas-fired natural draft	≥87.9 kW and <732.7 kW[d]	77% E_t	
		≥732.7 kW[a]	77% E_t	
	Oil-fired[e]	<87.9 kW	80% E_t	10 CFR Part 430
		≥87.9 kW and <732.7 kW[d]	81% E_t	10 CFR Part 431
		≥732.7 kW[a]	81% E_t	

a. These requirements apply to boilers with rated input of 2,344 kW or less that are not packaged boilers, and to all packaged boilers. Minimum efficiency requirements for boilers cover all capacities of packaged boilers.
b. E_c = thermal efficiency (100% less flue losses). See reference document for detailed information.
c. E_t = thermal efficiency. See reference document for detailed information.
d. Maximum capacity - minimum and maximum ratings as provided for and allowed by the unit's controls.
e. Includes oil fired (residual).
f. Systems shall be designed with lower operating return hot water temperatures (<55 C) and use hot water reset to take advantage of the higher efficiencies of condensing boilers.
g. Section 11 contains a complete specification of the referenced test procedure, including the referenced year version of the test procedure.

TABLE C-8 (Supersedes Table 6.8.1G in ANSI/ASHRAE/IES Standard 90.1)
Performance Requirements for Heat Rejection Equipment (SI)

Equipment Type	Total System Heat Rejection Capacity at Rated Conditions	Rating Standard	Rating Conditions	Performance Required[a,b]
Open-loop propeller or axial fan cooling towers[a]	All	CTI ATC-105 and CTI STD-201	35°C entering water 29.4°C leaving water 23.9°C wb entering air	>3.38 L/s kW
Closed-loop propeller or axial fan cooling towers[b]	All	CTI ATC-105 and CTI STD-201	38.9°C entering water 32.2°C leaving water 23.9°C wb entering air	>1.27 L/s kW
Open-loop centrifugal fan cooling towers[a]	All	CTI ATC-105 and CTI STD-201	35°C entering water 29.4°C leaving water 23.9°C wb entering air	>1.86 L/s kW
Closed-loop centrifugal fan cooling towers[b]	All	CTI ATC-105 and CTI STD-201	38.9°C entering water 32.2°C leaving water 23.9°C wb entering air	>0.68 L/s kW
Air-cooled condensers	All	ARI 460		Not applicable, air cooled condenser shall be matched to the HVAC system and rated per Table C-3

a. For purposes of this table, open circuit cooling tower performance is defined as the water flow rating of the tower at the thermal rating condition listed in this table divided by the fan nameplate rated motor nameplate power.

b. For purposes of this table, closed circuit cooling tower performance is defined as the process water flow rating of the tower at the thermal rating condition listed in this table divided by the sum of the fan motor nameplate power and the integral spray pump motor nameplate power.

TABLE C-9 (Supersedes Table 6.8.2A in ANSI/ASHRAE/IES Standard 90.1)
Minimum Duct Insulation R-Value[a] Cooling and Heating Only Supply Ducts and Return Ducts (SI)

Climate Zone	Duct Location						
	Exterior	Ventilated Attic	Unvented Attic Above Insulated Ceiling	Unvented Attic with Roof Insulation[a]	Unconditioned Space[b]	Indirectly Conditioned Space[c]	Buried
				Heating Ducts Only			
1, 2	None	None	None	None	None	None	None
3	R-1.06	None	None	None	R-1.06	None	None
4	R-1.06	None	None	None	R-1.06	None	None
5	R-1.41	R-1.06	None	None	R 1.06	None	R-1.06
6	R-1.41	R-1.41	R-1.06	None	R 1.06	None	R-1.06
7	R-1.76	R-1.41	R-1.41	None	R-1.06	None	R-1.06
8	R-1.76	R-10	R-1.41	None	R-1.41	None	R-1.41
				Cooling Only Ducts			
1	R-1.06	R-1.41	R-10	R-1.06	R-1.06	None	R-1.06
2	R-1.06	R-1.41	R-10	R-1.06	R-1.06	None	R-1.06
3	R-1.06	R-1.41	R-1.41	R-1.06	R-0.62	None	None
4	R-0.62	R-1.06	R-1.41	R-0.62	R-0.62	None	None
5, 6	R-0.62	R-0.62	R-1.06	R-0.62	R-0.62	None	None
7, 8	R-1.9	R-0.62	R-0.62	R-0.62	R-0.62	None	None
				Return Ducts			
1 to 8	R-1.06	R-1.06	R-1.06	None	None	None	None

a Insulation R-values, measured in $m^2 \cdot k/kW$, are for the insulation as installed and do not include film resistance. The required minimum thicknesses do not consider water vapor transmission and possible surface condensation. Where exterior *walls* are used as plenum *walls*, *wall* insulation shall be as required by the most restrictive condition of this table or Section 7.4.2. Insulation resistance measured on a horizontal plane in accordance with ASTM C518 at a mean temperature of 23.8 C at the installed thickness.
b Includes crawl spaces, both ventilated and non-ventilated.
c Includes return air plenums with or without exposed *roofs* above.

TABLE C-10 (Supersedes Table 6.8.2GB in ANSI/ASHRAE/IES Standard 90.1)
Minimum Duct Insulation R-Value[a], Combined Heating and Cooling Supply Ducts and Return Ducts (SI)

Climate Zone	Duct Location						
	Exterior	Ventilated Attic	Unvented Attic Above Insulated Ceiling	Unvented Attic w/ *Roof* Insulation[a]	Unconditioned Space[b]	Indirectly Conditioned Space[c]	Buried
				Supply Ducts			
1	R-1.41	R-1.41	R-1.76	R-1.06	R-1.06	None	R-1.06
2	R-1.41	R-1.41	R-1.41	R-1.06	R-1.41	None	R-1.06
3	R-1.41	R-1.41	R-1.41	R-1.06	R-1.41	None	R-1.06
4	R-1.41	R-1.41	R-1.41	R-1.06	R-1.41	None	R-1.06
5	R-1.41	R-1.41	R-1.41	R-0.62	R-1.41	None	R-1.06
6	R-1.76	R-1.41	R-1.41	R-0.62	R-1.41	None	R-1.06
7	R-1.76	R-1.41	R-1.41	R-0.62	R-1.41	None	R-1.06
8	R-1.76	R-1.94	R-1.94	R-0.62	R-1.41	None	R-1.41
				Return Ducts			
1 to 8	R-1.06	R-1.06	R-1.06	None	None	None	None

a Insulation R-values, measured in $m^2 \cdot k/kW$, are for the insulation as installed and do not include film resistance. The required minimum thicknesses do not consider water vapor transmission and possible surface condensation. Where exterior *walls* are used as plenum *walls*, *wall* insulation shall be as required by the most restrictive condition of this table or Section 7.4.2. Insulation resistance measured on a horizontal plane in accordance with ASTM C518 at a mean temperature of 23.8 C at the installed thickness."

b Includes crawl spaces, both ventilated and non-ventilated.

c Includes return air plenums with or without exposed *roofs* above.

TABLE C-11 (Supersedes Table 7.8 in ANSI/ASHRAE/IES Standard 90.1)
Performance Requirements for Water Heating Equipment (SI)

Equipment Type	Size Category (Input)	Subcategory or Rating Condition	Performance Required [a]	Test Procedure [b]
Electric water heaters	12 kW	Resistance > 75.7L	$EF \ge 0.97-0.00132V$	DOE 10 CFR Part 430
	>12 kW	Resistance > 75.7L	$SL \le 20 + 35\sqrt{V}$, W	ANSI Z21.10.3
	All	Heat Pump	$EF \ge 2.0$	DOE 10 CFR Part 430
Gas storage water heaters	<22.98 kW	Resistance > 75.7L	$EF \ge 0.67$	DOE 10 CFR Part 430
	>22.98 kW	<309.75 W/L	$E_t \ge 80\%$ and $SL \le (Q/800 + 110\sqrt{V})$, W	ANSI Z21.10.3
Gas instantaneous water heaters	>14.66 kW and <58.62 kW	>309.75 W/L and <7 L	$EF \ge 0.82$	DOE 10 CFR Part 430
	>58.62 kW[c]	>309.75 W/L and <37.5 L	$E_t \ge 80\%$	ANSI Z21.10.3
	>58.62 kW	>309.75 W/L and <37.5 L	$E_t \ge 80\%$ and $SL \le (Q/800 + 110\sqrt{V})$, W	
Oil storage water heaters	<30.78 kW	Resistance >75.7L	$EF \ge 0.59 - 0.0019V$	DOE 10 CFR Part 430
	>30.78 kW	<309.75 W/L	$E_t \ge 78\%$ and $SL \le (Q/800 + 110\sqrt{V})$, W	ANSI Z21.10.3
Oil instantaneous water heaters	<61.55 kW	≥309.75 W/L and <7.56 L	$EF \ge 0.59 - 0.0019V$	DOE 10 CFR Part 430
	>61.55 kW	≥309.75 W/L and <37.5 L	$E_t \ge 80\%$	ANSI Z21.10.3
	>61.55 kW	≥309.75 W/L and ≥37.5 L	$E_t \ge 78\%$ and $SL \le (Q/800 + 110\sqrt{V})$, W	
Hot-water supply boilers, gas and oil	61.55 kW and <3663.8 kW	≥309.75 W/L and <37.5 L	$E_t \ge 80\%$	ANSI Z21.10.3
Hot-water supply boilers, gas		≥309.75 W/L and ≥37.5 L	$E_t \ge 80\%$ and $SL \le (Q/800 + 110\sqrt{V})$, W	
Hot-water supply boilers, oil		≥309.75 W/L and ≥37.5 L	$E_t \ge 78\%$ and $SL \le (Q/800 + 110\sqrt{V})$, W	
Pool heaters, oil and gas	All		$E_t \ge 78\%$	ASHRAE 146
Heat pump pool heaters	All		≥4.0 COP	ASHRAE 146
Unfired storage tanks	All		≥R-12.5	(none)

a Energy factor (EF) and thermal efficiency (Et) are minimum requirements, while standby loss (SL) is maximum W based on a 21.1 C temperature difference between stored water and ambient requirements. In the EF equation, V is the rated volume in gallons. In the SL equation, V is the rated volume in gallons and Q is the nameplate input rate in kW

b Section 11 contains a complete specification, including the year version, of the referenced test procedure.

c Instantaneous water heaters with input rates below 58.62 kW shall comply with these requirements if the water heater is designed to heat water to temperatures 82.2°C or higher."

TABLE C-12 Minimum Nominal Efficiency for General Purpose Design A and Design B Motors[a] (SI)

	Minimum Nominal Full-Load Efficiency (%)[a]					
	Open Motors			Enclosed Motors		
Number of Poles ==>	2	4	6	2	4	6
Synchronous Speed (RPM) ==>	3600	1800	1200	3600	1800	1200
Motor Size (kW)						
0.7	77.0	85.5	82.5	77.0	85.5	82.5
1.1	84.0	86.5	86.5	84.0	86.5	87.5
1.5	85.5	86.5	87.5	85.5	86.5	88.5
2.2	85.5	89.5	88.5	86.5	89.5	89.5
3.7	86.5	89.5	89.5	88.5	89.5	89.5
5.6	88.5	91.0	90.2	89.5	91.7	91.0
7.5	89.5	91.7	91.7	90.2	91.7	91.0
11.2	90.2	93.0	91.7	91.0	92.4	91.7
14.9	91.0	93.0	92.4	91.0	93.0	91.7
18.7	91.7	93.6	93.0	91.7	93.6	93.0
22.4	91.7	94.1	93.6	91.7	93.6	93.0
29.8	92.4	94.1	94.1	92.4	94.1	94.1
37.3	93.0	94.5	94.1	93.0	94.5	94.1
44.8	93.6	95.0	94.5	93.6	95.0	94.5
56.0	93.6	95.0	94.5	93.6	95.4	94.5
74.6	93.6	95.4	95.0	94.1	95.4	95.0
93.3	94.1	95.4	95.0	95.0	95.4	95.0
111.9	94.1	95.8	95.4	95.0	95.8	95.8
149.2	95.0	95.8	95.4	95.4	96.2	95.8
186.5	95.0	95.8	95.4	95.8	96.2	95.8
223.8	95.4	95.8	95.4	95.8	96.2	95.8
261.1	95.4	95.8	95.4	95.8	96.2	95.8
298.4	95.8	95.8	95.8	95.8	96.2	95.8
335.7	95.8	96.2	96.2	95.8	96.2	95.8
373.0	95.8	96.2	96.2	95.8	96.2	95.8

a Nominal efficiencies shall be established in accordance with NEMA Standard MG1. Design A and Design B are National Electric Manufacturers Association (NEMA) design class designations for fixed frequency small and medium AC squirrel-cage induction motors.

TABLE C-13 Commercial Refrigerator & Freezers (SI)

Equipment Type	Application	Energy Use Limit (kW/h per day)
Refrigerators with solid doors	Holding temperature	2.831 V + 57.75
Refrigerators with transparent doors		3.40 V + 94.55
Freezers with solid doors		11.32 V + 39.07
Freezers with transparent doors		21.23 V + 116.07
Refrigerators/freezers with solid doors		the greater of 3.40 V + 94.55 or 19.82
Commercial Refrigerators	Pulldown	1.26 V + 99.37

V means the chiller or frozen compartment volume (Liters) as defined in the Association of Home Appliance Manufacturers Standard HRF1-1979

TABLE C-14 Commercial Clothes Washers (SI)

Product	MER	WF
All commercial clothes washers	48.7	30.3

MER = Modified Energy Factor, a combination of Energy Factor and MEF = Modified Energy Factor, a combination of Energy Factor and Remaining Moisture Content. MEF measures energy consumption of the total laundry cycle (washing and drying). It indicates how many liters of laundry can be washed and dried with one kWh of electricity; the higher the number, the greater the efficiency.

(This is a normative appendix and is part of this standard.)

NORMATIVE APPENDIX D— PERFORMANCE OPTION FOR ENERGY EFFICIENCY

D1. GENERAL

D.1.1 Performance Option Scope. *Building projects* complying with Section 7.5, the "Performance Option," shall comply with the requirements in Normative Appendix G of ANSI/ASHRAE/IES Standard 90.1 with the following modifications and additions. When a requirement is provided in this appendix, it supersedes the requirement in ANSI/ASHRAE/IES Standard 90.1. This appendix shall be used both for *building projects* demonstrating compliance with the requirements of this standard and for *building projects* demonstrating performance that substantially exceeds the requirements of this standard. Where stated in Normative Appendix G of ANSI/ASHRAE/IES Standard 90.1, the rating authority or program evaluator shall be the *AHJ.*

Note to Adopting Authority: ASHRAE Standing Standard Project Committee 189.1 recommends that a compliance shell implementing the rules of a compliance supplement that controls inputs to and reports outputs from the required computer analysis program be adopted for the purposes of easier use and simpler compliance.

D1.1.1 Performance Option (Section G1.2 of ANSI/ASHRAE/IES Standard 90.1). In addition to the requirements in Section G1.2 of ANSI/ASHRAE/IES Standard 90.1, all requirements in Sections 5.3, 6.3, 7.3, 8.3, and 9.3 shall be met.

D1.1.2 Trade-Off Limits (Section G1.3 of ANSI/ASHRAE/IES Standard 90.1). In addition to the requirements in Section G1.3 of ANSI/ASHRAE/IES Standard 90.1, future building components shall meet all requirements in Section 7.4.

D.1.1.3 Documentation Requirements (Section G1.4 of ANSI/ASHRAE/IES Standard 90.1).

a. In addition to the requirements in Section G1.4(d) of ANSI/ASHRAE/IES Standard 90.1, the documentation list shall include compliance with the requirements in Section 7.3.
b. In addition to the requirements in Section G1.4(e) of ANSI/ASHRAE/IES Standard 90.1, the documentation list shall identify aspects that are less stringent than the requirements in Section 7.4.

D1.1.4 Energy Rates (Section G2.4 of ANSI/ASHRAE/IES Standard 90.1). In addition to the requirements in Section G2.4 of ANSI/ASHRAE/IES Standard 90.1, when the total modeled annual on-site renewable energy generated by the *proposed design* exceeds that generated by the *baseline building design*, the difference in the annual on-site generated renewable energy between the *baseline building performance* and the *proposed building performance* shall be based on the energy source used as the backup energy source in the *proposed design* or on the use of electricity if no backup energy source has been specified.

D1.1.5 Baseline HVAC System Type and Description (Section G3.1.1 of ANSI/ASHRAE/IES Standard 90.1). The hood or hood section modeled according to Exception (d) to Section G3.1.1 of ANSI/ASHRAE/IES Standard 90.1 shall also meet the requirements of Section 7.4.3.9.

D1.1.6 Ventilation (Section G3.1.2.6 of ANSI/ASHRAE/IES Standard 90.1).

a. Exception (a) to Section G3.1.2.6 of ANSI/ASHRAE/IES Standard 90.1 shall be used only where *DCV* is not required by Section 7.4.3.2.
b. Exception (c) to Section G3.1.2.6 of ANSI/ASHRAE/IES Standard 90.1 shall not apply.

D1.1.7 Economizers (Section G3.1.2.7 of ANSI/ASHRAE/IES Standard 90.1).

a. Outdoor air economizers shall be included in the baseline systems identified in Section G3.1.2.7 of ANSI/ASHRAE/IES Standard 90.1 for the climate zones and capacities specified in Table 7.4.3.4A.
b. Exception (a) to Section G3.1.2.7 of ANSI/ASHRAE/IES Standard 90.1 shall not apply.

D1.1.8 System Fan Power (Section G3.1.2.10 of ANSI/ASHRAE/IES Standard 90.1).

a. System fan brake horsepower shall be 10% less than the values calculated using Table G3.1.2.9 of ANSI/ASHRAE/IES Standard 90.1.
b. Fan motor efficiency shall meet the requirements of Section 7.4.7.1.

D1.1.9 Exhaust Air Energy Recovery (Section G3.1.2.11 of ANSI/ASHRAE/IES Standard 90.1). Exhaust air energy recovery shall be modeled in the *baseline building design* as specified in Section 7.4.3.8.

D1.1.10 VAV Minimum Flow Setpoints (Section G3.1.3.13 of ANSI/ASHRAE/IES Standard 90.1). Zone minimum airflow setpoints shall be modeled as specified in Section 7.4.3.5.

D1.1.11 Building Performance Calculations (Table G3.1 of ANSI/ASHRAE/IES Standard 90.1). In addition to Table G3.1 of ANSI/ASHRAE/IES Standard 90.1, the *baseline building design* and *proposed design* shall comply with all modifications and additions in Table D1.1. All references to "Table G3.1" in Table D1.1 refer to Table G3.1 of Appendix G of ANSI/ASHRAE/IES Standard 90.1.

Table D1.1 Modifications and Additions to Table G3.1 of Appendix G in ANSI/ASHRAE/IES Standard 90.1

Proposed Building Performance	Baseline Building Performance
1. Design Model	
No modifications	No modifications
2. Additions and Alterations	
In addition to the requirements in Table G3.1 (2.a), work to be performed in the excluded parts of the building shall comply with Sections 7.3 and 7.4.	No modifications
3. Space Use Classification	
No modifications	No modifications
4. Schedules	
No modifications	No modifications
5. Building Envelope	
Exception (c) of Table G3.1 (5) shall be replaced with the following: The exterior roof surface shall be modeled using the solar reflectance and thermal emittance determined in accordance with Sections 5.3.2.3 and 5.3.2.4. Where test data are unavailable, the roof surface shall be modeled with a reflectance of 0.30 and a thermal emittance of 0.90.	1. In addition to the requirements in Table G3.1 (5), the *baseline building design* shall comply with Section 7.4.2. 2. If the *proposed design* does not comply with Section 7.4.2.9, then the fenestration area in the *baseline building design* shall be uniformly reduced until it complies. This adjustment is not required to be made when rotating the building as required in Table G3.1 (5.a). 3. In addition to the requirements in Table G3.1 (5.d) and (5.e), roof surfaces shall comply with Section 5.3.2.3.
6. Lighting	
1. In addition to the requirements in Table G3.1 (6.c), when lighting neither exists nor is specified, lighting power shall comply with Section 7.4.6. 2. When taking credit for daylight controls under Table G3.1 (6.f), credit may be taken only for lighting controls that are not required by Section 7.4.6. Credit for daylighting controls is allowed to be taken up to a distance of 2.5 times window head height where all lighting more than one window head height from the perimeter (head height is the distance from the floor to the top of the glazing) is automatically controlled separately from lighting within one window head height of the perimeter.	In addition to the requirements in Table G3.1 (6), lighting power shall comply with Section 7.4.6. Automatic and manual controls shall be modeled as required in Section 7.4.6.
7. Thermal Blocks—HVAC Zones Designed	
No modifications	No modifications
8. Thermal Blocks—HVAC Zones Not Designed	
No modifications	No modifications
9. Thermal Blocks—Multifamily Residential Buildings	
No modifications	No modifications
10. HVAC Systems	
No modifications	In addition to the requirements in Table G3.1 (10), the *baseline building design* shall comply with all requirements in Section 7.4.3.

Table D1.1 Modifications and Additions to Table G3.1 of Appendix G in ANSI/ASHRAE/IES Standard 90.1 *(Continued)*

Proposed Building Performance	Baseline Building Performance
11. Service Hot-Water Systems	
In addition to the requirements in Table G3.1 (11), service hot-water usage is allowed to be lower in the *proposed design* than in the *baseline building design* if service hot-water use can be demonstrated to be less than that resulting from compliance with Sections 6.3.2, 6.4.2, and 6.4.3.	1. In addition to the requirements in Table G3.1 (11.b) and (11.c), service hot-water systems shall meet the requirements of Sections 7.4.4, 7.4.7.2, and 7.4.7.3. 2. In addition to the requirements in Table G3.1 (11.f), the *baseline building design* shall meet the requirements of Section 7.4.7.2. If a condenser heat recovery system meeting the requirements described in Section 7.4.7.2 cannot be modeled, the requirement for including such a system in the actual building shall be met as a prescriptive requirement and no heat-recovery system shall be included in the *proposed design* or *baseline building design*. 3. In addition to the requirements in Table G3.1 (11.i), the *baseline building design* shall meet the requirements of Sections 6.3.2 and 6.4.3.
12. Receptacle and Other Loads	
No modifications	In addition to the requirements in Table G3.1 (12), the *baseline building design* must meet the requirements in Section 7.4.7.
13. Modeling Limitations to the Simulation Program	
No modifications	No modifications
14. Exterior Conditions	
No modifications	No modifications
15. Renewable Energy Systems	
1. Purchase of off-site renewable energy shall not be modeled in the *proposed design*. 2. The annual energy production of any *on-site renewable energy systems* in the *proposed design* shall be subtracted from the *proposed building performance*.	The *baseline building design* shall have an *on-site renewable energy system* that complies with the annual energy production specified in Section 7. This annual energy production shall be subtracted from the *baseline building performance*. No exceptions shall apply.

(This is a normative appendix and is part of this standard.)

NORMATIVE APPENDIX E— IAQ LIMIT REQUIREMENTS FOR OFFICE FURNITURE SYSTEMS AND SEATING

E1. IAQ LIMIT REQUIREMENTS

Installed office furniture system workstations and seating units shall comply with both the requirements of Sections E1.1 and E1.2.

E1.1 At least 95% of the total number of installed office furniture system workstations and at least 95% of the total number of seating units installed shall comply with either of the following criteria at 168 hours:

a. Emissions concentration limits as shown in Table E1.1 and defined in Section 4.2.1 of ANSI/BIFMA X7.1

b. Emission factors as shown in Table E1.2 and defined in Section 7.6.1 of BIFMA e3.

E1.2 At least 50% of the total number of installed office furniture system workstations and at least 50% of the total number of seating units installed shall not exceed the individual volatile organic chemical (VOC) concentration limits listed in Table E1.3 at 336 hours (14 days) or sooner when determined in accordance with the ANSI/BIFMA M7.1.

When the emission factor at 336 hours is determined using the power-law defined in ANSI/BIFMA M7.1 (Sections 10.4 and 10.5), emission factors with $-0.20 < b < 0.20$ shall be reported as constant.

Small chamber testing of component pieces of workstations per the ANSI/BIFMA M7.1 shall be allowed, provided that there is third-party oversight in selecting representative components and in applying the calculations in ANSI/BIFMA M7.1 to estimate the emission factor of a product.

TABLE E1.1 Workstation Systems and Seating Office Emissions Concentration Limits

Chemical Contaminant	Workstation Emission Limits	Seating Emission Limits
$TVOC_{toluene}$	≤0.5 mg/m^3	≤0.25 mg/m^3
Formaldehyde	≤50 ppb	≤25 ppb
Total aldehydes	≤100 ppb	≤50 ppb
4-Phenylcyclohexene	≤0.0065 mg/m^3	≤0.00325 mg/m^3

TABLE E1.2 Individual Furniture Components Maximum Emission Factors

	Open Plan Workstation	Private Office Workstation
Formaldehyde, μg/m^2·h	42.3	85.1
TVOC, μg/m^2·h	345	694
Total aldehyde, μmol/m^2·h	2.8	5.7
4-Phenylcyclohexene, μg/m^2·h	4.5	9.0

TABLE E1.3 Individual Volatile Organic Chemical (VOC) Concentration Limits

				Workstation	Seating	Individual Components	
Compound Name	CASRN	MW	CREL	Maximum Allowable Conc., μg/m^3	Maximum Allowable Conc., μg/m^3	Open Plan Maximum Allowable Emission Factor, μg/m^2·h	Private Office Maximum Allowable Emission Factor, μg/m^2·h
Ethylbenzene	100-41-4	106.2	Y	1000	500	689	1392
Styrene	100-42-5	104.2	Y	450	225	310	627
p-Xylene	106-42-3	106.2	Y	350	175	241	487
1,4-Dichlorobenzene	106-46-7	147	Y	400	200	276	557
Epichlorohydrin	106-89-8	92.52	Y	1.5	0.75	1.0	2.1
Ethylene Glycol	107-21-1	62.1	Y	200	100	138	278
1-Methoxy-2-propanol (Propylene glycol monomethyl ether)	107-98-2	90.12	Y	3500	1750	2413	4874
Vinyl Acetate	108-05-4	86.1	Y	100	50	68.9	139
m-Xylene	108-38-3	106.2	Y	350	175	241	487
Toluene	108-88-3	92.1	Y	150	75	103	209
Chlorobenzene	108-90-7	112.56	Y	500	250	345	696
Phenol	108-95-2	94.1	Y	100	50	68.9	139
2-Methoxyethanol	109-86-4	76.1	Y	30	15	21	42
Ethylene glycol monomethyl ether acetate	110-49-6	118.13	Y	45	22.5	31	63
n-Hexane	110-54-3	86.2	Y	3500	1750	2413	4874
2-Ethoxyethanol	110-80-5	90.1	Y	35	17.5	24	49
2-Ethoxyethyl acetate	111-15-9	132.2	Y	150	75	103	209
1,4-Dioxane	123-91-1	88.1	Y	1500	750	1034	2089
Tetrachloroethylene	127-18-4	165.8	Y	17.5	8.75	12.1	24.4
Formaldehyde	50-00-0	30.1	Y	16.5	8.25	11	23
Isopropanol	67-63-0	60.1	Y	3500	1750	2413	4874
Chloroform	67-66-3	119.4	Y	150	75	103	209
N,N-Dimethyl Formamide	68-12-2	73.09	Y	40	20	28	56
Benzene	71-43-2	78.1	Y	30	15	21	42
1,1,1-Trichloroethane	71-55-6	133.4	Y	500	250	345	696
Acetaldehyde	75-07-0	44.1	Y	9	4.5	6	13
Methylene Chloride	75-09-2	84.9	Y	200	100	138	278
Carbon Disulfide	75-15-0	76.14	Y	400	200	276	557
Trichloroethylene	79-01-6	131.4	Y	300	150	207	418
1-Methyl-2-Pyrrolidinone	872-50-4	99.13	N	160	80	110	223
Naphthalene	91-20-3	128.2	Y	4.5	2.25	3	6
o-Xylene	95-47-6	106.2	Y	350	175	241	487

(This is a normative appendix and is part of this standard.)

NORMATIVE APPENDIX F— BUILDING CONCENTRATIONS

F1. BUILDING CONCENTRATIONS

Building concentrations shall be estimated based on the following parameters and criteria:

a. Laboratory-measured VOC emission factors and actual surface area of all materials as described in (b) below.
b. At minimum, those materials listed in Section 8.5.2 (a) through (g) to be installed shall be modeled.
c. The actual building parameters for volume, average weekly minimum ventilation rate, and ventilated volume fraction for the building being modeled shall be used.
d. Standard building scenarios or modeling from similar buildings shall not be allowed.
e. Average weekly minimum air change rates shall be calculated based on the *minimum outdoor airflow rates* and hours of operation for the specific building being modeled.
f. Steady-state conditions with respect to emission rates and building ventilation may be assumed.
g. Zero *outdoor air* concentrations, perfect mixing within the building, and no net losses of VOCs from air due to other effects such as irreversible or net sorption on surfaces (i.e., net sink effects) and chemical reactions may be assumed.
h. All assumptions shall be clearly stated in the design documents.
i. The estimated building concentration, C_{Bi} (µg/m^3), of each target VOC shall be calculated using Equation 2 of CDPH/EHLB/Standard Method V1.1 (commonly referred to as California Section 01350), as shown below. Estimated building concentrations of individual target VOCs with multiple sources shall be added to establish a single total estimated building concentration for individual target VOCs.

$$C_{Bi} = (EF_{Ai} \times A_B) / (V_B \times a_B \times 0.9)$$

where

EF_{Ai} = area specific emission rate or emission factor at 96 hours after placing a test specimen in the chamber (14 days total exposure time), µg/m^2·h

A_B = exposed surface area of the installed material in the building, m^2

V_B = building volume, m^3

a_B = average weekly minimum air change rate, 1/h

(This appendix is not part of this standard. It is merely informative and does not contain requirements necessary for conformance to the standard. It has not been processed according to the ANSI requirements for a standard and may contain material that has not been subject to public review or a consensus process. Unresolved objectors on informative material are not offered the right to appeal at ASHRAE or ANSI.)

INFORMATIVE APPENDIX G—INFORMATIVE REFERENCES

This appendix contains informative references for the convenience of users of this standard and to acknowledge source documents when appropriate. Section numbers indicate where the reference occurs in this document.

Reference	Title	Section
American Institute of Architects (AIA) **1735 New York Avenue NW** **Washington, DC 20006, United States** **1-800-AIA-3837 or 202-626-7300; www.aia.org**		
AIA National/AIA California Council	Integrated Project Delivery: A Guide, v. 1 - 2007	Appendix H
American Institute of Steel Construction **One East Wacker Drive, Suite 700** **Chicago, Illinois 60601, United States** **1-312-670-2400; www.aisc.org**		
Brochure	Steel Takes LEED® With Recycled Content	9.4.1.1
American Society of Heating, Refrigerating and Air-Conditioning Engineers (ASHRAE) **1791 Tullie Circle NE** **Atlanta, GA 30329, United States** **1-404-636-8400; www.ashrae.org**		
ASHRAE Guideline 0-2005	The Commissioning Process	10.3.1.1
ASHRAE Guideline 1.1-2007	HVAC&R Technical Requirements to Support the Commissioning Process	10.3.1.1
ASHRAE Guideline 4-2008	Preparation of Operating and Maintenance Documentation for Building Systems	10.3.1.1
ASHRAE Standard 62.1-2010 with Appendix B	Ventilation for Acceptable Indoor Air Quality	Table 10.3.1.4
ASHRAE Handbook	Fundamentals – 2009	Appendix D
ASHRAE Handbook	HVAC Applications – 2011	Appendix H
ASTM International **100 Barr Harbor Dr.** **West Conshohocken, PA 19428-2959, United States** **1-610-832-9585; www.astm.org**		
ASTM D6329-98(2008)	Standard Guide for Developing Methodology for Evaluating the Ability of Indoor Materials to Support Microbial Growth Using Static Environmental Chambers	8.4.2
State of California, Department of General Services, Procurement Division **Ziggurat Building** **707 Third Street** **West Sacramento, CA 95605-2811, United States** **1-916-376-5000**		
RFP DGS-56275	Section 5.7: Indoor Air Quality Requirements for Open Office Panel Systems,	Appendix E
California Environmental Protection Agency, Office of Environmental Health Hazard Assessment **Post Office Box 4010** **Sacramento, CA 95812-4010, United States** **1-916-324-7572; www.oehha.ca.gov**		
http://www.oehha.org/air/allrels.html	Air Toxics Hot Spots Program Risk Assessment Guidelines. Technical Support Document for the Derivation of Noncancer Reference Exposure Levels	8.4.2, 8.5.2, Appendix E

Reference	Title	Section
Canadian Standards Association (CSA) **5060 Spectrum Way, Suite 100** **Mississauga, Ontario, L4W 5N6, Canada** **1-800-463-6727 and 1-416-747-4000; www.csa.ca**		
CSA S478-95 (R2007)	Guideline on Durability for Buildings	9.4.1, 10.3.2.3
Carpet and Rug Institute **730 College Drive** **Dalton, Georgia 30720, United States** **1-706-278-3176; www.carpet-rug.org**		
		8.4.2.3
Cool Roof Rating Council **1610 Harrison Street** **Oakland, California 94612, United States** **1-510-482-4421; www.coolroofs.org**		
CCRC-1-2008	Cool Roof Council Product Rating Program	5.3.2.4
Forest Stewardship Council (FSC) **1155 30th Street NW, Suite 300** **Washington, DC 20007, United States** **1-202-342-0413; www.fsc.org**		
		9.4.1.3.1
GREENGUARD Environmental Institute **2211 Newmarket Parkway, #110** **Marietta, GA 30067, United States** **1-678-444-4055; www.greenguard.org**		
		8.4.2, 8.5.2, Appendix E
Institute of Transportation Engineers **1099 14th Street NW, Suite 300 West** **Washington, DC 20005-3438, United States** **1-202-289-0222; www.ite.org**		
4th Edition, 2004	Parking Generation	10.3.6
Market Transformation to Sustainability (MTS) **1511 Wisconsin Avenue, N.W.** **Washington, D.C. 20007, United States** **1-202-338-3131; www.sustainableproducts.com**		
MTS 1.0 WSIP Guide – 2007	Whole Systems Integrated Process Guide for Sustainable Buildings and Communities	Appendix H
National Institute of Building Sciences (NIBS) **1090 Vermont Avenue, NW, Suite 700** **Washington, DC 20005-4905, United States** **1-202-289-7800; www.nibs.org**		
NIBS Guideline 3-2006	Exterior Enclosure Technical Requirements for the Commissioning Process	10.3.1.1
National Renewable Energy Laboratory (NREL) **1617 Cole Blvd.** **Golden, CO 80401-3393, United States** **1-303-275-3000; www.nrel.gov**		
NREL/TP-550-38617	Source Energy and Emissions Factors for Energy Use in Buildings	Table 7.5.3.1
Resilient Floor Covering Institute **115 Broad Street, Suite 201** **LaGrange, GA 30240, United States** **1-706-882-3833; www.rfci.com**		
		8.4.2.3

Reference	Title	Section
Sheet Metal and Air Conditioning Contractors National Association (SMACNA) 4201 Lafayette Center Drive Chantilly, VA 20151, Unites States 1-703-803-2980		
ANSI/SMACNA 008-2008	IAQ Guidelines for Occupied Buildings under Construction	10.3.6
Steel Recycling Institute 680 Andersen Drive Pittsburgh, PA 15220, United States 1-412-922-2772; www.recycle-steel.org		
Brochure	Steel Takes LEED® With Recycled Content	9.4.1.1
Sustainable Forestry Initiative, Inc. (SFI) 1600 Wilson Blvd, Suite 810 Arlington, VA 22209, United States 1-703-875-9500; www.sfiprogram.org		
		9.4.1.3.1
U.S. Department of Health and Human Services Agency for Toxic Substances and Disease Registry (ATSDR) 4770 Buford Hwy NE Atlanta, GA 30341, United States 1-800-232-4636; www.atsdr.cdc.gov		
www.atsdr.cdc.gov/mrls	Minimal Risk Levels (MRLs)	Table 10.3.1.4
United States Environmental Protection Agency (EPA) 1200 Pennsylvania Ave NW Washington, DC 20460, United States 1-888-782-7937 and 1- 202-775-6650; www.energystar.gov		
	Portfolio Manager	10.3.3.3.1
United States Department of Energy (DOE) Washington, DC 20585, United States 1-202-586-5000; www.energyplus.gov		
	EnergyPlus (or predecessors BLAST or DOE-2)	Appendix D
United States General Services Administration (GSA) 1800 F Street, NW Washington, DC 20405, United States 1-800-488-3111 and 1-202-501-1100; www.gsa.gov		
U.S. GSA – 2005	The Building Commissioning Guide	10.3.1

(This appendix is not part of this standard. It is merely informative and does not contain requirements necessary for conformance to the standard. It has not been processed according to the ANSI requirements for a standard and may contain material that has not been subject to public review or a consensus process. Unresolved objectors on informative material are not offered the right to appeal at ASHRAE or ANSI.)

INFORMATIVE APPENDIX H— INTEGRATED DESIGN

H1. INTEGRATED DESIGN PROCESS/ INTEGRATED PROJECT DELIVERY

Integrated design, and related concepts such as *integrated project delivery* and integrative design, requires early stakeholder collaboration to enable stronger, more balanced design solutions in all aspects of a project through the sharing of knowledge and expertise among project team members. This *integrated design process* is in contrast to traditional methods, where there is a limited utilization of the skills and knowledge of all stakeholders in the development of design solutions. An *integrated design process* enables the construction of high-performance green buildings that consume fewer resources and achieve better comfort and functionality. A goal of integrated processes is to better enable the construction of high-performance green buildings that consume fewer resources and achieve better comfort and functionality, as well as increased predictability of project outcomes early on.

Integrated design facilitates higher building performance by bringing major issues and key participants into the project early in the design process. For the most part, the opportunities for creatively addressing solutions occur very early in the design process. The complex interactions of sophisticated building systems require early coordination in order to maximize effectiveness and output of such systems. Early team building and goal setting may also reduce total project costs. This collaborative process can inform building form, envelope, and mechanical, electrical, plumbing and other systems. The later in the design process that systems are introduced to the project, the more expensive the implementation of such systems will be. Use of building information technologies can also be a valuable asset in increasing predictability of outcomes earlier in the project and is recommended for all integrated teams.

An iterative design process is intended to take full advantage of the collective knowledge and skills of the design team. A linear process approaches each problem sequentially. In contrast, an integrated process approaches each problem with input from the different viewpoints of the participants and the issues they represent, circling back after each design decision to collectively evaluate the impact on all stakeholders. This process acknowledges the complex interdependency of all building systems and their relationship to resource consumption and occupant well being.

There are several existing, and currently evolving, models for collaboration which can be considered: for example, the *ASHRAE Handbook—HVAC Applications*, Chapter 57; the MTS 1.0 WSIP Guide, *Whole Systems Integrated Process Guide for Sustainable Buildings and Communities*; and *Integrated Project Delivery: A Guide*, by the AIA and AIA California Council.

Project specific integrated design and/or *integrated project delivery* processes should be determined with full participation of the stakeholder team. What works for one project may not prove the best approach for the next. Additionally, the team should collectively identify the performance standards and the associated metrics by which the project success will be judged. Design charrettes of varying duration may be an effective tool to consider, though ultimately it is the responsibility of the stakeholder team to determine the process that will best fit any specific problem or project.

H1.1 Design Charrette. The following outlines one type of design charrette process that has resulted in successful integrated design.

At the initial stages of building design, a charrette process can be initiated and the members of the process should include all the stakeholders.

H1.1.1 Charrette Process. Experienced personnel representing each specialty should participate in the charrette process. A discussion of all the systems and all the items that affect the *integrated design* should be discussed. Stakeholders should be able to decide and vote on the best integrated system.

The integrative team process should entail the following steps of design optimization:

a. The original goals and budget of the project should be revisited to see whether the overall intentions of the project are intact.
b. The project should be compared against this standard or at least one existing green rating system.
c. Each of the building and *site* components should be scrutinized to help ensure natural systems for energy conservation, lighting, ventilation, and passive heating and cooling are maximized before mechanical systems are engaged.
d. The appropriateness and integration logic of the building's primary systems should be confirmed.
e. The impact of the design on the *site* and its larger context should be evaluated, including the environmental impact on a life-cycle cost basis.
f. Building information modeling (BIM) software, design tools, and the experience of the design team should be used if practical to help optimize the design.
g. All members of the design team should be included when making design decisions.
h. Commissioning and consideration of future operation and maintenance (O&M) requirements should be included within the design optimization process.

H1.1.2 Design Charrette Matrix. At the end of the charrette process, a matrix for each proposed building scheme can be developed and evaluated to summarize the impact on the *site*, water, energy, materials, and indoor environmental quality and to help lead to a decision as to the best integrated system. The matrix contains cells indicating the high-performance value, grading a particular building system to its appropriate high-performance criteria. Each high-performance value is qualitatively rated from 1 to 10, with 1 being the lowest (minimal energy savings, low air quality, low water efficiency, high cost) and 10 being the highest (high energy savings, high air quality, high water efficiency, low cost). The average of the high-performance values for each building system is the aggregate index. Selection of the best system should be based upon a comparison of these aggregate indices for each matrix.

Scheme #1—with Atrium, maximum exposure on the south, three-story office building.

Building System	High Performance Criteria						
	Site	IAQ	IEQ	Energy	Comm. M&V	Initial Cost	O & M
Arch	8	7	6	1	6	1	6
HVAC	–	5	6	2	6	2	7
Plumbing	N/A	–	–	–	–	2	7
Structural	–	–	–	–	–	2	
Aggregate index	8	6	6	1.5	6	2	6.8

Result:
Least numbers under energy and cost column defines consumption of substantial energy with high initial cost.

Scheme #2—without Atrium, three-story, minimum exposure on the south and west side.

Building System	High Performance Criteria						
	Site	IAQ	IEQ	Energy	Comm. M&V	Initial Cost	O & M
Arch	6	7	7	7	7	7	6
HVAC	N/A	5	7	7	7	7	7
Plumbing	N/A	–	–	–	7	7	7
Structural	–	–	–	–	–		
Aggregate index	6	6	7	7	7	7	6.8

Result:
High numbers on all columns indicate the building is conceived optimally.

Figure H1.2 Sample Charrette Design Matrices

(This appendix is not part of this standard. It is merely informative and does not contain requirements necessary for conformance to the standard. It has not been processed according to the ANSI requirements for a standard and may contain material that has not been subject to public review or a consensus process. Unresolved objectors on informative material are not offered the right to appeal at ASHRAE or ANSI.)

INFORMATIVE APPENDIX I— ADDENDA DESCRIPTION INFORMATION

ANSI/ASHRAE/USGBC/IES Standard 189.1-2011 incorporates ANSI/ASHRAE/USGBC/IES Standard 189.1-2009 and Addenda a, b, c, d, e, f, g, h, i, j, k, l, m, o, p, q, r, u, v and w to ANSI/ASHRAE/USGBC/IES Standard 189.1-2009. Table I-1 lists each addendum and describes the way in which the standard is affected by the change. It also lists the ASHRAE, and ANSI approval dates for each addendum.

TABLE I-1 Addenda to ANSI/ASHRAE/USGBC/IES Standard 189.1-2009

Addendum	Section(s) Affected	Description of Changes*	ASHRAE Standards Committee Approval	Cosponsor Approval	ASHRAE BOD Approval	ANSI Approval
a	3.2 Definitions	This addendum provides daylighting criteria in 189.1 that is consistent with the criteria in Standard 90.1-2007 with published addenda.	June 26, 2010	June 5, 2010	June 30, 2010	July 1, 2010
b	8.5.1 Daylighting Simulation	This addendum modifies daylighting definitions and criteria so the standard is consistent with what will be published in Standard 90.1-2010.	June 26, 2010	June 5, 2010	June 30, 2010	July 1, 2010
c	7.4.3 Heating Ventilating and Air Conditioning; 8.3.1 Indoor Air Quality; 10.3.1 Construction	This addendum groups outdoor air definitions for easier reference. It deletes (b) and (c) in 8.3.1.1 that were needed for the 30% more air requirement in previous public reviews of 189.1. The addendum does not change the requirements of the standard, because the definitions are reorganized and outdoor airflow rates are still permitted to be greater than the minimum requirements. The current definition of design outdoor airflow rate is being proposed for deletion, and the occurrences of this term in the standard are being modified to be clearer.	June 25, 2011	May 3, 2011	June 30, 2009	June 30, 2009
d	8.4 Prescriptive Option; 8.5 Performance Option; Appendix F Building Concentrations	Standard 189.1-2009 references CA/DHS/EHLB/R-174 (commonly referred to as California Section 01350). In February of 2010, this document was updated, and this addendum proposes to modify the standard to update the reference to CDPH/EHLB/Standard Method V1.1 (commonly referred to as California Section 01350).	June 25, 2011	May 3, 2011	June 30, 2009	June 30, 2009
e	Appendix E IAQ Limit Requirements for Office Furniture Systems and Seating	This addendum clarifies the requirements of E1.1, IAQ Limit Requirements. It corrects cited references in Section 11 and adds one reference as it relates to the clarifications made in Appendix E, IAQ Limit Requirements for Office Furniture Systems and Seating. All requirements remain the same.	June 25, 2011	May 3, 2011	June 30, 2009	June 30, 2009

TABLE I-1 Addenda to ANSI/ASHRAE/USGBC/IES Standard 189.1-2009 *(Continued)*

Addendum	Section(s) Affected	Description of Changes*	ASHRAE Standards Committee Approval	Cosponsor Approval	ASHRAE BOD Approval	ANSI Approval
f	7.3.2 On-Site Renewable Energy Systems; 7.4 Prescriptive Option	This addendum changes to the renewable energy requirements in Standard 189.1-2009: • In prescriptive section 7.4.1.1, On-Site Renewable Energy Systems, changes are made to the on-site renewable to be based on roof area rather than conditioned space. This results in a higher requirement for buildings more than one story than for one-story buildings. • In mandatory section 7.3.2, On-Site Renewable Energy Systems, changes make the metric units consistent with prescriptive section. An exception is added for buildings that meet the prescriptive section. • In NAECA section 7.4.3.1, Minimum Equipment Efficiencies, changes are made to be consistent with other sections.	June 25, 2011	May 3, 2011	June 30, 2009	June 30, 2009
g	5.3.3 Reduction of Light Pollution; 7.4.2 Building Envelope; 7.4.6 Lighting; 7.4.7 Other Equipment; 8.3.1 Indoor Air Quality	This addendum proposes to update the references to Standards 62.1-2010, 55-2010, and 90.1-2010 and ENERGY STAR.	September, 27, 2011	October 28, 2011	November 1, 2011	November 2, 2011
h	7.4.3 Heating Ventilating and Air Conditioning	This addendum proposes to update Section 7.4.3 (Heating, Ventilating, and Air Conditioning) of Standard 189.1-2009 to 90.1-2010.	June 25, 2011	May 3, 2011	June 30, 2009	June 30, 2009
i	7.4.6.1 (Lighting Power Allowance)	This addendum adds a requirement in section 7.4.6.1 (Lighting Power Allowance) that the exterior LPDs allowed by 189.1 be some percentage of those allowed by 90.1-2010.	September, 27, 2011	October 28, 2011	November 1, 2011	November 2, 2011
j	7.5 Energy Performance Path; Appendix D Performance Option for Energy Efficiency	This addendum proposes to update Section 7.5 (Energy Performance Path) and Appendix D (Performance Option for Energy Efficiency) of 189.1-2009. This addendum will result in a reference to Appendix G of 90.1-2010 (Performance Rating Method).	June 25, 2011	May 3, 2011	June 30, 2009	June 30, 2009
k	5.3.1 Allowable Sites; 5.3.2 Mitigation of Heat Island Effect; 5.4.1 Site Development	This updates portions of section 5 (Site Sustainability) to improve clarity, to improve requirements related to tree-growth rate, and to add a mandatory requirement related to invasive plants.	June 25, 2011	May 3, 2011	June 30, 2009	June 30, 2009
l	5.3.2.1 Site Hardscape	This updates portions of Section 5 (Site Sustainability) to improve clarity related to heat island reduction provisions, treating porous pavers and open graded aggregate as separate category from paving materials.	June 25, 2011	May 3, 2011	June 30, 2009	June 30, 2009
m	6.3.2 Building Water Use Reduction	This addendum clarifies the climates where condensate collection would be required for air conditioning units by exempting dry climates where little if any condensate would be expected.	September, 27, 2011	October 28, 2011	November 1, 2011	November 2, 2011

TABLE I-1 Addenda to ANSI/ASHRAE/USGBC/IES Standard 189.1-2009 *(Continued)*

Addendum	Section(s) Affected	Description of Changes*	ASHRAE Standards Committee Approval	Cosponsor Approval	ASHRAE BOD Approval	ANSI Approval
o	5.3 Mandatory Provision	This adds a mandatory requirement to section 5.3 (Site Sustainability) to introduce pedestrian friendly environments through the use of designated walkways.	June 25, 2011	May 3, 2011	June 30, 2009	June 30, 2009
p	5.4.1.1 Effective Pervious Area for All Sites	This addendum clarifies pervious area exceptions for brownfield sites in Section 5.4.1.1.	September, 27, 2011	October 28, 2011	November 1, 2011	November 2, 2011
q	5.3.3 Reduction of Light Pollution	This addendum provides exceptions for safety and functionality, defines hazardous areas and will better align Standard 189.1 with 90.1.	September, 27, 2011	October 28, 2011	November 1, 2011	November 2, 2011
r	7.4.6 Lighting	This addendum modifies the interior LPD requirement by adding multipliers in Standard 189.1. This addresses some issues related to the update to 90.1-2010.	September, 27, 2011	October 28, 2011	November 1, 2011	November 2, 2011
u	7.4.6.Lighting	This addendum adds a requirement for automatic controls to assure that lighted signs which are bright enough to be visible during daytime hours are operated during nighttime hours with a 65% reduction in power (reduce power to 35% of full power).	September, 27, 2011	October 28, 2011	November 1, 2011	November 2, 2011
v	7.4.3 Heating Ventilating and Air Conditioning	This addendum adds a minimum time period to automatic HVAC and lighting control requirements for guest rooms.	September, 27, 2011	October 28, 2011	November 1, 2011	November 2, 2011
w	7.4.2 Building Envelope; 8.4.1 Daylighting by Sidelighting; 10.3.1 Construction	This addendum modifies sections 7, 8, and 10 by adding exceptions to include automatically controlled shading devices and dynamic glazing for fenestration.	September, 27, 2011	October 28, 2011	November 1, 2011	November 2, 2011

* These descriptions may not be complete and are provided for information only.

NOTICE

INSTRUCTIONS FOR SUBMITTING A PROPOSED CHANGE TO THIS STANDARD UNDER CONTINUOUS MAINTENANCE

This standard is maintained under continuous maintenance procedures by a Standing Standard Project Committee (SSPC) for which the Standards Committee has established a documented program for regular publication of addenda or revisions, including procedures for timely, documented, consensus action on requests for change to any part of the standard. SSPC consideration will be given to proposed changes within 13 months of receipt by the manager of standards (MOS).

Proposed changes must be submitted to the MOS in the latest published format available from the MOS. However, the MOS may accept proposed changes in an earlier published format if the MOS concludes that the differences are immaterial to the proposed change submittal. If the MOS concludes that a current form must be utilized, the proposer may be given up to 20 additional days to resubmit the proposed changes in the current format.

ELECTRONIC PREPARATION/SUBMISSION OF FORM FOR PROPOSING CHANGES

An electronic version of each change, which must comply with the instructions in the Notice and the Form, is the preferred form of submittal to ASHRAE Headquarters at the address shown below. The electronic format facilitates both paper-based and computer-based processing. Submittal in paper form is acceptable. The following instructions apply to change proposals submitted in electronic form.

Use the appropriate file format for your word processor and save the file in either a recent version of Microsoft Word (preferred) or another commonly used word-processing program. Please save each change proposal file with a different name (for example, "prop01.doc," "prop02.doc," etc.). If supplemental background documents to support changes submitted are included, it is preferred that they also be in electronic form as word-processed or scanned documents.

ASHRAE will accept the following as equivalent to the signature required on the change submittal form to convey non-exclusive copyright:

Files attached to an e-mail:	Electronic signature on change submittal form (as a picture; *.tif, or *.wpg).
Files on a CD:	Electronic signature on change submittal form (as a picture; *.tif, or *.wpg) or a letter with submitter's signature accompanying the CD or sent by facsimile (single letter may cover all of proponent's proposed changes).

Submit an e-mail or a CD containing the change proposal files to:
Manager of Standards
ASHRAE
1791 Tullie Circle, NE
Atlanta, GA 30329-2305
E-mail: change.proposal@ashrae.org
(Alternatively, mail paper versions to ASHRAE address or fax to 404-321-5478.)

The form and instructions for electronic submittal may be obtained from the Standards section of ASHRAE's Home Page, www.ashrae.org, or by contacting a Standards Secretary, 1791 Tullie Circle, NE, Atlanta, GA 30329-2305.
Phone: 404-636-8400. Fax: 404-321-5478. E-mail: standards.section@ashrae.org.

FORM FOR SUBMITTAL OF PROPOSED CHANGE TO AN ASHRAE STANDARD UNDER CONTINUOUS MAINTENANCE

NOTE: Use a separate form for each comment. Submittals (Microsoft Word preferred) may be attached to e-mail (preferred), submitted on a CD, or submitted in paper by mail or fax to ASHRAE, Manager of Standards, 1791 Tullie Circle, NE, Atlanta, GA 30329-2305. E-mail: change.proposal@ashrae.org. Fax: +1-404/321-5478.

1. Submitter:

Affiliation:

Address: City: State: Zip: Country:

Telephone: Fax: E-Mail:

I hereby grant ASHRAE the non-exclusive royalty rights, including non-exclusive rights in copyright, in my proposals. I understand that I acquire no rights in publication of the standard in which my proposals in this or other analogous form is used. I hereby attest that I have the authority and am empowered to grant this copyright release.

Submitter's signature: ________________________________ Date: ____________________

All electronic submittals must have the following statement completed:

I *(insert name)* ________________________________, through this electronic signature, hereby grant ASHRAE the non-exclusive royalty rights, including non-exclusive rights in copyright, in my proposals. I understand that I acquire no rights in publication of the standard in which my proposals in this or other analogous form is used. I hereby attest that I have the authority and am empowered to grant this copyright release.

2. Number and year of standard:

3. Page number and clause (section), subclause, or paragraph number:

4. I propose to: *(check one)*

[ ] Change to read as follows
[ ] Add new text as follows
[ ] Delete and substitute as follows
[ ] Delete without substitution

Use underscores to show material to be added (added) and strike through material to be deleted (~~deleted~~). Use additional pages if needed.

5. Proposed change:

6. Reason and substantiation:

7. Will the proposed change increase the cost of engineering or construction? If yes, provide a brief explanation as to why the increase is justified.

[ ] Check if additional pages are attached. Number of additional pages: _______

[ ] Check if attachments or referenced materials cited in this proposal accompany this proposed change. Please verify that all attachments and references are relevant, current, and clearly labeled to avoid processing and review delays. *Please list your attachments here:*

Rev. 3-9-2007

POLICY STATEMENT DEFINING ASHRAE'S CONCERN FOR THE ENVIRONMENTAL IMPACT OF ITS ACTIVITIES

ASHRAE is concerned with the impact of its members' activities on both the indoor and outdoor environment. ASHRAE's members will strive to minimize any possible deleterious effect on the indoor and outdoor environment of the systems and components in their responsibility while maximizing the beneficial effects these systems provide, consistent with accepted standards and the practical state of the art.

ASHRAE's short-range goal is to ensure that the systems and components within its scope do not impact the indoor and outdoor environment to a greater extent than specified by the standards and guidelines as established by itself and other responsible bodies.

As an ongoing goal, ASHRAE will, through its Standards Committee and extensive technical committee structure, continue to generate up-to-date standards and guidelines where appropriate and adopt, recommend, and promote those new and revised standards developed by other responsible organizations.

Through its *Handbook*, appropriate chapters will contain up-to-date standards and design considerations as the material is systematically revised.

ASHRAE will take the lead with respect to dissemination of environmental information of its primary interest and will seek out and disseminate information from other responsible organizations that is pertinent, as guides to updating standards and guidelines.

The effects of the design and selection of equipment and systems will be considered within the scope of the system's intended use and expected misuse. The disposal of hazardous materials, if any, will also be considered.

ASHRAE's primary concern for environmental impact will be at the site where equipment within ASHRAE's scope operates. However, energy source selection and the possible environmental impact due to the energy source and energy transportation will be considered where possible. Recommendations concerning energy source selection should be made by its members.

ASHRAE · 1791 Tullie Circle NE · Atlanta, GA 30329 · www.ashrae.org

About ASHRAE

ASHRAE, founded in 1894, is an international organization of some 50,000 members. ASHRAE fulfills its mission of advancing heating, ventilation, air conditioning, and refrigeration to serve humanity and promote a sustainable world through research, standards writing, publishing, and continuing education.

For more information or to become a member of ASHRAE, visit www.ashrae.org.

To stay current with this and other ASHRAE standards and guidelines, visit www.ashrae.org/standards.

—·—

ASHRAE also offers its standards and guidelines on CD-ROM or via an online-access subscription that provides automatic updates as well as historical versions of these publications. For more information, visit the Standards and Guidelines section of the ASHRAE Online Store at www.ashrae.org/bookstore.

IMPORTANT NOTICES ABOUT THIS STANDARD

To ensure that you have all of the approved addenda, errata, and interpretations for this standard, visit www.ashrae.org/standards to download them free of charge.

Addenda, errata, and interpretations for ASHRAE standards and guidelines will no longer be distributed with copies of the standards and guidelines. ASHRAE provides these addenda, errata, and interpretations only in electronic form in order to promote more sustainable use of resources.

Product code: 86604 3/12
Errata noted in the list dated 3/6/12 have been corrected.